SCIENCE, SE

SCIENCE, SEEDS AND CYBORGS

BIOTECHNOLOGY AND THE APPROPRIATION OF LIFE

FINN BOWRING

VERSO

London • New York

First published by Verso 2003

1 3 5 7 9 10 8 6 4 2

Verso
UK: 6 Meard Street, London W1F 0EG
USA: 180 Varick Street, New York, NY 10014–4606

www.versobooks.com

Verso is the imprint of New Left Books

ISBN 1–85984–687–4

British Library Cataloguing in Publication Data
A catalogue record for this book is available from the British Library

Library of Congress Cataloging-in-Publication Data
Bowring, Finn, 1969–
Science, seeds and cyborgs : biotechnology and the appropriation of life / Finn Bowring.
p. cm.
Includes index.
ISBN 1–85984–687–4
1. Genetic engineering–social aspects. 2. Biotechnology–Social aspects. I. Title.

TP2486 .B695 2003
306.4′6—dc21 2002032950

Typeset in Dante by M Rules
Printed by Biddles, UK

For Ann, who loves life

Contents

Preface

Genetic engineering is a technology which, if our worst fears are justified, may imperil the future of humankind. As we have accustomed ourselves to the risks of nuclear meltdown, global warming, and the destruction and pollution of vital natural resources, today we must also consider the prospect of a future world populated by organisms containing genetic sequences that have never naturally occurred during millions of years of evolution. Despite the industrial metaphor, genetic engineering differs from the mechanical assembling of objects because it is a deliberate and always partially blind intervention into, and modification of, a self-acting biological system. Because the intervening scientist must surrender his or her efforts to the autonomous and inscrutable workings of the organism, genetic modification is always an experiment. There can be no rehearsal of this procedure, no 'models' which may be innocently tested, corrected and revised, no trials without commitment. If the test of a genetically engineered organism is to yield valid results, it must operate on the real thing. 'The experiment *is* the real deed, and the real deed is the experiment.'[1]

As ecologists have repeatedly pointed out, while the faulty products of mechanical engineering can be recalled and corrected, the results of the genetic manipulation of organisms may be irrevocable and irreversible. Since living organisms propagate and reproduce, intervening in the genome of a single organism may mean intervening in countless future generations across a time span that exceeds our responsible intentions and power of control. By changing genetic sequences that have remained stable for hundreds of thousands of years, scientists risk disrupting the evolutionary barriers to the arbitrary spread of DNA, allowing novel combinations of genetic material to bring about new and deadly viruses, aggressive epidemics of weeds or bacteria, and an ecosystem whose future functioning can no longer be safely predicted through extrapolating from the experiences of the past. With the new biotechnologies, 'the cumulative self-propagation of the technological change of the world constantly overtakes the conditions of its contributing acts and moves through none but unprecedented situations, for which the lessons of experience are powerless'.[2]

The dissolving boundary between intervention and experimentation occurs not just in the realm of agricultural biotechnology, but is evident everywhere scientists have won the right to do something badly – take the invariably unsuccessful 'treatment' of infertility with IVF – in order to accumulate knowledge through their mistakes. This fact alone must prompt us to ask political and ethical questions that too often remain absent from the debate over genetic engineering. Whose interests are being served by the expansion of the biotech industry? And what image of ourselves, of humankind, animates and is animated by the accomplishments of biotech science? There is abundant evidence that biotechnology is a tool of power which, in both its intended and unintended consequences, rarely lends itself to fair, responsible, and democratic control.

Why is it, then, that so many scientists prepared to think creatively about the future of biotechnology are so resolutely unwilling to question the moral and political legitimacy of this future? How is it that Princeton professor of molecular biology, Lee Silver, can spend three hundred pages documenting the current trajectory of reproductive technology, can indulge in futuristic imaginings stretching several centuries from today, can engage in numerous 'thought experiments' and offer vivid clinical and domestic scenarios generously embellished for the least scientific of readers, and can even allow himself a later-edition afterword in which he registers his readers' puzzlement at his lack of political commitment – how can this distinguished geneticist and enthusiastic writer do all this and yet *still* appeal to a spirit of pragmatic realism which sees any attempt to oppose the market-led advance of technology as futile idealism, while professing himself unqualified to judge the social desirability of the future he so caringly describes?[3]

Like many scientists writing on this issue, Silver predicts that the extensive genetic engineering of human beings will become a commonplace reality – not just because reproductive biotechnology is advancing at an unstoppable pace, but also because no ethical precedents will be established in the course of this advance. 'If it is within the rights of parents to spend $100,000 for an exclusive private school education,' Silver asks, 'why is it not also within their rights to spend the same amount of money to make sure that a child inherits a particular set of their genes?'[4] Arguments like this, which abound in the literature on the new reproductive technologies, reason as if the social, technological and scientific precedents that have already been set, the liberties won, the risks run, the inequalities tolerated, were not themselves the result of a historically contingent and always potentially unstable balance of forces, of conflicting interests and powers. They reason as if anyone who has accepted the market as the principal mechanism for the distribution of resources is disqualified from identifying limits to the operation of that market, and must therefore commit themselves to the manipulation and commodification of everything – including bodies and persons. They also fail to recognise that the legitimacy of new events does not rest exhaustively on people's habituation to old versions of those events, that it is easier, both intellectually and legislatively, to oppose the future than it is to change the past, and that future wrongs – even if only more of the same – can, with a little political leadership, awaken people's opposition to wrongs that they have hitherto thoughtlessly accepted.

The refusal of scientists to make a moral and political judgement over, and to take responsibility for, the new biotechnologies, both legitimises and is legitimised by the self-expanding dynamic of these technologies, their cumulative results and their opaque, long-term effects. Not only is responsibility difficult to locate amidst the unintended and unpredictable consequences of genetic engineering, but the biotech industry may well be fashioning a world so dangerous, complex and uncontrollable that our species' modest capacity for responsibility – and with it the anguish of guilt and suffering which marks the alienation of this responsibility – may itself become a problem, a deficiency to be corrected by the same technology and expertise.

The *re-engineering of human nature* thus completes the dream of the biotech revolutionaries, which is to fashion a world so barren and inhospitable to the human body and spirit, so denuded of recognisable intentions and values – of things that belong, make sense, and are true to themselves – that the mere *potentiality* of human beings for responsible and meaningful conduct is no longer sufficient to justify the preservation of the species. Instead, human nature must be upgraded and improved – even transcended, in some accounts – to adapt us to a world which we cannot recognise as our own. Emboldened by reductionist theories of genetic determinism, the goal of the biotech programme is, by turning fertility, abundance, health and vitality into socially defined, artlessly fabricated and always renewable commodities, to render 'unnatural' the biological rhythms, limits and needs of living organisms – whether crops, animals, or humans themselves. Masquerading as the triumph of freedom over necessity, this vision requires the wholesale adaptation of those organisms to mechanistic values and imperatives – to the mathematical principles of speed, exactitude, processing power, causation and universal exchangeability – which bare scare resemblance to the world as we experience it, to the truly human world.

The spectre raised by the biotechnology paradigm thus ultimately converges with the fetishised image of the cyborg, in which the mechanical enhancement of human powers and capacities is accomplished by neurological implants and silicon prostheses. With this vision of an improved human nature in mind, we should consider Hans Jonas's rejoinder to those who believe humanity can become the object of its own instrumental interventions in the world. Though there is no doubt that our social and cultural environments enrich or degrade our humanity, what is human in our being, Jonas insists, has always been there, in both our glory and our baseness:

> All such utopian dreams must be countered by the fact that 'man' has always been present with everything in him that should be avoided and all that cannot be surpassed. From this we become aware of what is worthwhile in man, *that* there is something worthwhile about him and that our existence is worthy of a future – of an always new chance to develop our *potentiality* for the Good. And all we can attempt to do is to assure that this *potentiality* continues to exist.[5]

Definitive of humans' unique potentiality is the capacity for responsibility and, in Jonas's view, having this capacity 'obligates us to perpetuate its presence in the world'. This effectively means preventing the conditions arising in which the capacity and desire for responsibility is seen as a nuisance, as the unwanted and dispensable legacy of an out-of-date morality which places unjustified restrictions on the formal calculation of costs and benefits, the steady expansion of the commodity economy, and the forward march of science. *Taking care* of ourselves and our world cannot be entrusted to genetic engineers, corporate accountants or high-tech machines, for there can be no biochemical production of, commercial substitute for, or cybernetic alternative to, human meaning. Demonstrating this fact is perhaps the biggest political and philosophical challenge of our time.

Topics!
1. Biological reductionism
2. Powers
3. Meaning: how we treat animals applies to how we treat ourselves

Introduction

This book presents a critical analysis of the ecological, social and ethical implications of the revolution in biotechnology. Its purpose is not to venerate the hallowed convention of academic impartiality, but to sharpen the curiosity and concern over this subject which I know I share with many people. The book covers several different fields of knowledge, ranging from molecular biology to political economy and ethics. It examines the impact of biotechnology in agriculture, animal experimentation, medical science and human reproduction. In doing so it brings together terminology and discourses from disciplines that do not always talk kindly to one another, though as an amateur in all of them I have done my best to mediate. It seeks to understand what the new biotechnologies will mean to the human species, in terms of our relationship to each other, our perception of the non-human environment, and our sense of our own selves.

The arguments of the following chapters revolve around three central themes. The first theme concerns the theoretical *reductionism* which underpins the efforts and ambitions of many genetic scientists. The idea that the functioning of organisms can be distilled to discrete and transferable units of information is the dominant fiction which underpins and legitimises the practice of genetic engineering. Chapter 1 addresses this issue by discussing the role that DNA plays in the functioning and reproduction of life forms. While this chapter offers a detailed introduction to molecular biology, as well as a description of the technical processes involved in genetic engineering, it also seeks to expose the fallacy that DNA is the self-contained and primordial origin or cause of organisms' traits and behaviours. It argues instead that we should recognise the intelligence of organisms as total systems, as living wholes with the limited but often remarkable capacity to adapt to challenging environments by modifying their internal processes and functioning – even at the genetic level.

If we simplify and misrepresent the true molecular functioning of organisms, then our efforts to modify that functioning by altering genetic sequences are likely to carry a large burden of uncertainty. This is the central subject of chapter 2, which concentrates on the ecological risks of genetically engineered crops. Although the biotech agribusiness lobby has promoted genetic engineering as an ecologically benign alternative to chemical-intensive farming, the evidence suggests that genetically modified crops represent a new and more worrying threat to the delicate equilibrium between humanity and nature. By disturbing established ecosystems and introducing into organisms characteristics they could never have acquired through evolution, genetic engineering poses a serious risk to sustainable agriculture and the need for a safe and reliable food system.

Although I return to the problem of genetic reductionism in later chapters, chapter 3 introduces a second theme, which is the issue of *power*. Here I draw attention to the concentration of economic power in the biotech industry in the hands of a small number of giant multinationals, and the corresponding influence which these companies have in shaping economic trends, government policies, public debate, and indeed scientific research on biotechnology. I look at the conflicts of interest which result from this dramatic accumulation of economic muscle, and review several case studies, many from the medical field, which demonstrate how science is being corrupted by its growing dependence on big business.

Chapter 4 extends this discussion by examining the legal changes, made in Europe and North America over the last twenty years, which have succoured the growth of the 'life science' companies. It focuses, in particular, on the extension of patent law to allow the private ownership of living organisms and their components. It also considers the implications of the new biotechnologies for people in the poorer parts of the world, and suggests – though undoubtedly there will be well-publicised exceptions – that genetic engineering has little potential to eliminate poverty, and that modern biotechnology is likely to consolidate rather than correct the global maldistribution of wealth.

Am I guilty here of replacing genetic determinism with technological determinism? I hope not. The problem, certainly, is that genetic engineering, to the extent that it works, represents an altogether new and more powerful means of removing the capacity for production and reproduction of living things from the environments and communities in which they are historically embedded. Just as the forcible separation of the peasantry from the land was the material precondition for the exploitation of productive labour, so the separation of genes from the life forms to which they belong is the technological means by which the productive powers of specific organisms and habitats are being privatised and redeployed in the commercial laboratories and territories of the wealthy. Whether or not the new biotechnologies, in a fairer society, could be a means of democratising access to the earth's resources may remain an open question. But let us at least recognise that, in the thoroughly unequal world in which we live, a powerful technology like genetic engineering is almost predestined to become a technology of power.

The theme of power, like the question of biological reductionism, is revisited in subsequent chapters of this book. In chapter 5, however, I begin to develop a third theme, which is the question of *meaning*. I am concerned here not so much with studying, as many sociologists have done, the unspoken assumptions and prejudices embedded in the supposedly objective discourse of genetic science, as with asking how the practice of genetic engineering may change our ethical understanding of ourselves and our ideals of the human good. This chapter focuses on the genetic engineering of animals, and reviews some of the ethical debates that have addressed this practice. Here I suggest that the treatment of sentient living beings as artefacts designed for the satisfaction of human wants is detrimental to humans' own vital sensibilities, and that the scientific project to redesign nature according to functional criteria will progressively erode the scope for and substance of humans' moral existence.

Chapter 6 moves away from the subject of plant and animal biotechnology, and looks at the possible impact of the mapping of the human genome on medical care and human health. Here I return to the problem of genetic reductionism, and show how this exerts a pervasive influence on research and debate about human illness. I challenge this reductionism by drawing attention to the enormous genetic diversity in human populations, and by emphasising how the health implications of specific genetic sequences cannot be properly assessed without studying the interaction of those sequences with different genetic and external human environments. I spend considerable time discussing a range of well-documented 'genetic disorders', demonstrating in the process that a consistent and unilateral relationship between genetic mutation and biological malfunction is virtually impossible to detect. Returning to the issue of power, I agree with the many critics of the new genetic science that the growing geneticisation of illness serves an ideological function, which is to distract attention from the social causes of human suffering and to justify the shaping of public health policies by the biotech-pharmaceutical industry.

Chapter 7 takes a closer look at the forms of clinical treatment being developed by medical geneticists, dwelling in particular on a form of human genetic engineering called 'gene therapy'. The chapter highlights some of the safety concerns which this experimental practice has raised, and asks whether the moral boundary generously accepted by its many proponents – between medical ('corrective') and non-medical ('enhancement') applications of genetic engineering – can realistically be sustained. I suggest that the relevant precedents indicate that it cannot, and that the economic incentives for biotech-pharmaceutical companies to medicalise every deviation from a manufactured model of health and normalcy will be the driving force behind its collapse. I also suggest that the modest success of human genetic modification, combined with the spectacular growth of forms of laboratory-assisted reproduction in recent years, is creating a technocratic attitude to the manipulation of life which endangers our moral relationship to future generations.

The 'slippery slope' from the medicalisation of life to its industrial manufacture is the subject of chapter 8. Here I argue that a second moral boundary – that which separates the genetic manipulation of body cells (generally approved), from genetic alterations which may be transmitted to non-consenting future offspring, thus modifying the human gene pool (generally deplored) – is also likely to be eroded by the expansion of the biotech industry. I also discuss the prospects for new forms of regenerative medicine based on the revolutionary use of stem cells, arguing that current research aimed at producing such cells from medically cloned embryos (so-called 'therapeutic cloning') will inevitably lead to the crossing of a third moral boundary, and to the birth of a cloned human being.

I should point out here that my argument that we are on a slippery slope to the genetic manipulation of human beings is not as controversial as it was when I first contemplated writing this book. Enthusiasts for human genetic engineering like Lee Silver and Gregory Stock now make exactly the same claim, though for them the inexorable expansion of biotechnology is proof that anxious hand-wringing over the pace of scientific advance is futile: 'if biological manipulation is indeed a slippery slope, then we are already sliding down that slope now and may as well enjoy the ride.'[1] Against this technocratic fanaticism, Francis Fukuyama bravely counsels that government regulation, if strengthened, could indeed function to protect common-sense moral distinctions such as that between therapeutic and enhancement applications of biotechnology.[2] I admire his pragmatism, but without a serious curtailing of the marketing power of the biotech industry – which as Fukuyama notes has doubled in size in the US since 1993 – the prospects remain bleak in my analysis.

Chapter 9 returns to the issue of power, addressing the social and political implications of genetic testing. Examining the likelihood of discrimination by insurance companies, and the way screening may accelerate the geneticisation of workplace illness as well as the medicalisation of childhood, I suggest this practice is likely to consolidate the injustices faced by people who are vulnerable, socially disadvantaged, or simply different. I also note that the growing commodification of human bodies, organs, tissues, cells and genes is extending the socially divisive impact of the market as well as demeaning the physical integrity of human selfhood.

The commodification of life is an issue further explored in chapter 10, which tackles the most controversial but popular development in human biotechnology: the industrialisation of human conception. I spend the first half of this chapter discussing the rights and wrongs of paid surrogacy, the legalisation of which seems to me to mark an unprecedented historical step towards the cultural acceptance of human life as a manufactured commodity – a step which is lengthened, with cold and rarefied logic, by the invention of the artificial womb. The rest of the chapter addresses the threat of eugenics, and the use of cloning and genetic engineering to produce more desirable offspring. *Power* re-emerges as the central theme of this discussion – particularly the absolute power enjoyed by the programming adult over its genetically designed child. Genetic determinist I am not, however, for my argument is that this power is really exercised not through the irresistible commands of a genetic programme, but rather through the lived relationship between the designing parents and the pre-designed child – a relationship in which the former will be naturally inclined to consolidate their genetic investment by strengthening their power over the child's personal development and choices.

I cannot help but think here of Sharon Duchesneau and Candace McCullough, two American lesbian women with congenital deafness who in

March 2002 revealed to the world's press that Duchesneau had recently given birth to the couple's second deaf child by deliberately using sperm from a deaf friend. Though this was effectively a DIY pregnancy, it re-ignited public debate about the right of adults to use the new reproductive technologies to predetermine the genetic capacities of their offspring. While the original *Washington Post* article[3] offered a lengthy and sensitive insight into why deaf people view the inability to hear, which has given rise to an alternative language and culture, as 'different not deficient', many liberal commentators rose to condemn the deliberate creation of children with four rather than the full five senses. Who indeed had anticipated that the passion for 'designer children' would result in the limitation rather than enhancement of normal human faculties?

Yet the difference between genetic limitation and genetic enhancement may not be so significant, and may in fact distract us from the deeper ethical problem. In a world of diverging norms and cultural fundamentalisms, eugenic choices may derive from the values of besieged communities as well as from the dominant technocratic culture of capitalism. The most important issue that arises from this 'communitarian eugenics' may not be whether adults have the right to reduce the sensory capacities of their offspring, but whether by doing this they are seeking to 'naturalise' – and thereby preempt resistance to – the expectation that their offspring commit themselves to the reproduction of the parents' *culture*. The conviction shared by Duchesneau and McCullough that deafness is more a culture than a disability, suggests that, regardless of the biological limitations imposed on these children, their parents may be unwilling to allow them to claim or fight for membership of the non-deaf community, to develop exclusive attachments with hearing people, and to assert a cultural identity which does not rest on their disability (and its creative compensations). Today's obsession with cultural separatism and 'difference' may thus find in the new eugenic technologies a means of transfiguring culture into biology, the result of which may be a radical subversion of what Habermas calls the 'ethical self-understanding of the species', and with it the invalidation of the modern humanist principle of *universalism*.

That said, technocratic rationality certainly stands in the ascendancy, and is likely to remain so as long as eugenic decisions are mediated and facilitated by organised science. This is the theme of the final chapter, in which I conclude by arguing that it is the image of the cyborg which above all articulates the logic and goals of the biotechnology revolution. According to robotics expert Hans Moravec, for example, human genetic engineering will logically progress to more powerful forms of bioengineering, in which miniaturised computer technologies replace, accelerate and upgrade the carbon-based molecules of the human body. Proposing instead that it is meaningless to divide humans' rational insight and power from their biological embodiment, I suggest that there is a deadly contradiction in the argument that the human species can and should become the direct object of its own transformative action, because it presupposes a human consciousness separate and abstracted from the nature – always imperfect and ambiguous – which is its own presence in the world.

If the source of this contradiction – that we are in fact human in our frailty and suffering, and natural in our consciousness, reasoning and doubt – cannot be surmounted in theory, then the cyborg promises to transcend it in practice. For what the cyborg *is* in the enchanted eyes of its enthusiasts is precisely what we are *becoming* due to the cultural impact of biotech science. When the advocates of human self-modification challenge their critics with the observation that 'we are already cyborgs anyway', their claim has greater significance than they might think. Long before we began to repair our malfunctioning bodies with prostheses and implants, we were already *thinking like machines*, measuring the costs and benefits of our existence, formalising the qualities of our experience and meticulously calculating the definition of the good. Transforming humans into cyborgs may not therefore be the consummate act of invention, in which the post-human being is finally brought into existence by its *homo sapiens* predecessors. For the cyborg is indeed already with us – is in fact *within* us – in the form of our growing conviction that human lives cannot, with any measurable certainty, be rendered more *meaningful*, but can be corrected and improved by modifying their internal functioning.

So this, it seems to me, is the ultimate goal of the biotech revolution: to mechanise human experience so thoroughly that it actually *makes no sense* to think of ourselves as philosophers and questioners – and indeed to think *as* philosophers and questioners, as anguished searchers, sufferers and producers of meaning. Gregory Stock, who believes, incidentally, that genetic engineering will triumph over the limitations of artificial intelligence and robotics, is kind enough to concede that this is in fact the true hope of the biotech revolutionaries: 'The more we succeed in modifying our biology and that of other animals, the more we will see it as something malleable that we can adjust and improve, and the more we will come to assess germline therapies *on the basis of risk and reward rather than philosophical meaning*.'[4]

1

The Revolution in Molecular Biology

Genetics is the study of biological production and reproduction, of how the physical characteristics of organisms are created and multiplied, produced and reproduced. The genetic theory of reproduction – or 'heredity' – that prevails today, dates back to the 1850s, when a Moravian abbot, Gregor Mendel, began breeding experiments with varieties of garden pea. Mendel found that the most distinct characteristics of different pea plants – such as the length of the stem or the form and colour of the seed – did not result, when the plants were mated, in offspring which blended those characteristics to form intermediary features (such as a stem of intermediate length, or a seed that was half-way between wrinkled and smooth). Instead the offspring exhibited, for each particular trait considered, the visible characteristic of only one parent. When identical offspring were mated, however, the trait of the forgotten parent frequently reappeared – with startling mathematical regularity – in the countenance of the next generation. By concentrating on the transmission of distinct and relatively rigid traits (and by ignoring the more ambiguous results of experiments with other plants), Mendel had discovered what came to be thought of as the atomic foundation of biology; namely, that traits are transmitted and expressed through self-contained units of inheritance.

Mendel's research was neglected during its author's lifetime, but his writings were revisited at the turn of the century when his theory of the segregation of parental traits converged with the discovery of paired chromosomes. It was another fifty years, however, before the search for the actual material of these units of biological inheritance came to sudden fruition. Francis Crick, who had drifted from an aborted career in physics to studying crystallography at Cambridge University, had started to work informally on the postdoctoral research project of a young American molecular biologist, James Watson. Both were interested in solving the riddle, generated thanks to the work of Oswald Avery and his colleagues in the early 1940s, of how the four chemical components of DNA – not, as previously assumed, the many thousands of different proteins in an organism – could serve as the medium for the inheritance of the staggering variety of traits observed in a living organism.

In 1953, after only a few months of collaboration, Crick and Watson advanced a theory that would eventually spark the biotechnology revolution. The structure of DNA, they argued, was far more complex than had previously been thought, for its molecular configuration represented the functioning of a kind of language or 'code'. This code, they confidently proposed, was the same for virtually all living organisms. Whether they are bacteria, insects, fungi, plants, animals or *homo sapiens*, they all speak with the same chemical language. Understanding this 'Esperanto of biology', took another decade, but by that time the foundations of a new science and technology had been firmly laid.

The Biology of the Cell

The smallest living unit of an organism is a cell. These are microscopic entities, and there may be thousands of billions of them in a single organism. With the exception of primitive organisms like bacteria, cells are comprised of three elemental structures: an outside wall or membrane; an inner cell fluid, known as the cytoplasm, where proteins are produced; and within that fluid a nucleus surrounded by its own porous membrane.

Cells have different shapes, sizes and behaviours which reflect their different functions. If we include the scores of different neurons that have been identified in the human brain, an adult human is made up of around 260 known cell types. The common weed, *Arabidopsis thaliana*, on the other hand, has fifty. By dividing and multiplying, cells may stack up to form tissue (such as skin, flesh or fruit), organs (such as kidneys or roots) or structures (such as bones or plant stems). Certain cells never divide – human nerve cells, for example, which are responsible solely for communication, last a lifetime without replacement. Other cells – such as human red blood cells, which also never divide – have particularly short life-spans, and have to be continually replaced by new ones (in this case produced by the precursor cells of the bone marrow). Cells may play a defensive role (such as white blood cells which produce antibodies, or the cells which enable nettles or jelly fish to sting), while others may be responsible for producing enzymes (to facilitate digestion, for example), fibrous proteins (collagen to build hair, elastin for ligaments), transport vehicles (such as oxygen-carrying haemoglobin), hormones (such as insulin) for metabolism, or storage space (for fat, for example).

The exact function and behaviour of particular cells is ultimately determined by the types of protein the organism requires them to produce, carry or excrete. Indeed, every metabolic process – whether this is growth, photosynthesis, locomotion or thinking – involves the formation and catalytic interaction of proteins. How particular cells are able to yield the precise proteins enabling them to perform their specific functions was one of the puzzles facing molecular biologists in the 1950s, and their answer to this puzzle is crucial to understanding the influence of today's dominant genetic paradigm.

DNA, Mitosis and Meiosis

Enclosed within the nucleus of every cell is DNA, or deoxyribonucleic acid. When a cell is not dividing, DNA exists in the form of a diffuse and tangled network of filaments called 'chromonemata'. With the onset of cell division, these filaments coil up and condense into the separate molecular structures known as chromosomes, a Greek word meaning 'coloured bodies', chosen because when stained with a dye they become visible through a light microscope. When magnified, chromosomes appear to be short rods or thick pieces of thread of uneven width. They are easiest to see when they have condensed and started to replicate, and when the cell they inhabit is about to split into daughter cells each with a full set of chromosomes. For this reason chromosomes tend to be represented in pictures and photographs as various X shapes, where two copies of the chromosome are still attached at a central point (called a centromere).

The number and length of chromosomes in a cell varies according to the particular species. In sexually reproducing organisms the chromosomes exist in homologous (i.e. virtually identical) pairs, one copy being inherited from the male parent and one from the female parent. Maize plants, for example, have ten pairs of chromosomes, tomatoes have twelve, potatoes have twenty-four, and fruit flies have four. In humans there are twenty-two matched pairs of 'autosomes' (numbered one to twenty-two), and – as in all mammals – one pair of 'sex chromosomes', which as the term implies determines whether the person is male or female. The chromosomes that comprise this latter pair exist in two different forms, which are denoted 'X' and 'Y'. A person who inherits one copy of each sex chromosome is male, while the cells of a female carry two copies of the X chromosome. Since women do not have a Y chromosome, sexual reproduction cannot produce offspring with a pair of Y chromosomes.

Examined more closely, a chromosome proves to be two intertwined strands of DNA, tightly coiled like a spring around a central scaffold made of a special class of proteins known as histones. These two strands are composed of chemical subunits called nucleotides. The nucleotides are bound together in a linear chain by consecutive units of alternating sugar (deoxyribose) and phosphate. From the inner edge of this sugar-phosphate backbone protrude 'bases', which take the form of one of four nitrogenous chemicals. The bases on the two intertwined strands complement each other in such a way that, with the help of hydrogen bonds, they each loosely attach to their opposing base, forming what is sometimes thought of as the rungs of a twisted ladder, or the steps of a spiral staircase (see Figure 1). The four chemical forms of the bases are adenine, cytosine, guanine and thymine. These are represented by the letters A, C, G and T. The pairing of opposite bases to form the rungs that join the two complementary threads of DNA follows a strict pattern: A pairs with T, and G with C.

When a cell divides, as it must do if growth is to occur and dying tissue is to be replaced, each chromosome in the nucleus of the cell must be replicated. This is made possible because the two intertwined strands of DNA can be uncoupled. Each 'double helix' is unzipped to create separate threads, and new matching strands are assembled adjacent to them (Figure 1). The sequence of bases on each strand thus functions as a template, a mould whose cast is formed through the universal base-pairing principle, with each cast in turn separating and functioning as a new template. This was why the discovery of the structure of DNA by Crick and Watson was so important: it explained how hereditary information could be copied from cell to cell, and how errors ('mutations') in the replication of DNA – omitted or inserted letters, duplicated or inverted sequences – could occur.

The exception to this regular pattern of cell division and replication (called 'mitosis'), is the form of division by 'meiosis', which is a prerequisite for sexual reproduction. The non-reproductive or 'somatic' cells of a sexually reproducing organism are described as 'diploid', because they carry two complete sets of chromosomes, one set inherited from each parent. If the germ cells (egg and sperm cells) were also diploid, the offspring produced by the unification of these reproductive cells would have twice as many chromosomes as its parents, with the number of chromosomes doubling in each subsequent generation. Meiosis is thus the process by which, to take humans as an example, specialised cells in the testis and ovaries divide to produce germ cells (or 'gametes') which are 'haploid' – possessing a *single* set of chromosomes. For each reproductive cell this involves, as a precursor to the final formation of the gametes, the more or less random exchange and recombination of common sections of the paired paternal and maternal chromosomes. Following this shuffling of inherited genetic material between homologous chromosomes,

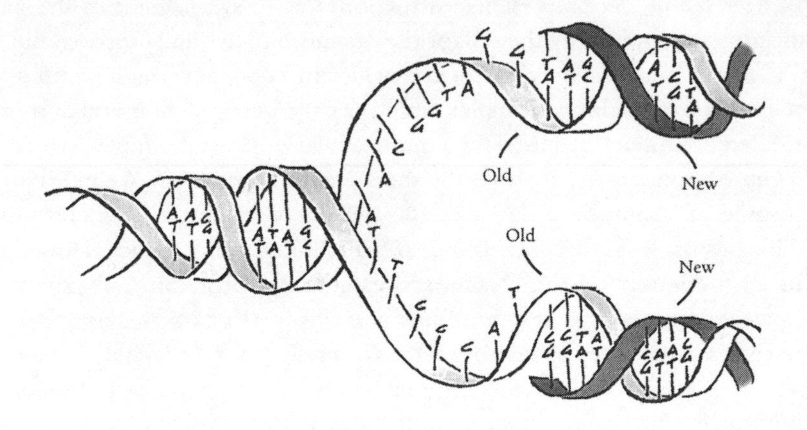

Figure 1 DNA replicating itself. A simplified representation of semiconservative replication of DNA, in which each strand of the original molecule acts as a template for the synthesis of a new complementary DNA molecule, following the rules of complementary base pairing: adenine (A) to thymine (T), and guanine (G) to cytosine (C). Two strands of DNA are thus obtained from one, identical to one another and to the parent molecule. (By Nick Thorkelson. Reprinted by permission of the publisher from *The Century of the Gene* by Evelyn Fox Keller, p. 24, Cambridge, Mass.: Harvard University Press, Copyright © 2000 by the President and Fellows of Harvard College.)

the latter then separate and migrate to form the nuclei of two haploid germ cells – each of which is now genetically unique. The fertilisation of the egg then brings the two sets of chromosomes back together again, as the resulting embryo inherits half its chromosomes from its mother and half from its father. The differences in these two sets of the human genome are likely to be tiny – around 0.1 per cent – and it is the interaction of these differences which in humans is thought to determine such things as skin pigment, height, eye colour, and other features of the person's visible characteristics or 'phenotype'.

The Synthesis of Proteins

Until now I have been describing how DNA is *replicated* when cells divide and multiply, and when organisms reproduce. But what of the actual function of DNA in the cell itself? Crick and Watson suggested that the chemical bases on the DNA strands which make up the chromosomes are also arranged according to the rules of a code, and that the particular arrangement of bases on any given stretch of DNA represents instructions for the synthesising of the specific proteins required by the cell for the organism to live and function. But if A, C, G and T are the only letters in this genetic code, how can it be used to spell out the multiple and complex directions, the hundred thousand or more of different proteins, required for a multicellular organism to function?

One obvious answer to this is the sheer length of the DNA. A single chromosome can be made up of over 100 million bases (letters), whereas the whole human genome contains over three billion letters of DNA. However, not all the letters on the chromosomes code for proteins. Only specific sequences do. It is these specific sequences, which are of course particular to the chromosome they appear on, which are referred to as 'genes'. This is a term which was actually coined in 1909 to describe the hypothetical units of inheritance which had yet to be identified (since the bulk of the chromosome is protein, genes were also originally assumed to be proteins).

Before the first rough draft of the human genome was announced in February 2001, it was generally believed that humans carried around 100,000 genes in their cells, and some estimates put the figure as high as 150,000. These protein coding sequences were known to account for at best only five per cent of total DNA, however, with the remaining ninety-five per cent or more described as jumbled and repetitive strips of seemingly unintelligible letters whose existence is still something of a mystery. Many researchers believe this excess DNA is meaningless junk (hence the term 'junk DNA' or, alternatively, 'selfish DNA', because its only function seems to be to replicate itself), a redundant, fossilised legacy of our distant evolutionary ancestors. Others have suggested that it may serve a 'management' function in organising the specific placement of genes along the chromosomes, that it is a protective form of stuffing or padding, or that it functions to plump up the size of the nucleus so that it can produce more proteins.

Each gene is thus thought of as a coded 'recipe' for the production of a particular protein. The code is written across a linear stretch – often thousands of bases long – of three letter words, each one formed out of the four letter alphabet of A, C, G and T. Almost every three letter word, or 'codon', specifies the production of one of twenty amino acids (there are sixty-four possible triplets of letters, which means that a number of different codons signify the same amino acid, while one codon designates the 'start' of a sequence to be translated, and three other codons all signify the end of the 'reading frame'). Amino acids are the subunits that make up proteins. A single cell may contain thousands of different proteins, and many more are likely to be present in the organism as a whole. The shape and function of a protein depends on the particular amino acids which constitute it, the specific order in which they are strung together, and the length of the total chain. A typical protein is made up of a sequence of several hundred amino acid units, each one corresponding to a triplet of chemical bases. Although the amino acids are formed, like the sequences of DNA from which they are translated, in a linear chain, the chain coils and pleats itself to give each protein a unique and stable shape.

Transcription, Translation, and Protein Folding

The translation of DNA instructions into proteins is a little more complicated than this, however, because it is a process which requires several intermediaries. To begin with, the letter sequences of a gene must be *transcribed* to form a copy, using the same base-pairing principle that operates in the division and replication of cells. The parent strands of DNA are unwound by a class of enzymes called DNA helicases, while other proteins and enzymes function to stabilise the unwound stretch, preventing it from binding back together again and helping to reduce tension in the strands. Since the two strands are complimentary to each other, only one strand – called the 'sense' strand – contains the 'message' for protein synthesis. This means, ironically, that it is the other strand – the 'anti-sense' strand – which serves as the template for the copy of the sequence, since this copy is created by the now-familiar base-pairing principle, but must resemble the 'sense' strand in its chemical structure.

When DNA is replicated in the process of ordinary cell division, the task of reading and copying the uncoupled DNA strands is performed by an enzyme called DNA polymerase. When DNA is copied (transcribed) as a precursor to protein synthesis, on the other hand, this role is played by RNA polymerase, and its product is not complementary DNA, but rather a slightly different chemical called RNA, or ribonucleic acid. RNA is found in the cytoplasm of cells, and has the same structure as DNA with two exceptions: it exists in single rather than double strands, and it uses the nitrogenous base uracil (U) in place of thymine (T).

A complimentary RNA copy of the anti-sense strand of the genetic sequence is thus made (with U replacing T), and this is then displaced when the two strands of the double helix bind back together again. This 'messenger' RNA (mRNA) then leaves the nucleus and enters the cytoplasm. There it is met by ribosomes, complex molecular organs which select amino acids and knit them together in the order indicated by the mRNA. This final stage of protein synthesis is called *translation* (see Figure 2).

Understanding the finer details of how DNA replicates itself and produces proteins is not so important to the rest of this book. What is important, however, is that we are aware of the numerous agents – proteins, enzymes, catalysts, and so on – which are involved in these processes, since they demonstrate, contrary to popular interpretation, that the 'gene' is not an independent cause or origin of biological traits. (Even the celebrated pairing of nitrogenous bases to form DNA's double helix is, for example, the continually accomplished *product* of large aggregates of enzymes – indeed, all DNA is effectively a product of the copying mechanisms of cells.) It should also be noted at this stage that not all genes are recipes for proteins. Some genes must code for a form of RNA that is the key ingredient of ribosomes (ribosomal RNA), while others must yield a type of RNA (called transfer RNA) which brings amino acids to the ribosomal complexes and helps assemble the proteins. Other vital DNA sequences do not code for any amino acids, but instead function as 'docking' locations to which catalytic enzymes attach themselves in order to initiate or suppress transcription.

To further complicate matters, not every protein is the product of a single gene, for some proteins are assembled from several different sequences, and some sequences can contribute to the formation of different proteins. This is

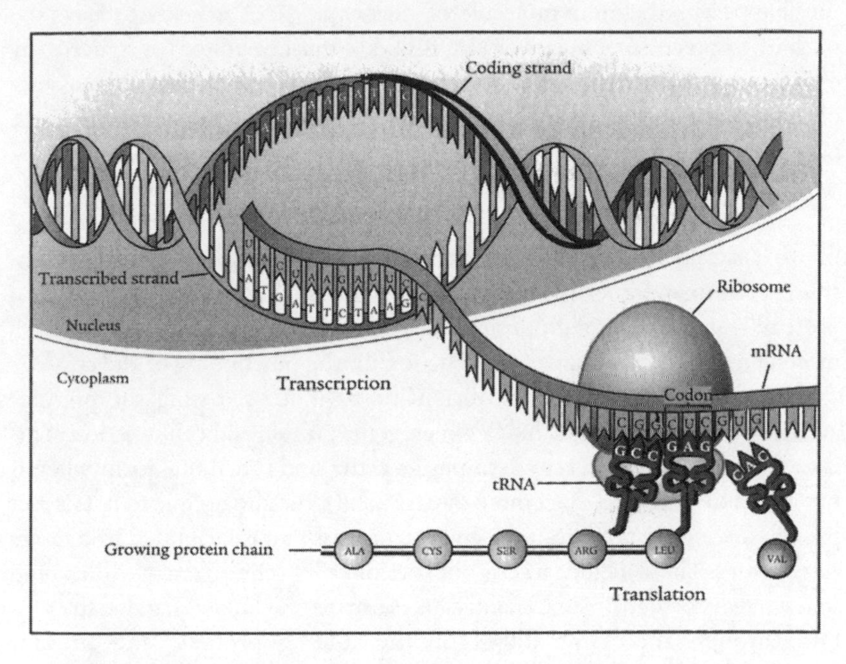

Figure 2 The Central Dogma: transcription and translation. To make a protein molecule, the DNA double helix separates at the site of a gene, and transcribing enzymes (not shown) copy the bottom strand of nucleotides into a complementary mRNA strand. Where there is a G in the DNA, a C appears in mRNA; where there is a C in DNA, a G appears in mRNA; where there is a T, an A appears. However, an A in DNA appears as a U instead of a T in mRNA. Consequently, the upper or coding strand of the DNA has the same sequence as the mRNA except that T is present in DNA and U in mRNA. After the mRNA is transported from the nucleus, it joins ribosomes in the cytoplasm, where it is translated. Each codon (or triplet of bases) in the mRNA is complementary to a specific transfer RNA (tRNA), and each tRNA carries a specific amino acid to add to the growing protein chain. In this example, the amino acids arginine, leucine, and valine are being added to the chain, in the order dictated by the codons in the mRNA. When the chain is completed, it will fall off the mRNA-ribosome complex and become a functioning protein molecule. (*By Kathy Stern.* Reprinted by permission of the publisher from *Genes, Blood, and Courage: A Boy Called Immortal Sword* by David G. Nathan, p. 157, Cambridge, Mass.: The Belknap Press of Harvard University Press, Copyright © 1995 by David G. Nathan.)

possible because before a molecule of messenger RNA is ready to leave the nucleus, its precursor – nuclear RNA (nRNA) – must be edited (or 'spliced') by enzymes which delete non-coding segments of it (called 'introns'), and reassemble the remaining meaningful segments ('exons') in a shorter strip. Since non-coding or 'junk' DNA is scattered within as well as between specific genes, only a fraction of primary nRNA transcripts survives this editing procedure and progresses to the cytoplasm. The routine editing of nRNA and the prevalence of what is called 'alternative splicing' – it is now thought that almost two-thirds of human genes may be used to synthesise more than one type of protein[1] – shows what a gross simplification it is to depict the gene as a primordial recipe or commanding instruction for the functioning of the cell.[2]

There is, moreover, another scientific basis for rejecting the popular account of a linear relationship between gene, protein and cell function. This is the fact that a protein only assumes its active and functional form when it has folded to create the exact three-dimensional structure unique to it. Correct protein folding is a temperature-sensitive process that is facilitated by a range of enzymes. These include a class of proteins called 'chaperonins' which bind to the newly formed protein chain, catalyse structural linkages, guide the protein through successive stages of the folding process, and prevent partially-folded proteins from attaching to each other in sticky aggregates. The carbohydrate components of glycoproteins – proteins which are modified, post-translation, by the attachment of sugar molecules, and which form important constituents of cell membranes as well as lubricant body fluids – are also known to play crucial roles in protein folding.

Early attempts by Ely Lilly and other biotech-pharmaceutical companies to synthesise medically valuable proteins, such as insulin and growth hormone, by inserting human genes into bacteria, foundered precisely because the amino-acid chains, though synthesised in full accordance with their DNA templates, regularly failed to fold into functional structures in the new cellular environment. Instead the properly expressed proteins formed sticky, insoluble deposits ('inclusion bodies') which accumulated in the cytoplasm of the bacterial cells.[3] To address this problem scientists had to develop counter-strategies, modifying the solution in which the bacteria were cultured, reducing the rate at which the foreign genes were expressed in bacteria (when over-produced, the likelihood increases that partially-folded proteins will stick to each other and form aggregates before they have achieved their final stable structure), and even manipulating the host cells to over-produce specific chaperonins.[4]

Research into protein folding has also accelerated with the discovery that the accumulation of misfolded protein deposits is common to several degenerative diseases, including Alzheimer's and Huntington's, as well as cystic fibrosis, a type of inherited emphysema, and probably many cancers. Prions, which are the infectious agents that cause transmissible spongiform encephalopathies such as scrapie, and 'Mad Cow' and Kreutzfeldt-Jacob diseases, are also misfolded proteins. The discovery of prions is important for another reason, since the existence of prions clearly invalidates the theory that the instructions for the functioning of all living things is encoded in nucleic acid. Prions are pure protein, containing neither DNA nor RNA. Yet they can still infect hosts and 'replicate' themselves, essentially by inducing identical amino-acid chains to copy their own misfolded shape, thus allowing them to aggregate and form insoluble plaques (called 'amyloid fibrils'), normally in the brain.[5]

I shall return to the limitations of the DNA-centric model of organisms later in this chapter, after looking more thoroughly at the science of genetic engineering. Before this, there is one obvious question regarding the functioning of genes that needs to be answered. Given that virtually every cell in a single organism has the same DNA content in its nucleus, how do different cells produce different proteins in order to perform different functions? The answer is that at any one time only a select number of genes in a given cell are being used to make the proteins for which they are the template. Whereas these genes are said to be in an 'on' or 'expressed' state, the rest of the genes in the cell are silent, or switched off. The expression pattern of a cell's DNA, in other words, varies according to the developmental stage and location of the cell in the organism as a whole. Human genes that code for insulin or saliva, for example, will not be switched on in bone cells or heart tissue. Leaf cells on a plant will not produce pollen or nectar, and youthful shoots will not yield fruit. Control over and co-ordination of gene expression in complex organisms involves the gatekeeping role played by a large class of DNA sequences called 'promoters'. Together with other regulatory regions known as 'operators', 'enhancers' and 'silencers', these control sequences respond to their tissue-specific environment by signalling to another class of proteins – called 'transcription factors' – when to bind to the DNA in order to guide and activate (or repress) transcription. Despite the fact that they themselves do not 'code for' any amino acids, promoters are absolutely vital to tissue-specific gene expression. They are consequently also the key to manipulating gene expression. A gene which is normally only expressed in the pancreas, for example, does so thanks to the pancreas-sensitive control sequences adjacent to it. A geneticist wishing to make the pancreas produce a different protein would then have to join together the gene for the new protein with the pancreas-sensitive promoter region.

Genetic Engineering: A New Industry is Born

Genetic engineering, or 'recombinant DNA technology', is the term used to describe the practice of manipulating the expression of proteins in an organism by changing the sequence of DNA in its cells. It is a relatively young technology, whose birth dates back to 1973 and the work of two American biochemists, Stanley Cohen and Herbert Boyer, who were the first to successfully manipulate the genome of a bacterium (and who profited handsomely from the patenting of their techniques in the 1980s). The academic research of Cohen and Boyer established the foundations of the modern biotechnology industry, which, in a series of joint ventures and bold government programmes, spread during the 1980s and '90s from the US not just to Western Europe and Japan, but also to developing countries like India, China, Thailand and Indonesia, and to Latin-American and Caribbean countries such as Mexico, Argentina and Cuba.[6]

The first successes of the fledgling industry centred on the manipulation of bacteria for medical purposes. By the middle of the 1980s, bacteria had been engineered to synthesise a number of pharmaceutically important proteins, including human insulin, human growth hormone, hepatitis B vaccine, human blood clotting Factor VIII, and several types of human interferon. Meanwhile in the agricultural domain, genetic engineering, by adding extra genes, silencing existing genes, or altering the DNA pattern of coding and promoting sequences, brought with it the possibility of deliberately changing the type, quantity and timing of a plant's protein production in order to create traits which could not be satisfactorily generated by traditional methods of selective breeding.

By 2000, according to the annual report of the International Service for the Acquisition of Agri-biotech Applications (ISAAA), 44.2 million hectares of genetically modified crops had been sown throughout the world, covering a total area around twice the size of the UK, and making up sixteen per cent of all crops grown globally. Nearly seventy per cent of these transgenic crops lay in the US, with Argentina (23 per cent), Canada (7 per cent), and China (1 per cent) accounting for most of what remains.[7] After India decided in March

2002 to end its moratorium on the commercial planting of genetically modified cotton – a crop which covers more acreage in India than in any other country – the global spread of transgenic plants was predicted to accelerate.[8]

In terms of commercial revenue and land use, by far the most successful application of genetic engineering in agriculture to date is the creation of crops which are poisonous to pests or tolerant of herbicides – products the implications of which we shall look at in greater depth in the next two chapters.[9] Although research into herbicide-resistant crops was being funded by all the major chemical companies by the mid-1980s, it was not until the second half of the 1990s that the results of this research began to reach the market. Before this point, agricultural biotechnology had taken an important step forward in the West when genetic engineering was used to slow down the softening of fruit and vegetables, thus allowing these crops to be mechanically picked and transported with reduced risk of damage and a longer shelf-life. This technique was developed by the California-based company Calgene Inc., and it resulted in the Flavr Savr tomato, which in 1994 became the first genetically modified food to reach a Western market (the second was a virus-resistant squash).[10] In 1997 Calgene was bought up by Monsanto who, in the hope of reversing the tomato's poor reception amongst suspicious American consumers, quickly changed the fruit's brand name to the more traditional-sounding 'McGregor'.[11]

The same technique was used by Zeneca Plant Science (a subsidiary of UK-based Zeneca Group PLC, which subsequently merged with Astra AB of Sweden to form AstraZeneca), who chose not to compete with Calgene in the US fresh tomato market, but aimed their product instead at the European market for processed tomato. This project – premised on the discovery that tomatoes low in polygalacturonase make thicker and more viscous sauces and pastes – turned out to be a more astute commercial strategy, not least because, by modifying a product already sold as a processed commodity, the manufacturers minimised consumers' concerns about 'interfering with nature'. Originally developed in the more lax regulatory environment of the US, Zeneca's tomato purée reached supermarket shelves in Britain in 1996, where it was sold by both Safeway and Sainsbury under their own respective labels. Despite combined sales of over two millions items, the product was removed from the shelves in 1998 due to consumer pressure.[12] AstraZeneca is now considering using this same technique to slow down the production of the ripening gas ethylene in fruit such as bananas, strawberries, melons and peaches.[13]

Since virtually all living things share the same genetic 'code', proponents of genetic engineering also believe they can effectively transfer advantageous physical characteristics from one species to another. In the race to produce a salmon that can be farmed in cold northern waters, for example, a Californian company called DNA Plant Technology has transplanted into the genome of salmon a gene which, by producing a protein which inhibits the growth of ice crystals in the blood and tissue cells, naturally enables arctic flounder and other polar fish to survive in freezing temperatures. The same 'antifreeze gene' is also being used in experiments to produce frost-resistant plants, fruit and vegetables. Scientists have also engineered tobacco plants which contain the DNA of fireflies, showing how the genes that code for self-illumination could be used to 'mark' transgenic organisms and keep track of them. The same principle is being used in trials for the world's first release of a genetically engineered insect – moths that carry a gene which, when activated in the next generation of cotton-feeding bollworm larvae, will kill the larvae before they damage the crops. In the US trial the gene will be replaced by a dummy gene from a jellyfish which will cause the caterpillars to glow under a special light, thus allowing scientists to predict the likely dispersion and transfer of the deadly gene.[14] Scientists are also working on the possibility of breeding out mosquitoes' capacity to carry the malaria-causing parasite by introducing harmless genetically engineered mosquitoes with enhanced reproductive fitness.[15]

Amongst other more recent accomplishments is the production of silk through the insertion of spider silk genes into mammalian cell cultures by the Canadian company Nexia Biotechnologies, and the genetic engineering of rubber trees to yield the human blood protein albumin in their sap. In the first case, separate cultures of bovine mammary cells and hamster kidney cells were persuaded to secrete soluble silk proteins outside of the cells, where they were collected and spun to form nano-fibres with enormous tensile strength and elasticity. Patented under the trademark 'BioSteel', the development of this product, which Nexia hopes will eventually be harvested from the milk of genetically engineered goats, is being closely monitored by the military, medical and aerospace industries.[16] In the second experiment, Malaysian scientists, by transferring a human gene to rubber trees and other plants, successfully persuaded them to produce in their sap a seventy per cent concentration of human albumin, the main protein in blood plasma and a vital resource for medical transfusions.[17]

The Role of Promoters

One technical problem with transferring genes from one organism to another, however, concerns the controlling function of the promoters. Adding a new genetic sequence to an organism will not normally be enough to switch the foreign gene on. This requires suitable control sequences recognisable to the host cell, and which indicate to the cell where the new gene to be transcribed is, as well as when – and ideally how much – transcription is to occur. In agricultural genetic engineering, this control sequence normally derives from viruses which naturally contain promoters powerful enough to subvert the plant cells' metabolic machinery and command it to read the viral genes. The Flavr Savr tomato, for example, contained a widely-used promoter from the cauliflower mosaic virus.

In animals a more sophisticated means of activating foreign genes is usually necessary. Some biotech firms, for instance, have sought to genetically manipulate livestock so that they generate in their milk medically important proteins, such as alpha-1-antitrypsin (AAT). AAT is a human serum glycoprotein which is misfolded and consequently deficient in many emphysema sufferers, and extra quantities of AAT are also thought to alleviate the respiratory problems of people with cystic fibrosis. In one of the earliest attempts to realise this goal, scientists at the Roslin Institute in Scotland removed from the genome of a sheep cell the promoter for the milk protein ß-lactoglobulin, then attached it to a strip of human DNA coding for AAT. The new DNA construct was then injected into the nuclei of artificially inseminated sheep eggs (single cell embryos), which were then brought to term in the wombs of surrogate mothers.[18] Equipped with the human AAT gene and ovine milk promoter, the transgenic animals, when mature, were able to produce high levels of AAT in their mammary glands (both sheep and humans normally produce their own version of this protein in the liver, and current medical demand is met by purification from human plasma). The sheep's milk must then be treated to isolate the human protein in a process which the Roslin Institute's commercial partner, PPL Therapeutics, is currently developing on a large scale. In addition to its Scottish flock, the company is rearing hundreds of transgenic sheep in New Zealand, and has signed a deal with the German pharmaceuticals group Bayer worth £27 million in the development, trials and marketing of a drug which was originally expected to reach the market in 2004. The same company has also agreed terms with a New Zealand research institute, Celentis, in a project centred on the genetic manipulation and cloning of cows designed to secrete pharmaceutical products in their milk.[19]

Injecting DNA into single-cell embryos remains, however, an inefficient and unpredictable method for creating transgenic animals, and one which does not guarantee high levels of transgene expression nor germ-line transmission to the progeny of the transgenic sheep. Because of the high costs and low success rates of this procedure, PPL Therapeutics has also played a key role in financing the development of nuclear transfer techniques – or cloning – at the Roslin Institute, in experiments which led to the famous creation of Dolly the sheep.[20] Although Ian Wilmut and his colleagues at the Institute had already cloned two sheep, Megan and Morag, from embryo cells which had been starved to a quiescent state,[21] the arrival of Dolly in July 1996 was interpreted as a major breakthrough because the DNA cloned in this case came from a cell line derived from the frozen mammary tissue of a deceased *adult* ewe.[22] The same technique was subsequently employed to produce cloned sheep from foetal skin cells (fibroblasts) which had been genetically manipulated *in vitro* to contain the AAT gene and ß-lactoglobulin promoter.[23] The transgenic cells were fused with sheep eggs from which the genetic material had been removed, and then allowed to develop into embryos, and subsequently to a number of live-born lambs (three of which survived beyond six months, and one of which, when treated with hormones, lactated with AAT in its milk). It remains to be seen whether the cloning of genetically altered cells in this way will become a more viable commercial alternative to the established practice of genetically engineering animals by injecting foreign DNA directly into the nuclei of single-cell embryos.

The Role of Vectors

Effectively changing the genetic material of organisms is a process which has several stages. First the promoter and the foreign gene ('transgene') must be spliced from their organisms and then combined. This is accomplished using a family of enzymes which are produced by bacteria as a form of defence against invading DNA. Called 'restriction endonucleases', these enzymes each recognise a specific sequence of DNA which they always cut in the same place. By utilising different endonucleases, scientists can cut away targeted strips of DNA. Once the desired DNA sequences have been selected, cut and combined, the new genetic material has to be multiplied (i.e. cloned) to produce quantities large enough to work with. Not only are the failure rates in genetic engineering staggeringly high, but the cloning of genes is also an essential element of molecular analysis and genetic medical diagnosis.

Before the invention of the polymerase chain reaction in the late 1980s, DNA could not be cloned in its naked state, but had to ride piggy back on fast-multiplying cells like bacteria. The most commonly used 'vector' for this process was a non-pathogenic strain of the *Escherichia coli* bacterium. Like most bacteria, *E. coli* carry an additional loop of double-stranded DNA called a plasmid. Plasmids have a small number of genes which are normally inessential to bacterial metabolism, and they can replicate independently of the main bacterial chromosome (bacteria are single-cell organisms without a nucleus and which, like viruses, have only one chromosome). Although they often fuse with the main chromosomal loop of bacterial DNA, plasmids are also capable of passing from one bacterial cell to another, and in the course of their travels may pick up and carry with them neighbouring bacterial genes. In this fashion they play a central role in the exchange of genetic material between different bacteria, including the transfer of antibiotic resistance.

Once the genetic scientist has separated the plasmid from a bacterial cell, a specific segment of the plasmid can be cut away with a restriction enzyme and replaced with the new DNA sequence using another enzyme which acts as a glue. At this point so-called 'marker genes' will be added as well. A common choice here, used in Calgene's tomatoes for example, is another gene taken from *E. coli* which codes for resistance to the antibiotic kanamycin. The mended plasmids are then reintroduced to the bacteria, either by electrically charging them to fire through the walls of the bacteria cells, or by making the walls of those cells more permeable to the plasmids by the application of a calcium salt. Once inside a bacterial cell, the plasmid rapidly multiplies, and its clones are carried into the bacterium's daughter cells. The bacterial cells are then exposed to the appropriate antibiotic (or herbicide or toxin, depending on the choice of marker gene), thus allowing the identification, by a process of elimination, of those cells which have successfully incorporated the new DNA.

In some cases genetically modified bacteria may be used as transgenic organisms in their own right, functioning as the 'protein factories' which were the first commercial success stories of recombinant DNA technology. *E. coli*, for example, has been genetically engineered to produce insulin for diabetics, human growth hormone for growth-deficient children, bovine growth hormone to increase cows' milk production, and anti-haemophilic globulin used in the treatment of haemophiliacs. Another example is the production, since 1996, of riboflavin (vitamin B2) from genetically engineered *Bacillus subtilis*, and indeed a wide range of enzymes used today in processed food are derived from the same genetic technology. These include chymosin – the main enzyme in rennet (traditionally obtained from the stomachs of slaughtered infant calves) – which is today produced for consumers of 'vegetarian' cheese from a yeast which has had the gene for calf chymosin inserted into it (this was in 1990 the first product of recombinant DNA technology to reach the US food supply). Bacteria can now also be genetically engineered to degrade toxic waste, and may in the future become a common anti-pollution tool.

In other cases bacteria are used as vectors for carrying new genes into the cells of plants or animals, which are later harvested for proteins, used for experimentation and study, or consumed in the form of genetically modified foods. Today one of the fastest-growing areas in plant biotechnology is research into the production of edible fruit and vegetables with pharmaceutical properties like cholesterol-lowering enzymes, antibacterial peptides that protect against tooth decay, vaccines against viruses like hepatitis B, or with enhanced nutritional qualities. The successful development of genetically engineered vaccine-producing fruit and vegetables could be a huge public relations victory for the biotech industry, since the low cost of their production (less than a penny a dose) combined with the removal of the need for refrigeration, could make them affordable to Third World health providers.[24]

In the early 1980s the first successful attempts to modify the genetic structure of plants used a natural soil-inhabiting bacterium called *Agrobacterium tumefaciens*. In the wild, this bacterium carries a plasmid which produces a tumorous swelling or 'gall' in the crown of plants. It does this by inserting part of its DNA into the chromosomes of the invaded plant cell, and forcing the cell to synthesise from this DNA a metabolite that feeds the rapid proliferation of the bacteria, as well as plant hormones which render the cell cancerous. The *A. tumefaciens* plasmid is thus unique in its ability to routinely exchange genes between bacteria and higher organisms (called 'eukaryotes') whose DNA is contained in the nuclei of their cells. When a mutant form of the plasmid was found which did not cause cancerous cell growth, the bacterium was adopted and customised as a central tool for plant biotechnology. Once combined with the new foreign genes, the bacterium is co-cultured with embryonic tissue cells of the target plant, allowed to infect those cells, and in the process smuggle the new genes into the plant's genome.

However, since most narrow-leafed plants – such as cereals, legumes and grasses – are not natural hosts of this bacterium, alternative methods also had to be developed in plant biotechnology. One such approach is called 'biolistics', which involves firing into plant cells tiny particles of gold or tungsten which have been coated with the recombinant DNA. Many of the cells are destroyed or missed, but some of them incorporate the foreign DNA (often multiple copies of it) into their chromosomes – its point of integration being random.

One other type of vector growing in popularity and use in the genetic modification of plants is the virus. Viral vectors are utilised mainly because they are cheap and potent. Plant viruses are pathogens of their host plants, which may range from one or two different species to several hundred, but they are normally harmless to herbivorous animals. Because plants tend to be protected by rigid cell walls, plant viruses are primarily spread in the wild through the wounds made by the bites of virus-carrying insects. Once a virus has been genetically engineered to express a novel protein, it is made to infect the cells of the plant – normally via a purpose-made wound – and enter the cytoplasm of those cells. The genomes of many viruses are made of a single strand of RNA. Although some viruses (such as retroviruses) function by converting this into double-stranded DNA and inserting the latter into the host's chromosomes,[25] most simply occupy the cell's cytoplasm, replicating themselves and synthesising their own proteins until the cell disintegrates and the mature viruses are released.

Because there may be many more copies of fast-replicating viral RNA in a plant cell than plant DNA, and because the level of expression of viral RNA is likely to be high, this method can yield a soluble foreign protein – such as a vaccine, hormone or antibiotic – at concentrations of up to twenty-five per cent of total leaf protein.[26] In addition, plant viruses have been used to inoculate commercial plants against viral attacks. Though scientists remain somewhat puzzled as to the mechanisms which allow this to work, the process involves transferring to the plant genome the gene that codes for the protein coat of the enemy virus. Insect viruses have also been genetically modified and tested as rapid-acting biological pesticides potentially capable of replacing their chemical-intensive rivals.[27]

The Myth of Seamless Continuity

With only three decades of research and experimentation behind it, genetic engineering remains, in comparison to the thousands-of-years-old techniques of animal and plant husbandry, an industry in its early infancy. Keen to attract investment in what is still a highly experimental, research-intensive science, the biotech companies have worked hard to present an unthreatening account of their work and aspirations to the general public, promoting the genetic engineering of plants as potentially the safest, cheapest and most environmentally progressive source of the kinds of high-yielding, disease-resistant, hardy and nutritious crops needed to feed the world's burgeoning population. Central to this marketing strategy is the dissemination of the view that genetic engineering is only a more controlled, speedy and precise version of the processes of selective breeding practised by human beings since they abandoned hunter–gatherer modes of subsistence some 10,000 years ago.

This view is well represented in a report on biosafety by the US National Research Council, which concluded that 'no conceptual distinction exists between genetic modification of plants and micro-organisms by classical methods or by molecular techniques that modify DNA and transfer genes'.[28] Mae-Wan Ho recalls a public debate in 1997 between herself and a prominent spokesperson for the biotech industry, in which the latter even went so far as to refer to the natural products of traditional breeding methods as 'transgenics'.[29]

The terms 'genetic modification' and 'GM', crucial to the protest vocabulary of the environmentalist movement, have also been dismantled and redefined in a similar way. For example, the eminent biologist Richard Dawkins, in a broadside aimed at the romantic anti-scientism of Prince Charles, observed that '[a]lmost every morsel of our food is genetically modified – admittedly by artificial selection not artificial mutation'.[30] His view is shared by US rice-specialist Susan McCouch: 'If you look even briefly at the history of plant breeding, then you know that every crop we eat today is GM. Human beings have imposed selection on them all.'[31]

While it is true that farmers have always selected and combined plants and animals in order to preserve and reproduce their most desirable traits, the idea that there is a seamless continuity between conventional breeding and genetic engineering is, for a number of reasons, difficult to sustain. To begin with, conventional breeding is limited to sexually compatible species. Though there are a number of recorded exceptions,[32] when different but related species (usually from the same genus) do interbreed, nature normally intervenes to inhibit the sexual reproduction and proliferation of the hybrid. In most cases, due to the mixing of species-specific chromosomes, the offspring of different species either die before they can reproduce or are sterile. If a mare is crossed with a donkey, for instance, any offspring that result are sterile mules, whilst the breeding in captivity of the Central American platyfish with a different species from the same genus, the swordtail, yields hybrids in which the distinctive skin spots inherited from the former become malignant and usually lethal melanomas.[33] In other cases the number of hybrids is contained by novel characteristics which make them particularly vulnerable to predators (when the *Heliconius erato* and *Heliconius himera* butterflies mate in the Ecuadorian Andes, for example, their offspring lack the colouring patterns normally recognised by predators as signifying indigestible food[34]).

A second difference between selective breeding and genetic modification is that conventional breeding is normally used to select and reproduce plants or animals whose desirable characteristics result from particular variants (called 'alleles') of common species-specific genes. This almost inevitably brings with it some loss or reduction in the organism's capacity to survive in the wild, partly because the breeder is powerless to prevent the carrier of the desired trait from passing on through sexual reproduction other undesirable or maladaptive genes, and partly because agronomically important traits may themselves interfere with an organism's natural adaptability. Breeding a plant to germinate earlier, for example, may be beneficial to well-protected and irrigated crops, but could be fatal to untended plants which germinate with the first spring rain and are then exposed to a late frost or early dry summer. Maize bred to produce ears of tightly compact grains is well-adapted to the storage needs of human consumers, but would be unlikely to survive without cultivation because of the difficulty of dispersing its own seeds.

In normal cases of plant and animal breeding there is therefore, as Philip Regal puts it,[35] a trade off between fitness for nature and fitness for human need, the most important consequence of which is that classical breeding methods tend to result in plants and animals that depend for their survival on continual human stewardship – including the regular reinvigoration of domesticated crops and animals by back-crossing them with wild relatives. The same is not necessarily true of genetically engineered organisms which, since selective genes can be transferred on their own, may be armed with adaptive traits giving them a significant ecological advantage over their wild competitors.

A third difference between breeding and genetic engineering centres on the way gene transfers are mediated by vectors and promoters derived from plasmids, bacteria, and viruses – and in some cases from a combination of all three. The result is novel transgenic products with genetic material from a range of unrelated organisms, many of which are naturally virulent, pathogenic, and often highly mobile. Indeed, since these vectors are selected and engineered precisely for their capacity to cross species boundaries and to overcome the natural defence mechanisms of cells which normally destroy or inactivate invading genes, they carry with them, in the eyes of many ecologists, an unprecedented risk of environmental pollution.

Finally, for all its claims to safety, efficiency and precision, genetic engineering frequently involves the random incorporation of foreign genes into the host organism's genome, the full and long-term effects of which are impossible to predict. Such disruption of the structural ordering of DNA is extremely rare in sexual reproduction, which carefully preserves the sequence integrity of the chromosomal DNA. Transgenic organisms thus end up with entirely novel DNA sequences which may disturb the regulation and expression of indigenous genes, disrupt the organism's metabolism, and leave it weak, disfigured or disease-prone. Where promoter sequences are inserted, they may not only influence the expression of the transferred gene, but may also activate other neighbouring genes in the chromosome. If produced in excessive quantities or if expressed in the wrong part of the organism, both indigenous and alien proteins may be toxic or allergenic to the organism itself, or to other organisms in the food chain.[36]

Genome Plasticity and the Decentred Gene

I shall consider in greater depth in the next chapter the ecological and health implications of genetically engineered crops. Before doing so I want to return to the science of molecular biology and look more critically at the conventional understanding of cellular and genetic functioning as I have presented it here. This is a worthwhile exercise, because the most persuasive ecological critiques of genetically modified organisms tend to be founded on the claim that the science which underpins genetic biotechnology is flawed. A particular source of controversy for genetic science is what many critics believe is its tendency towards biological reductionism. This tendency is founded on the premise that organisms can best be understood by reducing them to their most elementary parts – ideally parts which are chemically or structurally equivalent, and therefore interchangeable. 'The ultimate aim of the modern movement in biology is in fact to explain *all* biology in terms of physics and chemistry', wrote Francis Crick in 1966. 'Thus eventually one may hope to have the whole of biology "explained" in terms of the level below it, and so on right down to the atomic level'.[37]

This search for the atomic origins of life was apparently completed with the discovery of the chemical structure and physical functioning of DNA – the so-called 'self-duplicating molecule' and the 'universal code of life'. Since every living organism is, in this reductive account, simply the product of a particular variation in the assemblage of universal genetic building blocks, there can be no natural barrier preventing the transfer of genes across species. Living beings, in a perverse triumph of functionalist reason, are merely sophisticated machines like computers, operating according to a genetic programme which, once decoded, can be rewritten to perform new functions. 'Biotech is a subset of information technology', explains Robert B. Shapiro, the former Chairman and CEO of Monsanto. 'It's a way of encoding information in nucleic acids as opposed to encoding it in charged silicon'.[38] 'If you want to think about life', echoes the arch-priest of ultra-Darwinism, Richard Dawkins, 'don't think about vibrant throbbing gels and oozes, think about information technology'.[39]

The practice of genetic engineering is thus founded on the belief that the physiological functioning of organisms is an effect of or sequel to a germinal cause, and that this cause, once it has been identified and circumscribed as the primordial fount and progenitor of everything else, can with the right technology and expertise be rearranged to produce alternative effects. This particular form of biological reductionism plays an ideological function whose effects we shall explore in the following chapters. By ideology, I mean a coherent set of ideas which, by distorting or misrepresenting reality, serves to legitimise or conceal the power of dominant groups.

Genetic science performs an ideological function by enabling those who manipulate the genomes of living things to portray those things as manufactured artefacts beholden to their creators, devoid of independent (and unpredictable) agency, and warranting legal protection as patented inventions. The term 'genetic engineering' alone confers on the actions of the molecular biologist an appearance of mathematical precision and mechanical predictability which, as we shall see in the next chapter, is a long way from the truth.[40] First we need to remind ourselves why the conventional account, popularised by Richard Dawkins amongst others, of genes as fixed and determining units of information, as the master molecules which, by programming the cell's productive machinery, 'construct for themselves containers, vehicles for their continued existence',[41] is a serious misrepresentation of reality.[42]

To begin with, Dawkins' depiction of genes as 'selfish' units of 'digital information' whose purpose is to produce phenotypes – the expressed characteristics of organisms – capable of ensuring the replication and thus survival of that information into the next generation, is inconsistent with the fact that genes, if they are to be replicated in offspring, must 'co-operate' to ensure that the *organism as a whole* develops to the age of sexual reproduction. As Steven Rose puts it, 'the survival of any gene to the point at which the organism is mature enough to reproduce depends upon the "goodwill" of other genes'.[43] Reducing the organism to the level of 'selfish genes', in other words, makes it impossible to explain the growth and functioning of the full biological system, which relies on the co-ordinated activities of thousands of different genes and scores of specialised cell types.

Moving the point of analysis away from the gene to the organism as a whole is also necessary if we are to explain the existence of apparently superfluous expression processes, as well as the employment by cells of multiple overlapping pathways of protein synthesis. Many thousands of human 'genes', for example, are now known to produce types of 'noncoding RNA' – RNA which is not translated into amino acids – while others appear to express amino acid chains that have no regular function and are destroyed. Genes also demonstrate a peculiar flexibility of operation, performing different functions, expressing different traits, and having different effects, according to the regulatory signals they receive from their cellular environment. The same primary RNA transcript may be edited to express different proteins (in some organisms as many as several hundred), the transcripts of different genes may be cut and combined so that they yield hybrid proteins, and genes may even overlap so that the same stretches of DNA can be read in different ways. For this flexibility to work correctly, gene expression must rely on the regulatory activities of innumerable transcription factors, proteins which are themselves expressed by genes, hundreds of which have already been identified. More than this, proteins themselves, once synthesised, can undergo changes in their shape, function and activity, thus enabling feedback mechanisms to fine-tune the organism's adaptive development and functioning.

As Barry Commoner has argued, in one of the first direct critiques of the DNA-centric model, 'specificity ultimately embodied in DNA nucleotide sequence and in protein amino-acid sequence has a multi-molecular origin in a system which is characterised by circular feed-back relations'.[44] 'Thus transmission of biochemical specificity within the cell is fundamentally circular rather than linear and the total system rather than any single constituent is responsible for the biochemical specificity which gives rise to biological specificity.'[45]

The role of DNA is thus dynamic, delocalised, and entangled with surrounding networks and agents, including those of the external environment, the inter-cellular environment of the organism, and the metabolic environment of the specific cell. This was already implicit in the discovery by biologists in the 1960s of the vital role played by a number of previously undetected enzymes, a discovery which eventually exposed the first major flaw in the classical portrait of the regally detached gene commanding the cells to faithfully and obediently replicate it. The integrity of DNA and its miraculous stability through generations depends, it is now known, on the ingenuity of an elaborate system of enzymes involved in checking, editing and repairing damaged or miscopied DNA – in the absence of which 'one out of every hundred bases would be copied erroneously'. 'The stability of gene structure', Evelyn Fox Keller continues, 'thus appears not as a starting point *but as an end-product* – as the result of a highly orchestrated dynamic process requiring the participation of a large number of enzymes organised into complex metabolic networks that regulate and ensure both stability of the DNA molecule and its fidelity in replication'.[46]

What this means is that the gene, instead of being a fixed and primary unit of biological production and inheritance, is in a very practical sense actually *produced and defined by* the complex self-regulating system of the organism as a whole.

> To the extent that we can still think of the gene as a unit of function, that gene (we might call it the *functional gene*) can no longer be taken to be identical with the unit of transmission, that is, with the entity responsible for (or at least associated with) intergenerational memory. Indeed, the functional gene may have no fixity at all: its existence is often both transitory and contingent, depending critically on the functional dynamics of the entire organism.[47]

Once the gene is dethroned as the locus of causal agency and biological reproduction, and replaced instead by 'the functional dynamics of the entire organism', then the natural stability of the genome must also cease to be the sacred point of departure for molecular biology. Genes can, it is now known, change their chromosomal positions – and in doing so their functional effects – can rearrange their sequencing order, multiply themselves[48] or contract in number, and can be carried outside the organism into the wider environment.[49] Defective genes, including those of human beings, can also become corrected in derivative cells which regain their normal functioning, while organisms which have had vital DNA sequences deleted have displayed a surprising ability to preserve their phenotype.[50] This illustrates the way organisms are able to draw upon what is now known to be widespread functional redundancy of DNA sequences and expression processes, rearranging the metabolic pathways and networks of genes and proteins in order to coax from the genome an extraordinary degree of phenotypic variation.

Witness, for example, the recorded ability of male long-distance runners in their sixties to match the performance of twenty-three-year-old competitors, as the bodies of the older athletes modify the expression of important enzymes in such a way that an increased rate of oxidative metabolism compensates for their lower heart rate.[51] This illustrates what Steven Rose calls 'developmental plasticity' – 'the capacity of a living system to adapt to experience and environmental contingencies, and to compensate for deficiencies'[52] – or what is otherwise known as 'epigenesis': 'the mechanisms by which DNA is contextualised, controlled and regulated to produce changing patterns of gene expression in the face of changing environmental signals'.[53] The true genome, as Ralph Greenspan describes it, is 'a complex, emergent system made up of many non-identical components, with non-exclusive roles, non-exclusive relationships, several ways of producing any given output, and a great deal of slop along the way'.[54]

Adaptive Mutation

Perhaps more than anything else, it is the identification of processes of adaptive or 'directed' mutation which has challenged the limitations of the neo-Darwinian credo of 'passive' natural selection, the Central Dogma of unidirectional protein synthesis, and the definition of cells and organisms as vehicles for the replication of genes. In the neo-Darwinian picture painted by biologists like Richard Dawkins, evolutionary change is the outcome of a 'Blind Watchmaker', as adaptive genes are inherited from parents fortuitously equipped (due to replication errors or the effects of chemical fluctuations) with random mutations. Yet the experiments of John Cairns and his colleagues were among the first to suggest that genomes can acquire adaptive mutations by responding directly to the challenges of their environment – through a 'reversible process of trial and error', in which cells 'produce a highly variable set of mRNA molecules and then reverse-transcribe the one that made the best protein'.[55] Several enzymes have now been identified – Miroslav Radman calls them 'DNA mutases' – which actually function, when an organism is under threat, to produce replication errors at high rates. 'This inducible, genetically programmed process allows individual cells to mutate when their survival is threatened, thereby increasing the genetic diversity and adaptability (fitness) of the endangered population'.[56]

Non-random changes in DNA structure have also been observed in multicellular organisms, including the selective amplification (i.e. the autonomous multiplication) of developmentally important genes in amphibian eggs, as well as in the follicle cells surrounding the developing egg of the *Drosophila* fruit fly. As already noted in considering the role of transposons, the mammalian immune system is known to generate enormous antibody diversity 'through the rearrangement and joining of . . . germ-line segments, by the addition of nucleotides to them, and by hypermutation in some regions of the assembled genes'.[57] Adaptive changes observed in the chromosomes of plant cells exposed to herbicides, as well as in the genomes of starved bacteria and yeast cells, have readily contributed to scientists' understanding of how plants, insects and bacteria are able to mutate and develop genetic resistance to human-made pathogens like herbicides, insecticides and antibiotics, in a process described by James Shapiro as 'natural genetic engineering'.[58]

In mammals, one recognised origin of the resistance of cancer cells to chemotherapy treatment is the amplification of the gene that codes for the enzyme targeted by the cytotoxic drug methotrexate.[59] In humans, a similar process has been seen to occur with positive effect. In 1989 a father and son, members of a farming family regularly exposed to the organophosphorous insecticide parathion, were found to possess a 100-fold amplification in a gene that codes for an enzyme used in hydrolising the human neurotransmitter, acetylcholine. Not only is the enzyme specifically inhibited by organophosphates, but this family had inherited a mutated version of the gene for the enzyme, making them particularly vulnerable to insecticide poisoning. Through the adaptive amplification of the defective gene – thought to have occurred during spermatogenesis in the grandfather – the father and grandson appeared to have acquired compensatory genetic resistance to the chemical pathogen.[60] For biologists such as John Campbell, cases like these provide ample justification for the rejection of classical genetic reductionism:

> Classical genetics was predicated on the genotype having a sacred status. Genes were inviolable messages from ancient ancestors passed on faithfully from generation to generation, except for rare mutational corruptions. They were transcendental to the activities and metabolism of the organism carrying them. We now realise that this simply is not true. Instead cells actively manipulate the structure of their DNA molecules for both physiological and evolutionary reasons.[61]

Non-DNA Inheritance Systems

Research into epigenetic mutation and inheritance pathways has also shaken the hitherto impregnable belief that environmentally acquired characteristics are never passed on to the next generation, and that adaptive changes to biological systems are purely random in origin. The idea that environmentally induced changes could be transmitted through the germ-line was the original theory of Jean-Baptiste Lamarck, a French naturalist whose post-religious account of evolution predated Darwin's theory of natural selection by half a century. Lamarck believed, to repeat his most famous example, that the long necks of giraffes evolved from animals who repeatedly stretched to reach the highest leaves on the trees, and who passed on their elongated necks to their offspring.

This belief that adaptive changes in the physiology of adult organisms are inherited by their offspring appealed directly to those who thought that human nature, no less than the nature of animals and plants, could be modified and improved by changing the influences of their social environments. Indeed, Lamarck's theory of 'the inheritance of acquired characters' was ruthlessly adopted and promoted by the Soviet geneticist Trofim Lysenko during Stalin's reign, despite the fact that the theory had been widely discredited by the experiments of a German zoologist in the 1880s. Attempting to prove Lamarck right, August Weismann had cut off the tails of mice for twenty-two generations and was surprised to find that the amputations had no effect on the tail length of the mice's progeny. His research gave birth to the conviction that there was an insuperable barrier separating the reproductive pathway (the 'germ-line') of a multicellular organism from the mortal cells of its body or 'soma', and that heritable changes in the characteristics of organisms can only derive from random mutations in the DNA of germ cells or their precursors.

It now appears, however, that 'Weismann's barrier', as it came to be called, is not entirely impermeable. Agronomists in the UK, for example, recently reported the case of a naturally occurring mutant variety of the toadflax plant (*Linaria vulgaris*). In the mutant, the flowers of the plant exhibit radial rather than the normal bilateral symmetry. The source of this difference, which was inheritable, was identified not as a mutated gene, however, but as the *silencing* of the relevant gene.[62] This occurred due to a process called 'methylation', where methyl groups attach to specific DNA bases (normally cytosines) and prevent them from being transcribed and expressed. Methylation, which is still poorly understood by molecular biologists, is thought to play a role in stabilising functionally inactive genes, and is probably used by organisms to defend themselves against foreign DNA by inhibiting its expression. It may also be a response to changes in temperature.

While methyl groups are normally removed from the inherited genome during the early development of the embryo, this study showed that this is not necessarily the case, and that specific alterations in gene *expression* can indeed be inherited. In this case it was the methylated deactivation of the relevant gene, not the gene itself, which was the source of the unusual trait being transmitted from the plant to its progeny. The same epigenetic inheritance path has been identified in mammals. Researchers studying the variability of coat colour amongst stable genetic lines of transgenic mice, for example, found that the critical factor determining the colour of a mouse's offspring was the extent to which the gene for coat colour was methylated, and that these methylation patterns were again inherited.[63] In surveying the scientific literature in the mid-1990s, Jablonka and Lamb catalogued thirty-eight different eukaryotic (non-bacterial) organisms – ranging from protozoa, fungi and plants, through to fish, mice and humans – which have inherited changed patterns of cellular functioning that did not derive from random genetic mutation.[64]

Further evidence of epigenetic inheritance comes from the study of the related phenomenon of 'imprinting'.[65] This is the process whereby specific chromosomes, DNA sequences or genes appear to carry a 'memory' of the sex of the parent organism from which they derived, a memory which can affect the expression pattern of homologous genes. In female mammals, for example, one of the two X chromosomes in nearly all somatic cells is inactivated (only one X chromosome is required), and in many particular species the sexual derivation of the inactivated X chromosome is always the same (the mechanism of deactivation is normally methylation). Scientists have also found that mice embryos given two sets of chromosomes from parents of the same sex do not fully develop, and that the nature of the developmental defects that ensue varies according to whether the two chromosomes come from male or female parents. Identical chromosomes, in other words, seem to mean different things to cells and organisms according to the sex of the organism from which they derived.

Whether or not foreign genes are expressed in transgenic mice has also been found to be affected by the sex of the parent of the animal from which the transgene is inherited (the human oncogene *c-myc*, for example, normally only leads to tumours in mice when it is inherited from a male parent). In humans there are several cases of dominant genetic disorders which affect people differently according to whether the mutation is inherited from the mother or the father (people who inherit the genetic mutation that gives rise to Huntington's disease from their mother, for example, develop symptoms during middle age, whereas those who inherit the mutation from their father develop symptoms in adolescence).

What the study of methylation and imprinting seems to demonstrate is that DNA is not the only medium for the inheritance and determination of traits. Inheritance is not a closed system impervious to environmental influences and subject only to random variation, for the genome can be 'marked' in such a way that records the effects of external stimuli and the attempts of the organism, or its forbears, to adapt to them. 'Since the environment influences epigenetic marks, and epigenetic marks influence changes in DNA sequence, the role of the environment in evolution is not solely that of an agent of selection. It is also an agent of variation.'[66]

This also amounts to a refutation of the 'Central Dogma' originally advanced by Francis Crick in 1957. According to this formula, 'once "information" has passed into protein it *cannot get out again*. In more detail, the transfer of information from nucleic acid to nucleic acid, or from nucleic acid to protein may be possible, but transfer from protein to protein, or from protein to nucleic acid is impossible'.[67] Yet these studies show that 'information' can indeed pass from protein to protein (as when methylation enables novel expression patterns to be inherited independently of DNA mutations, and more directly in the corruption of protein-folding by prions as mentioned earlier), and from protein to DNA (as when the genome is autonomously rearranged to better equip the organism for a challenging environment). In fact, at a purely somatic level, we could say that there simply *has* to be a non-genetic system of inheritance or memory if, when a liver, skin or blood cell divides, the daughter cells, despite having a genome identical to that of the rest of the cells in the organism, are to 'know' what cell type, what expression pattern, is required of them. For differentiated cells to 'breed true', in other words, they have to 'remember' the phenotype of their progenitor.

Rethinking Genetic Engineering

This discussion of genome plasticity, directed mutation, and epigenetic inheritance systems is important because a strong source of justification for the genetic engineering of organisms is the neo-Darwinian orthodoxy that evolutionary change is driven by natural selection of advantageous genotypes, that these genotypes are themselves the product of entirely random changes (replication errors), and that genotypes express phenotypes in a linear and one-way process of production. It is because DNA represents, in this account, the absolute starting point, the germinal *de novo* template, for the reproduction and functioning of organic life, that the modification of DNA is a technically conceivable goal. And it is because 'natural' changes to the genome of organisms are, according to this same account, entirely blind and random – because DNA 'programmes' the behaviour of organisms but remains essentially ignorant of why it gives those particular instructions and of the possibility of giving more suitable ones – that the right to take charge of the otherwise arbitrary processes of genetic mutation and change has been so confidently assumed by molecular genetics.

It is clear, however, that instead of being a centralised programme whose code predetermines the developmental path and functioning of an organism, DNA is merely one dynamic element in the self-stabilising networks that make possible living beings, and that one effect of the circular feedback mechanisms that regulate these networks is the restructuring of the genome itself. If there really is such thing as a biological 'programme', Keller points out, then it 'consists of, and lives in, the interactive complex made up of genomic structures and the vast network of cellular machinery in which those structures are embedded. It may even be that this program is irreducible – in the sense, that is, that nothing less complex than the organism itself is able to do the job.'[68]

In 1984 Barbara McClintock anticipated future research which would 'determine the extent of knowledge the cell has of itself and how it utilises this knowledge in a "thoughtful" manner when challenged'.[69] The 'thoughtful cell' does indeed seem to be the inescapable truth uncovered by more recent research, and it is a truth which should make us think hard about the wisdom of 're-engineering' biological systems as if they were reprogrammable machines. If the reproduction and behaviour of organisms can no longer be credibly reduced to the 'self-duplicating molecule' of the gene, but must instead take account of multiple inheritance pathways, self-directed mutation, and mechanisms which allow organisms to transmit epigenetic adaptations to their offspring, then, as Commoner fatefully observed in 1968, 'any promise to control inheritance by chemical manipulation of DNA is likely to be illusory and, in certain social circumstances, harmful to the welfare of man'.[70]

> If the code theory were correct, then it should be possible, by chemical modification of DNA, to develop specific nucleotide sequences hitherto absent from living things; and, in transformation experiments, to achieve wholly new inheritable biological characteristics and new forms of organisms. In contrast, the view that inheritance is based on multi-molecular sources predicts severe limits on the biological effectiveness of artificial modifications of DNA nucleotide sequence.[71]

We shall consider in later chapters the deeper problems and possibilities which genetic engineering, and the reductionism that underpins it, pose for 'the welfare of man', focusing in particular on the choice we face between engineering the human organism better to suit a hostile environment, and transforming the environment to bring it within reach of the natural adaptive capacities of human beings. If trends in agricultural biotechnology are portents for developments in the field of human health and medicine, then the aggressive growth of herbicide-resistant crops – adapting plants to chemicals, rather than the reverse – offers a fairly clear indication of what the dominant course will be. With the agricultural sector in mind, we shall now look at the evidence in support of the view that the genetic engineering of plants and bacteria carries significant risks to health and the environment.

2

Agricultural Biotechnology and Ecological Risk

In the previous chapter I argued against the prevailing DNA-centric account of biological functioning. I emphasised the numerous agents and non-linear processes involved in the development, performance and reproduction of organisms, and suggested that living things should be considered far more 'thoughtful' and adaptive in their responses to environmental challenges than genetic reductionism gives them credit for. The fluid and adaptive genome is what enables organisms to fine-tune their functioning to harmonise their interactions with the surrounding environment. It is the view of many ecologists that the genetic engineering of agricultural crops, by disregarding the slow and limited adaptive capacities of organisms in favour of the mechanical transformation of plant genomes, represents an unprecedented threat to ecological stability and sustainable farming. Living organisms are, from this perspective, too complex, too 'intelligent', and too entangled with their habitats to be appropriated and controlled as if they were reprogrammable machines. Since they can migrate, mutate, replicate and cross-fertilise, the long-term behaviour of genetically engineered species is, moreover, unlikely to match the predictions of even the most skilled scientist.

Random Insertions and Gene Silencing

'Although genetic engineers can cut and splice DNA molecules with base-pair precision in the test-tube', John Fagan warns, 'when an altered DNA molecule is introduced into the genome of a living organism, the full range of its effects on the functioning of that organism cannot be controlled or predicted'.[1] One reason for this is the way the random insertion of novel genes can disturb vital DNA sequences, and cause the rearrangement or 'scrambling' of the endogenous genome.[2] Even where *unintended* genetic alterations are avoided, however, the complex networks involved in the organism's functioning may still be disrupted. Biologists recognise, for example, that the specific protein produced by a single gene may be essential to a variety of functions and traits that are vital to the organism (a phenomenon called 'pleiotropy'). They also recognise that the effects of some genes can override or mask the phenotype normally produced by other unrelated genes (a phenomenon called 'epistasis'), implying that the inactivation of one gene may have unpredicted effects, and that the consequences of changing or silencing a gene cannot always be extrapolated from the trait with which it is directly associated.

In one experimental attempt to enhance the coloration of petunia petals, scientists introduced an extra copy of the gene which codes for a key catalytic enzyme – chalcone synthase (CHS) – used in the biosynthesis of the compounds responsible for plant coloration. Unexpectedly, the effect was in fact the widespread *silencing* of both the exogenous and the endogenous versions of the CHS gene, with the result that the flowers of nearly half the transgenic plants lost their normal violet colouring and reverted to white.[3] A similar experiment on petunia coloration conducted by a different group of scientists had almost identical results,[4] whilst another piece of research on manipulating colour expression in the plant revealed how the hypermethylation of the promoter driving the foreign gene was what caused the weakening of the artificial colour, suggesting this was the plant's attempt to defend itself against genetic assault. Methylation was also seen to occur at any time during the life cycle of the plant, and the frequency of weakly pigmented plants increased from around five per cent of those grown in a greenhouse to about sixty per cent of open-field plants over the course of the season.[5] According to Steinbrecher, these results were accompanied by a number of other unexplained side effects on the plants, including increased growth, lowered fertility, and enhanced disease resistance.[6]

Luke Anderson refers to a trial of maize genetically engineered to be toxic to insects, the results of which included an unexpected reduction in yield by more than a quarter, and lower levels of copper in the leaves, stalks and grain. In another trial, designed to assess the safety of herbicide-resistant soybeans made by Monsanto, it was found that cows fed on the crops produced on average an extra eight per cent of milk fat per day. This was unexplained, but the product was passed as fit for animal consumption.[7]

When Lappé and Bailey analysed Monsanto's own data on test plots of soybeans engineered to be resistant to their herbicide, they not only found that some genetically engineered plants grew taller than their non-transgenic neighbours, but side-by-side yield comparisons also revealed that, when the herbicide was *not* applied, the transgenic crops suffered a yield deficit averaging nearly ten per cent.[8] After American farmers had complained of unexpected crop losses with the new Monsanto soybean, independent scientists also found that the transgenic crops were unusually vulnerable to hot Spring weather, with high soil temperatures causing a level of height, yield and weight deterioration far in excess of the damage suffered by non-transgenic beans.[9]

In 1997, the debut year for the commercial planting of Monsanto's herbicide-tolerant cotton, at least sixty different farmers throughout the mid-south region of the United States saw their crops fail. While the conventional cotton crop in the region continued to grow healthily, the bolls from tens of thousands of acres of transgenic cotton became deformed and detached themselves from the plant, the damage costing millions of dollars in lost earnings.[10] Most analysts attributed the disaster to instability caused by the random insertion of the bacterial nitrilase gene, which enables the cotton plants to detoxify the herbicide bromoxynil. A study has also shown that foreign DNA in transgenic oilseed rape can end up being silenced, perhaps by methylation, several generations down the germ line.[11]

Toxic Food Chain

One of the most contentious developments in plant biotechnology today is the growing use by biotech companies of a toxin produced by the common soil bacterium, *Bacillus thuringiensis* (Bt). Organic farmers have used the *kurstaki* strain of this bacterium, dispensed in the form of a topical spray, in moderation as a safe form of biological pest control for several decades. Unlike indiscriminate organophosphate insecticides, the toxins it contains are only poisonous to specific pests, and are usually targeted at species of *Lepidoptera*, such as the European corn borer and the spruce budworm. In the case of the corn borer, which is normally only a significant pest problem for maize growers every three to five years, the spray is not commonly used every year.[12] It is also quickly degraded by solar UV light.

In 1996, Novartis Seeds, then a US subsidiary of the Swiss agricultural biotechnology firm, began selling maize that had, under licence from Monsanto, been genetically engineered with a Bt gene to express a toxin in its tissue so as to make it lethal to corn borers. By 1998, up to 16 million of the 80 million acres of corn grown in the US carried a Bt gene, with at least seven varieties available for farmers to buy either from Monsanto, Novartis (now Syngenta), Mycogen, or AgrEvo. Crops of Bt cotton and potato were also expanding, with the concentration of Bt cotton in central and northern parts of Alabama, for example, reaching ninety-six per cent of total cotton acreage.[13]

The endotoxins produced by *B. thuringiensis* during sporulation are naturally expressed in an inactive form. Properly called 'protoxins', they become poisonous only when the crystallised protein molecules are broken down by specific enzymes produced in the highly alkaline midgut of the target insect. The toxins expressed in Bt crops, however, have been modified so that they are immediately active and soluble when they form in the tissue of the plant. Laboratory research has shown that the active form of these insecticidal toxins, as they are expressed in the flesh of transgenic plants, is not easily degraded by natural processes, and that composted plant flesh, in all of whose cells the gene and the toxin are present, may transfer both gene and toxin to soil bacteria and inert particles, thus posing a potential threat to the fertility of complex soil ecosystems and the numerous insects and microorganisms that help maintain them.[14]

Perhaps more worrying is the discovery that the Bt toxin, when synthesised in the cells of plants, has the capacity to pass on its effects through the food chain to valuable pest-eating insects, birds, rodents and amphibians, or to important pollinating insects such as bees. This danger was previously minimised in topical applications of the inactive toxin, since only specific insects produced the stomach enzymes which could activate it. A Swiss study found, however, that when green lacewings were fed with corn borers – their chief food source – which had themselves been reared on a diet of Bt maize, the lacewings suffered disrupted development and increased mortality. Further experiments by the same research group directly verified the toxic effects of the activated Bt protein on lacewings, proving that it was the surviving toxin and not the ailing corn borers which was the cause of mortality.[15] A similar study for the Scottish Crop Research Institute reported that ladybirds fed on aphids raised on potatoes that had been genetically engineered to produce an insecticidal lectin, laid fewer eggs and lived half as long.[16]

Concerns that genetically engineered insecticidal plants could end up disrupting wider ecosystems were fuelled in 1999, when the British science journal *Nature* published a letter from a team of Cornell University entomologists announcing the results of their research on Monarch butterfly larvae. When the larvae were fed their normal diet of milkweed leaves, but this time dusted with pollen from Bt-corn, half of them died after only four days.[17] In field conditions, the researchers noted, corn pollen is normally dispersed at least sixty meters from its source plant, and milkweed is commonly found in both corn fields and adjacent habitats. In a subsequent study at Iowa State University, Monarch larvae were allowed to feed on milkweed leaves taken from plants exposed to the natural dispersal of transgenic Bt-corn pollen. Despite the fact that the larvae were exposed to the pollen for only forty-eight hours, around a fifth of them died as a result of this diet, whereas all the larvae fed on pollen-free milkweed survived. Although Syngenta has been applauded for recently withdrawing the most toxic version of Bt-corn (so-called 'event 176'), both this product and a significantly less toxic variety ('event Bt11') were found by the Iowa researchers to cause similar rates of mortality.[18]

Ecologists have also voiced fears that, because farmers are helpless to moderate the rate at which the transferred Bt gene expresses the toxin in the host plant (depending on the specific promoter used, the toxin may also be expressed in the pollen, kernels, roots, and other non-green parts of the plant), the constant exposure of insects to high levels of the insecticide will accelerate the pressure for the selection and adaptive survival of Bt-resistant 'superpests'.[19] As a result, the less harmful Bt spray will become ineffective, and organic farmers may have to abandon it in favour of more dangerous chemical pesticides. This threat loomed large after 20,000 acres of Monsanto's Bt cotton growing in eastern Texas succumbed in 1996 to cotton bollworm – one of three insects it was specifically designed to kill.[20]

The risk of encouraging the evolution of Bt-resistant pests was recognised by the US Environmental Protection Agency (EPA), who predicted that most target insects could become resistant to the Bt toxin in three to five years.[21] In October 2001 the EPA nonetheless approved the re-registration of Bt-corn for another seven years, having simply recommended that farmers devote at least twenty per cent of their acreage to non-Bt crops, enabling these to function as 'refugia' for non-resistant insects which can then mate with resistant ones in order to breed out this adaptive trait.[22] Meanwhile, biotech companies are pressing on with the development of more powerful gene-based insecticides, with patents already granted for the venom-expressing genes of spiders, scorpions and snakes.[23]

The ecological danger of industrial agriculture producing ever-more robust predators and competitors is also compounded by the spread of crops which are genetically engineered to be resistant to herbicides, thus allowing farmers to use those weed-killers indiscriminately, confident they will not be damaging their own harvest. With nearly a third of the soya produced in the US in 1998 already genetically modified to tolerate glyphosate – the main toxic ingredient in Monsanto's best-selling Roundup herbicide – there is a considerable risk that weeds themselves will develop resistance to herbicides, either by accelerated natural selection due to increased glyphosate usage, or indirectly due to cross-pollination with genetically modified crops.[24] As a result, new and more powerful chemical herbicides are likely to be required.

Introgression and Super-Weeds

When pollen from cultivated crops spreads ('introgresses') to sexually compatible wild relatives the result is often fertile weedy hybrids containing genes from the cultivated plant. Any genetically engineered crops with close relatives that are common weeds – this applies to oat, barley, potato, carrot, lettuce, squash, sorghum, sunflower, alfalfa, some trees, rice, blackberry and raspberry (and in the animal kingdom there are corresponding problems with fish, shellfish and insects, which are all close to wild-types) – thus pose a risk of transgene introgression. Probably the greatest source of concern, however, is the genetic manipulation of oilseed rape (known as canola in North America), which is part of the mustard family and the source of rapeseed oil. The main variety, *Brassica napus*, has out-crossing rates of up to thirty per cent with related plants, and there is extensive evidence of gene flow from oilseed rape – a crop which Monsanto have widely marketed in herbicide-resistant forms – to wild and weedy relatives.[25]

British researchers studying the risk of gene flow from an oilseed rape crop engineered to be resistant to glufosinate, a weed killer currently owned and marketed by Aventis, found that seven per cent of seeds from a non-transgenic rape crop sited 400 metres away had acquired the herbicide-resistance gene through cross-pollination.[26] A similar study conducted at France's national agricultural research agency found that genes conferring glufosinate tolerance on transgenic oilseed rape persisted for several generations when the plant was hybridised with wild radishes.[27] In 1998, worrying evidence of transgenic gene flow was also discovered in non-experimental conditions. A field previously sown with transgenic glufosinate-resistant and conventionally-bred imidazoline-resistant canola in Alberta, Canada, was found to contain 'volunteer' (self-planted) varieties of the crop nearly half of which produced progeny with high rates of glyphosate resistance. Though the source of this genetic trait was assumed to be an adjacent field of glyphosate-resistant canola, one plant which produced glyphosate-resistant seedlings was found growing 500 metres from the relevant field boundary. A cause of even greater concern was the discovery that many of the seedlings had acquired the genes for multiple resistance to two of the three herbicides, and that two of the 924 seedlings which were screened were resistant to all three.[28]

Research by scientists at the University of Chicago also indicates that transgenic plants may actually have inherently higher rates of out-crossing. In this case genetically engineered mustard plants were found to be twenty times more likely to interbreed with related species than their non-transgenic counterparts, a level of promiscuity the researchers were unable to explain, and which is particularly worrying given that the gene in question – coding for resistance to the herbicide chlorosulfuron – has been introduced into several crops already.[29]

Since scientists have recorded cases of pollen being blown by the wind across distances of more than 600 kilometres,[30] the prevention of unintended gene flow through cross-pollination does not seem a particularly realistic goal. This became vividly apparent to the world's scientific community when in November 2001 *Nature* published an article by two California environmental scientists reporting the discovery of introgressed transgenic DNA in traditional maize landraces in a remote mountain region of Mexico. Despite the fact that Mexico has imposed a moratorium on the planting of transgenic maize since 1998, and the last known transgenic crops in the region were grown sixty miles from where the scientists collected their samples, five of the seven tested samples of native 'criollo' corn contained the widely used p-35S promoter from the cauliflower mosaic virus, two out of six tested samples contained the nopaline synthase terminator sequence from the *Agrobacterium tumefaciens* (used to signal the end of the foreign gene to be read), and one sample also contained a Bt-toxin gene (*cryIAb*). The scientists, whose findings were later said by the Mexican government to have underestimated the true scale of contamination,[31] also observed that the transgenic promoter had become integrated into the genome of the maize in a variety of different chromosomal locations, indicating how the artificial introduction of new genes can end up scrambling the DNA of resulting plants, with the disruption of normal cellular functioning a possible effect. The researchers concluded that 'there is a high level of gene flow from industrially produced maize towards populations of progenitor landraces', and that 'the transgenic DNA constructs are probably maintained in the population from one generation to the next'.[32]

As a consequence of unintended gene transfer like this – at least where the transgenic constructs introgress in one piece – genetically modified plants with superior growing capacities, with genetic resistance to insects, diseases, herbicides and other environmental stresses, or which have been engineered for improved photosynthesis, quick propagation, or a capacity to survive in harsh climates, may pass these traits on to related species, producing new strains of 'super-weeds'.[33] Around the world, a number of troublesome weeds have already formed as a result of gene flow and hybridisation from non-transgenic crops, including a wild radish in California that is the product of cultivated radish hybridising with jointed charlock, and the hybridising of maize in parts of Central America and Mexico with teosinte. In the latter case, the weedy variety thrives in the same conditions designed to nourish the cultivated crop, and is also difficult to identify visually and remove. When rice breeders in India tried to solve a similar problem by developing easily distinguishable purple varieties of rice, the relevant genes introgressed to neighbouring wild varieties, turning the weedy rice plants purple as well.[34]

Weed control costs a huge amount in economic terms (billions of dollars a year in the US), and additional resources are lost or expended due to the effects of weed proliferation. This includes low crop yield due to weed competition, the harbouring by weeds of toxic compounds, pests and plant pathogens, the contamination of harvested seeds with weed seeds, the disruption of transport systems, and the production of toxins or allergens harmful to animals and humans.

Field trials have already shown how herbicide-resistance in genetically engineered potato and oil-seed rape can spread to wild relatives within a single growing season.[35] Despite being developed under greenhouse conditions by Aventis in Germany, some sugar beet plants genetically modified to resist the herbicide glufosinate were found in European trials in 2000 to have acquired, from other beet plants engineered by the company, the gene for glyphosate resistance.[36] In the case of Calgene's transgenic tomatoes, concerns that the PG antisense transgene – a by-product of which (another example of pleiotropy) appears to be genetic resistance to two fruit-rotting fungi – may be transferred to weedy relatives, were raised in protest against the US government's proposal to exempt the tomato from regulation as an environmental risk.[37] With highly mobile creatures like fish the risks are even greater. Salmon which have been genetically modified to grow ten times faster than normal fish by producing large amounts of growth hormone, for example, could easily escape and breed with wild salmon, which would in turn gain a rapid evolutionary advantage over competitors.[38]

Horizontal Gene Transfer

Genetic material is not only spread 'vertically', through sexual reproduction, but can also be transferred 'horizontally', migrating from adult to adult using a variety of natural vectors, such as viruses, bacterial plasmids, and transposons.[39] Genetic engineering is itself an artificial form of horizontal gene transfer, but one which generates unease among many scientists because, by introducing into the gene pool modified vectors deliberately designed to carry DNA across species, it magnifies the opportunity for foreign genes to escape from their host organisms and to recombine to form sequences which would never occur naturally. The risk of recombination is also known to be greatly enhanced by the disruption caused by genetic engineering to DNA sequences that may have preserved their integrity for hundreds of thousands of years.

Microbiologists reported in 1998 that genetically engineered sugar beet could transfer DNA – including antibiotic resistant marker genes – to common soil bacteria,[40] which could then be picked up and spread by birds, insects, or animals. Bacteria – numerous strains of which are permanent inhabitants of animal and human guts – are well known to be capable of incorporating foreign DNA from their surroundings, whether this is plant debris, excretion, or intestinal tissue,[41] and bacteria that have been genetically 'crippled' to prevent them from surviving outside of controlled environments can re-acquire from other bacteria the genes necessary to continue multiplying.[42] Research has also found a high degree of cross-species mating amongst bacteria, which involves the transferring of bacterial plasmids as well as segments of chromosomal DNA.[43] In the case mentioned in the previous chapter, concerning the proposed US

- horiz transfer creating dangerous bacteria/viruses

field trial of pink bollworms engineered to carry a jellyfish gene expressing a green fluorescent protein, the release was delayed due to a legal challenge by scientists worried over the use in this experiment of a particularly promiscuous vector. This included a transposon which, though disabled, could re-acquire its mobility from transposon-bearing insect viruses, or be carried, along with any new genetic cargo, by the viruses themselves.[44]

Horizontal gene transfer and subsequent genetic recombination is now thought to be responsible for the emergence of virulent new strains of antibiotic-resistant bacteria, including the cholera outbreak in India and Bangladesh in 1992, a *Stretptococcus* epidemic in Tayside in 1993, and the strain of *E. coli* 0157 that killed 21 people in Scotland in 1996.[45] According to Ho and her colleagues, there is persuasive evidence to suggest that the rapid growth in genetic engineering over the last two decades – because it involves the creation and release of modified microorganisms designed to break down the barriers to non-sexual gene transfer – has played a significant role in accelerating the emergence of drug- and antibiotic-resistant infectious diseases.

Crops engineered with viral genes to be resistant to predatory viruses – much in the way that vaccines are used to inoculate people and animals against diseases – have also shown a capacity to transfer those genes to other viruses that the plant is exposed to, the potential result being recombinant strains of viruses capable of infecting a wider range of hosts.[46] The great majority of transgenic crops also contain a powerful promoter taken from the cauliflower mosaic virus (CaMV), a para-retrovirus related to human hepatitis B. According to Ho and other concerned microbiologists, this viral promoter has the capacity to reactivate dormant viruses and combine with others to produce virulent new diseases that could wipe out whole harvests. Given that the CaMV promoter is promiscuous enough to function in a range of different organisms, including all plants, green algae, yeast, *E. coli*, and even human cell lines, it may also be capable of causing the over-expression of genes in any of the different species to which it may be accidentally transferred, potentially precipitating cancerous cell growth.[47] And since a gene from the CaMV, once manually recombined with a baculovirus (an insect virus), has been successfully expressed in the cytoplasm of insect cells infected with the baculovirus vector,[48] it is not inconceivable that DNA from the widely used CaMV might at some time recombine with a human virus, like hepatitis B, resulting in a highly infectious human super-virus.

Risks to the Mammalian Food Chain

The full health implications of genetically engineered food will be considered shortly, but at this point we should at least expose the flaw in the industry-sponsored view that DNA is degraded by the heat processing used in the manufacture of animal feed (and indeed other processed food). To the surprise of the UK government's Advisory Committee on Animal Feeding Stuffs, its own independent study reported that 'DNA fragments large enough to contain potentially functional genes survived processing in many of the samples studied', thus giving credence to the view that foreign genes – such as those coding for antibiotic resistance – may enter the animal and human food chain.[49]

This may be one explanation for the discovery by independent scientists, asked to review a study by Aventis on its own herbicide-resistant maize, that the mortality rate amongst chickens fed the genetically engineered maize was double that of a control group.[50] With research also proving, contrary to popular scientific opinion, that viral DNA can survive passage through the gut of mice, penetrate the gut cells and enter the blood stream,[51] worries about the long-term effects of genetically manipulated organisms on consumers' health seem well justified. The nature of this risk also touches on the need for adequate segregation and labelling of genetically modified products,[52] as well as the difficulty of policing this division – as illustrated by two incidents widely reported in the British press in May 2000.

In the first incident, samples of honey bought from shops near a UK test site for genetically modified oilseed rape were found to be contaminated with genetically modified pollen.[53] In the second, it was discovered that nearly two per cent of the rape seeds sown in the UK in 1999 and subsequently harvested and sold to industry and food producers (amongst other uses, rape oil is used in the manufacture of margarine, vegetable oil and cosmetics), were of a genetically engineered (herbicide-resistant) variety not licensed for use in Britain.[54]

According to evidence presented to the House of Commons Agricultural Select Committee by the seed company, Advanta (a joint venture between AstraZeneca and Cosun of the Netherlands), the Canadian-grown rape seed had come from plants that had been cross fertilised by pollen from herbicide-tolerant varieties grown over four kilometres away from the non-transgenic crop. This is well beyond the distances recommended by the biotechnology industry and government scientists, and goes someway to explain a level of adulteration which now appears commonplace. According to an American company responsible for screening agricultural products, in 1999 '12 out of 20 random American consignments of conventional maize seed contained detectable traces of GM maize', with two samples containing nearly one per cent contamination. If this rate of contamination applies also to Britain's annual soybean imports, each year around 5000 tonnes of genetically engineered soybeans may be entering the animal and human food chain.[55]

A further concern to environmentalists is the development by biotechnology firms of crops genetically engineered to express pharmaceutical products in their flesh or seeds.[56] If this enterprise takes off, there may be a major problem segregating these crops – particularly when they are grown, as they surely must be, on an industrial scale – from wildlife. 'Soil insects and microorganisms, foraging and burrowing mammals, seed-eating birds, and a myriad of other non-target organisms will be exposed for the first time to vaccines, drugs, detergent enzymes, and other chemicals expressed in the engineered plants', warn Rissler and Mellon. 'Herbivores will consume the chemicals as they feed on plants. Soil microbes, insects and worms will be exposed as they degrade plant debris. Acquatic organisms will confront the drugs and chemicals washed into streams, lakes, and rivers from fields.'[57]

The Threat to Biodiversity

Biodiversity is essential to maintaining both the human carrying capacity and the overall stability of the ecosystem, and it is most prodigious in the developing world. One of biggest fears regarding the ecological consequences of recombinant DNA technology, and the genetic reductionism that underpins it, is that scientists' interest in manipulating and preserving functional genes is being pursued at the cost of a diminishing concern for the survival of the whole organisms or species in which valuable DNA exists. As Rifkin reports, current rates of genetic erosion in agriculture are already alarming:

> The US soy crop, which accounts for 75 per cent of the world's soy, is a monoculture that can be traced back to only six plants brought over from China . . . [O]f the seventy-five kinds of vegetables grown in the United States, 97 per cent of all the varieties have become extinct in less than eighty years . . . [I]n the United States just ten varieties of wheat account for most of the domestic harvest, while only six varieties of corn make up more than 71 per cent of the yearly crop. In India, farmers grew more than thirty thousand traditional varieties of rice just fifty years ago.

Now, ten modern varieties account for more than 75 per cent of the rice grown in that country.[58]

With ninety-five per cent of the global genetic diversity employed in agriculture since the beginning of the twentieth century already lost,[59] ecologists believe that it is vital for humanity to make an effort to preserve what remains. Yet further decline in biodiversity may well be an inevitable consequence of genetic engineering in agriculture. This is so not just because the DNA-centric paradigm and the technology of genetic manipulation are largely indifferent to species-diversity. It is also a trend reinforced by the way intellectual property law makes genetic uniformity a legal prerequisite for patent-protection of seeds, and by the way the cloning of plant varieties reduces the natural genetic variation of sexually reproducing organisms.[60] The relatively high costs of DNA technologies – and the corollary expense of extensive legal support – also means that the industry is dominated by a small number of powerful companies who are determined to maximise consumer dependency and who face little competitive pressure to satisfy smaller and more diverse markets.

Of course, it would be naïve to ignore potential theoretical contradictions in this ecological critique, for although it opposes the corporate manufacturing of standardised agricultural commodities, it also seeks to draw attention to the danger of genetic scientists producing unintended mutations and novel organisms with no evolutionary precedents. This apparent contradiction partly explains why some sections of the Left have embraced the new biotechnologies as a means for humans to resist uniformity and the disciplinary imposition of supposedly natural categories and boundaries. Thinkers such as Donna Haraway reject the distinction between nature and culture, 'insist on noise and advocate pollution', and, envisaging humans themselves 'to be cyborgs, hybrids, mosaics, chimeras', seek to construct a 'kind of postmodernist identity out of otherness, difference, and specificity'.[61]

From an ecological rather than philosophical perspective, however, and in the context of the non-human natural world, the crossing of the barriers between different species may cause a significant decline rather than growth in biological diversity and difference – especially when the transgressing organisms are equipped with new and aggressive traits. The consequences will be not endlessly unique hybrids but rather super-species capable of displacing a wider range of genotypes, and even of causing the extinction of different species affected by a radically changed ecosystem and disrupted food chain. Small populations of rare plants may, for example, be so overwhelmed by pollen from sexually compatible transgenic neighbours that they may be hybridised to the point of extinction, or indeed given disadvantageous traits which weaken their chances of survival.[62]

Apart from the threat posed to the aesthetic and functional (industrial, medical, scientific) value of ecological heterogeneity, loss of biodiversity makes cultivated crops extremely vulnerable to attacks by species-specific diseases and pests which, in contrast to the effects of crop-rotation and intercropping, grow in strength when guaranteed a regular and unchanging supply of food.[63] Crops engineered with a single gene to be pest- or disease-resistant are also likely to offer less stable protection than those which have been selected by successive generations of farmers for their polygenic ability to survive hostile conditions.

Biodiversity also ensures the survival of the genetic reservoirs of traits which farmers have always used to maintain and renew the vitality of domesticated plants and animals, and to modify them in response to changing environmental conditions. As Lappé and Bailey point out, numerous famines over the last few centuries can, notwithstanding contributory political factors, be traced back to excessive dependence on a narrow range of food sources.[64] The *Phytophthora infestans* fungus which precipitated the calamitous 1845–49 Irish Potato Famine, for example, and which also brought famine to the West Highland crofting communities of Scotland, spread with such devastation because of the genetic susceptibility of the two crop varieties from which most European potatoes descended. The same blight struck the Andes in South America – from where the European potato originated, and from where European potato farmers were subsequently restocked with resistant varieties – but the hundreds of different varieties grown in the region meant that food security was never threatened. More recently, the severity of the US corn blight epidemic of 1970, which destroyed at least twelve per cent of the nation's crop, has also been attributed to genetic uniformity, with eighty per cent of the hybrid corn grown in that year possessing an intentionally-bred trait called 'male sterility factor'. Introduced to prevent the self-fertilisation of the plants and aid hybridisation, the genetic basis to the trait also rendered the crop sensitive to the toxin produced by a rare strain of the Southern leaf blight fungus.[65] This was another clear case of pleiotropy.

The Ecological Roulette of Novel Organisms

A further risk associated with transgenic plants is that they will be released into environments which have not evolved any ecological constraints capable of containing them (or indeed into unstable social and political environments where long-term safety precautions – such as the establishment of refugia for insects – may be difficult to maintain). It is unlikely that genetically engineered crops which reach the global marketplace will have undergone extensive trials in all the countries, with their different climates and ecosystems, in which they are sold. Since drugs and chemicals that have been banned in the rich nations are routinely exported by Northern multinationals to the poorer countries of the world, the approval of transgenic plants by Western governments is in many cases likely to be interpreted by countries with underdeveloped and under-resourced regulatory systems as an unquestionable seal of safety.[66] The export of transgenic plants to novel ecosystems may well magnify the risks already associated with the introduction of any transgenic organism into an environment which has not evolved in concert with it.

It is true that some transgenic organisms contain genes designed to yield completely artificial, trivial or cosmetic traits. Traits such as improved taste, shape, colour, texture, or longer shelf-life, do not confer on the organism any adaptive advantage over its natural competitors, and may not therefore pose a serious threat to surrounding ecosystems. It is also true that the hosts of novel genes are themselves sometimes too ecologically incompetent and domesticated to proliferate in the wild (corn, for example, has been cultivated for hundreds of years in the US, but has never established itself as a wild plant, and indeed all crops which have no local wild relatives – such as corn or wheat in Britain – are unlikely to hybridise).

With eleven of the worst eighteen weeds in the world grown as cultivated crops in one country or another,[67] however, ecologists have good reason to be concerned that some transgenic organisms, aided by novel genetic material and possibly crossing with other related species, may become uncontrollable pests, predators or diseases.[68] This is certainly a risk for crops (such as rape, raspberry, radish, or blackberry) with close weedy relatives, which may only require one or two new genes conferring ecologically advantageous traits to transform them into persistent or invasive weeds.

The most frequently made analogy here is with the transportation of foreign species into hospitable environments which have not evolved appropriate mechanisms for keeping the growth of those species in check. Although non-native species account for more than ninety-eight per cent of America's food output – and an economic value of around $800 billion a year – the estimated 50,000 non-indigenous species thought to have been introduced to the US by humans are now the cause of major environmental problems, which cost more than $138 billion a year in lost resources and counter-measures. Over forty per cent of the species classified as threatened or endangered in the US are at risk primarily because of non-native organisms,[69] making non-native species the second-highest cause of biodiversity loss (after habitat destruction by humans). It should be remembered that many plant species (cheatgrass and purple loosestrife in the US, rhododendron in Britain) or microbial plant diseases (chestnut blight, Dutch elm disease) which were deliberately or inadvertently introduced to Western Europe and North America in the nineteenth century, have taken a hundred years or more to become serious threats to indigenous species and their ecosystems, suggesting that the benefits of a precautionary approach to regulating the introduction of novel organisms may not be verified during the office or even lifetime of the regulators themselves.

The kudzu vine, for example, which is native to Japan and China, was transplanted to North America in the late nineteenth century initially as an ornamental plant providing shade for southern homes. Subsequently used to combat soil erosion on steep banks, and also promoted as a forage crop, the vine soon spread out of control, infesting an estimated seven million acres in the south eastern US. Similarly, the very thirsty paper bark tree was imported from Australia nearly a century ago to help dry out Florida's Everglades swamps, but is now the state's most pernicious and prolific weed. In the south west of the US, the heavy-drinking salt cedar, originally a Mediterranean tree, contributes to water shortages by draining reservoirs, springs and waterways. Another example is large crabgrass, which was introduced to the US from Europe in the nineteenth century to produce grain, but which has persisted as a serious weed after it was replaced by corn and wheat. And Johnson grass, which was introduced from Africa as a forage crop around the same time, is now considered one of the world's ten worst weeds (it has also hybridised with sorghum crops, producing a vigorous weed in sorghum fields which is difficult to identify and eradicate).[70]

In the UK, some native animals – including the red squirrel, and Britain's only native crayfish – have been seriously threatened by aggressive competition from their imported North American cousins, while in Lake Victoria in East Africa half of the original 400 indigenous species of *haplochromis* fish have been rendered extinct due to the delayed effects of the introduction of the Nile perch by the British in the 1950s.[71] Back in the UK, the plant now causing greatest concern is knotweed, which was introduced to Britain from Japan in the early nineteenth century as an ornamental plant. Strong enough to push through concrete and road surfaces, the plant is today the country's most invasive weed, costing millions of pounds in short-term control measures. Government biologists are now deciding whether to risk new ecological problems by importing from Japan some of the natural pests – mainly beetles and fungi – which control the growth of knotweed in its native habitat.[72]

Risk-Assessment: A Flawed Orthodoxy

Most ecologists argue that field trials – the final stage of assessment before new plants are passed fit for release – are themselves a seriously deficient means of assessing the ecological risks of genetically modified crops, since they lack a meaningful geographical and temporal scale.[73] Most field tests involve releases on no more than ten acres of land, whereas commercial cultivation of staple crops would involve millions of acres in potentially diverse climates and habitats. Risks also tend to be formally evaluated in terms of probability of occurrence rather than magnitude and time-scale of effect, with the result that the very small chance of a devastating or cumulative effect is often deemed more acceptable than the risk of a less serious but more immediate and measurable threat.

The first government-approved release of a genetically engineered organism for field-testing in the United States illustrated, for many ecologists, the refusal of scientists to consider the long-term consequences of the products of recombinant DNA technology. This was the 'ice-minus' microbe, so called because the sequence of DNA in its cells associated with the formation of ice-crystals had been deleted. In its natural form the *Pseudomonas syringae* bacterium is common in temperate regions of the world, where its tendency to attach to plants and nucleate ice crystals in cold weather can cause widespread frost damage. Environmentalists pointed out, however, that the ice-making organism probably plays an important role in shaping precipitation patterns and climatic conditions. If the genetically engineered ice-minus microbe were commercially released over millions of acres of land, thus replacing the wild-type *P. syringae* in major agricultural regions of the world, catastrophic climate change, ecologists argued, could be a possible result.[74]

When the risks of a transgenic organism are described as being 'very small', in other words, a fundamental deception is being propagated since, in the absence of a full-scale release, the magnitude of risk is at best unknown and unpredictable. Yet this criticism is not necessarily a deterrent to scientists working in the industry, who are increasingly guided by the 'reactive' principle of risk assessment and regulation, which was dominant in the scientific community until the 1970s, and is today being reintroduced under the pressure of commercial lobbyists. The reactive principle discounts all but strictly foreseeable risks, and allows new technologies and scientific practices, once they have successfully passed through basic laboratory and field trials, to be tested *in situ* – in effect by monitoring their actual use. This is how it was discovered, years after their introduction, that organochlorine insecticides like DDT built up in the fatty tissue of insects and their predators, eventually affecting, often lethally, the animals at the top of the food chain. The rehabilitation of the reactive principle reflects the financial concerns and political influence of an industry keen to avoid the cost of extensive testing and subsequent product delays.

Toxic Tendencies: Implications for Human Health

In the light of these general risks associated with genetic engineering in agriculture, critics have consequently identified three main areas of concern for human health.

The first of these is the danger that allergens or toxins, old or new, will be produced in genetically engineered (or derivative) food products. This hazard derives from the random placing of foreign genes, the use of powerful promoters, the general disturbance which genetic engineering may cause to an organism's natural metabolic functioning, and the fact that genes may be used from organisms which have never before been part of the human diet. Unintended chemical changes are also more likely to emerge when the transgenic organism is under stress (as occurred with the first Monsanto soybean harvest mentioned earlier), which also means that it may be years before the deleterious effects of the plant's altered genome become evident. If completely novel allergens, immuno-irritants, or poisonous substances are generated in the transgenic organism, these may also pass undetected by safety assessors and be revealed only when they have begun to affect consumers' health.

This problem is particularly pertinent to allergies, which tend to develop through repeated exposure to allergenic substances – thus making short-term assessments of potential allergens an inadequate method of appraising food safety. It is now known, for example, that soya – a product which has for a long time been thought of as a healthy protein-rich alternative to dairy products, and one suitable for lactose- or casein-intolerant infants – has become one of the top ten foods most likely to cause an allergenic reaction.[75]

Well-known cases of the unintended production of harmful substances through genetic engineering include the production of dangerous levels of a mutagenic substance, methylglyoxal, in yeast genetically engineered for faster fermentation,[76] and the unintended production of nut allergens in soybeans engineered by Pioneer Hi-Bred International, who had transferred a gene from the Brazil nut in order to produce soybeans with higher levels of growth-enhancing amino-acids.[77]

In another case, a lawsuit was launched in April 2000 by lawyers representing the victim of a synthetic form of human insulin manufactured using genetically engineered microorganisms. A minority of the diabetics who use the synthetic insulin, both in North America and Europe, have complained of serious side effects, including personality changes and insensitivity to changing blood sugar levels. A report highlighting the problems faced by some users was earlier suppressed by the British Diabetic Association in 1999 for being 'too alarmist'.[78]

By far the most disturbing case on record, however, relates to the amino-acid food supplement, L-tryptophan. Although the product had been extracted from fermented natural bacteria by a number of Japanese firms for several years, in the late 1980s one such firm, the Japanese petrochemical company Showa Denko, decided to increase their productivity levels by switching to a genetically engineered bacterium. In the Autumn of 1989, only months after the new product had hit the market, thousands of people across the US and elsewhere fell ill, poisoned by toxins which caused raised numbers of white blood cells and severe muscle pain. The epidemic of eosinophilia-myalgia syndrome eventually left thirty-seven people dead and 1500 permanently disabled.[79]

Because the genetically engineered product was not labelled as such, it took months for the US authorities to trace the source of the poisonings. By the time Showa Denko were implicated they had destroyed all stocks of the transgenic bacterium, and claimed that the most likely cause was cost-cutting at the filtering stage. Since no toxins had been identified in the original, non-transgenic bacterium, however, and since cutting corners in the purification process was commonplace right across the industry, most scientists are agreed that it was genetic manipulation and the raised levels of tryptophan biosynthesis which generated the deadly poison.[80] Undisclosed compensation claims were settled out of court, allowing the company to escape legal censure and close the case. The failure in April 2000 of the European parliament to endorse proposals to make biotechnology firms legally responsible for any harmful effects of their products, suggests that European politicians and policy makers have learned little from this tragedy.[81]

One of the most important things to arise from the tryptophan case was the discovery that a protein expressed by the same genetic sequence may critically differ according to the organism in which that supposedly determining sequence is located. In this case the degree of chemical difference between the toxic and non-toxic products was thought to be less than 0.1 per cent. Yet contamination levels far in excess of this concentration must be identified if comparable products – riboflavin, vegetarian rennet, the numerous bacterially-derived enzymes used as preservatives and stabilisers in a whole range of processed foods (none of which must be labelled as deriving from transgenic organisms) – are to be recognised by most legislative bodies as unfit for consumption.

A final danger in this category relates to the engineering of plants with inbuilt insecticides, such as the Bt toxin, as described earlier. While the use of the Bt toxin has a good safety record when applied in topical form, scientists cannot infer from this the safety of plants which continually produce the toxin internally, and thus largely protected from the degrading effects of UV light, since this may result, over a long period, in the exposure of human beings to high levels of the toxin.[82] Nor should they assume that pesticidal proteins with proven safety records will behave identically when produced in a foreign organism. When Arpad Pusztai and his colleagues at the Rowett Institute in Scotland fed rats on potatoes which had been genetically engineered to produce an insecticidal toxin – the lectin *Galanthus nivalis* agglutinin (GNA) – originally produced by the snowdrop flower, the rats exhibited damage to their intestines and metabolism. The fact that less adverse effects were recorded among rats fed potatoes that had been externally spiked with the same snowdrop lectin – GNA extracted from snowdrops is widely believed to be harmless to mammals – suggested to Pusztai that it was the internal expression of the snowdrop transgene, or the effects of the re-engineering of the potato genome with a viral vector (the cucumber mosaic virus), which caused the pathology.[83]

Since the US Environmental Protection Agency has not approved the Bt toxin for human consumption, deeming it safe only for animal feed, the discovery in the US in September 2000 that packets of the corn snacks known as 'taco shells' contained traces of Bt corn, understandably provoked huge concern amongst consumer groups, a desperate attempt to recall nearly 300 different products, a raft of compensation claims against the French pharmaceutical giant Aventis SA who produced the 'Starlink' corn, and strained relations between the US trade officials and major corn importers such as Japan.[84] In the wake of this débâcle, Don Westfall, the vice-president of a Washington-based consulting company reporting to the food industry, inadvertently disclosed to the press the underlying strategy of the biotech food companies: 'The hope of the industry is that over time the market is so flooded that there's nothing you can do about it.'[85]

Deadly Microbes

The second area of concern for human health and well-being revolves around the dangers associated with the vectors and markers used in genetic engineering which some scientists, such as Mae-Wan Ho, believe may contribute to the growth of new and more virulent infectious diseases. The risk of reactivating dormant or crippled viruses through the use of the cauliflower mosaic virus promoter has already been mentioned. Many scientists are also concerned that the genetic engineering of viruses and bacteria will inevitably result in the accidental creation of virulent new microbes – such as the lethal mousepox virus unintentionally produced by Australian researchers attempting to create a contraceptive vaccine for feral mice[86] – and that such accidental discoveries may in turn increase the risks of and opportunities for bio-terrorism.

To the dangers of bio-terrorism we must also add the hazards associated with the widespread use in genetic engineering of antibiotic-resistant marker genes. As previously discussed, genes from bacteria that code for resistance to antibiotics are routinely used in genetic engineering as a means of identifying, by the application of antibiotics, those cells which have successfully assimilated the new DNA sequence. The evidence already mentioned that DNA is not always broken down in the gut, that it can enter the blood stream and internal organs of mammals, and that it can be transferred to bacteria inhabiting the bodies of animals and humans,[87] raises the small but potentially catastrophic risk that the presence of antibiotic marker genes in human food (Zeneca's tomato purée carries a widely-used marker for resistance to kanamycin, for example) and animal fodder (a gene conferring resistance to ampicillin is used in Novartis's Bt maize), will accelerate the already disastrous decline in the medical efficacy of antibiotics in the treatment of infectious diseases.[88]

The refusal by the European Union in 1999 to allow the sale of Monsanto's herbicide-tolerant and Bt cotton in Europe seems to have been informed by this concern, which was also expressed in a report by the UK Advisory Committee on Novel Foods and Processes (ACNFP). Both Bollgard and Roundup Ready cotton contain a gene – called *aad* – which confers resistance to the antibiotics streptomycin and spectinomycin. Streptomycin is largely used as a second-line treatment for tuberculosis. But according to Ho it is also the drug of choice for treating strains of gonorrhoea already resistant to penicillin and third-generation cephalosporins, especially during pregnancy.[89] With sixty per cent of the cotton crop harvested for human (seed oil) and animal (seed cake) consumption, and with cotton fibres often worn against vulnerable contact areas in the form of nappies, sanitary towels, tampons and bandages, the millions of hectares of genetically modified cotton already being grown in the US and China may therefore represent a significant risk to public health.

It is well known that the dramatic increase in virulent superbugs and antibiotic-resistant infections over the last two decades has its origins in the reckless use of antibiotics in medicine and farming, the industrialisation of which is faithfully paralleled by the mechanical mind-set that pervades the biotech industry. Since antibiotics were introduced in the 1940s, Ho recounts,

> Resistance to penicillin, ampicillin and anti-pseudomonas penicillins in *Staphylococcus aureus* went from approximately nil in 1952 to more than 95 per cent in 1992. By the eighties, *S. aureus* had also developed high levels of resistance to the synthetic penicillin methicillin and all other ß-lactams. The new fluoroquinolone anti-microbial, ciprofloxacin, was introduced in the mid-eighties; resistance to it had reached more than 80 per cent by 1992. A study carried out by the Centres for Disease Control showed that ciprofloxacin-resistance in *S. aureus* went from less than 5 per cent to more than 80 per cent within one year. Recent data shows that vancomycin-resistance in *Enterococci* in hospitals in San Francisco grew from 3 per cent in 1993 to 95 per cent in 1997. In Italy there was a twenty-fold increase in erythromycin-resistance in *Streptococcus* between 1993 and 1995.[90]

Horizontal gene flow is now recognised as a central vehicle for the generation of new strains of bacteria, and for the transfer of antibiotic resistance from one bacterium to another. If the instability and promiscuity of transgenic DNA does indeed accelerate the rate of cross-species gene transfer and conjugation, then deliberately arming plants with antibiotic-resistant genes is hardly compatible with current medical needs. It goes without saying that the biggest victims of this industrial nemesis will be the poor populations of the developing world.[91]

Pesticide Pollution

The third potential danger which genetic engineering in agriculture may pose to human health derives from the risk that potentially toxic pesticides will be used in even greater amounts as a consequence of the spread of transgenic herbicide-tolerant crops. We have already seen that, by the end of the 1990s, herbicide-tolerant plants constituted around three-quarters of the total global acreage given over to genetically engineered crops. Poisoning from chemical pesticides is estimated to affect twenty-five million agricultural workers world wide, most of whom live in the Third World.[92] In the UK, pesticide drift onto hedgerows and field margins is a key factor contributing to the dramatic decline in wild bird populations in recent years, while in the developing world pesticide drift, like the general decline in biodiversity, also poses a threat to those who depend on wild plants and 'weeds' for their own basic survival.[93]

Unsurprisingly, herbicide-resistant crops are being promoted by their manufacturers as a solution to excessive chemical contamination of rural land and waterways, mainly because they allow farmers to limit their spraying by targeting the most sensitive stages of weed growth. Interestingly, however, the evidence suggests the reverse is occurring. On a global scale, of course, total sales of herbicides are continuing to rise, mainly due to greater penetration of Third World markets. But the industry's claim to be reducing herbicide use per hectare is also beginning to look suspect. When United States Department of Agriculture (USDA) data from over 8,000 university-based soybean trials conducted between 1996–98 was examined by Charles Benbrook, he discovered that, as well as suffering diminished yields, farmers growing Monsanto's Roundup Ready transgenic soybeans used between two and five times more herbicides than those growing conventional soybean varieties.[94]

There may be several reasons for this trend towards greater herbicide use on herbicide-tolerant crops, including growing weed-resistance, adaptive changes in weed growing patterns, the emergence of more stubborn weeds in place of the destroyed ones, aggressive marketing, and the general invitation which 'tolerant' crops offer to farmers to apply their herbicide more liberally. With mounting evidence of herbicide drift from farms growing transgenic crops, the pressure on adjacent farmers to convert to herbicide-resistant varieties is also likely to increase. Indeed, with sales of Monsanto's Roundup – the world's biggest-selling herbicide – crucial to the company's success, an overall reduction in consumption levels would be alarming news for investors.

Roundup is typically forty per cent glyphosate, an organophosphate which at high levels causes symptoms of liver toxification in animals, and which is banned from commercial use in the UK. Its toxic effects are particularly pronounced in fish, and it can poison earthworms and mycorrhizal fungi, both of which play a role in recycling soil nutrients.[95] In 1999 a confidential European Union report revealed that glyphosate was poisonous to insects and spiders, many of which are natural predators of crop pests.[96] It was the third highest cause of pesticide-related illness in California in the early 1990s,[97] and its growing use has been linked by Swedish oncologists to the rapidly increasing incidence of non-Hodgkin's lymphoma since the late 1950s.[98] In an extensive report assembling the results of numerous laboratory and *in situ* studies of glyphosate toxicity, the documented findings included 'medium-term toxicity (salivary gland lesions), long-term toxicity (inflamed stomach linings), genetic damage (in human blood cells), effects on reproduction (reduced sperm counts in rats; increased frequency of abnormal sperm in rabbits), and carcinogenicity (increased frequency of liver tumours in male rats and thyroid cancer in female rats)'.[99]

Glyphosate-tolerant crops may therefore pose an unprecedented threat to animal and human health, and not just because they are likely to consolidate a trend which has seen glyphosate usage increase since the mid-1990s at a rate of between fifteen and twenty per cent per year.[100] The threat is particularly great because these crops, rather than picking up accidental residues from spray drift and soil contamination like their conventional predecessors, will be directly exposed to the herbicide by farmers who no longer have to discriminate between the crop and the weed. Why else did Monsanto, for all its claims to be championing the cause of pesticide reduction, successfully lobby the US Environmental Protection Agency to raise the maximum permitted level of glyphosate residues in soybeans by over 200 per cent (from six parts per million to twenty parts per million),[101] then persuade the British government in 1997, for the sake of the US export industry, to harmonise their safety standards with those of the US (which in the UK meant raising the permitted level of glyphosate residues by 20,000 per cent[102])?

The Organised Irresponsibility of Science

Raising questions like these exposes the limitations of a strictly scientific analysis of the ecological implications of genetic engineering, since these questions can only be satisfactorily answered if we take into account the social and economic context in which biotechnology and its products are being developed and sold. Only a moment's reflection is required, for example, to see that most of the ecological risks associated with genetically engineered crops are both nurtured and magnified by an economic culture of competitive short-termism which is inimical to responsible and sustainable farming. When weeds, insects or diseases develop resistance to human treatments, when toxic residues build up over decades in the animal and human food chain, when market pressures drive farmers to restrict the use of refugia and to embrace more chemical intensive and biologically invasive forms of agriculture – when these collectively harmful phenomena occur, blame and accountability seem impossible to ascribe, for it is precisely the displacement of social responsibility from the market arena which makes collective care for the environment so difficult to achieve.

Although responsibility may *seem* impossible to attribute, however, appearances can be deceptive. For what is unintended is not necessarily unforeseen or unforeseeable. Moreover, what *is* unforeseeable is not *whether* there will be risks associated with the new biotechnologies – the existence of these risks is today rarely denied – but rather how the risks should be assessed and managed. Should science be allowed to assess such risks by *taking them*? Or is this practice, which is exemplified in the 'trial release' of genetically modified organisms, merely an attempt to deepen the system of what Beck felicitously calls 'organised irresponsibility'?[103] By permitting increasing chemical pollution, the dispersal and migration of novel genetic sequences, the crossing and confusion of the boundaries between different organisms, between different species, and between the natural and the artificial, the logic here seems to be to make risk and transgression such a ubiquitous feature of modern life that identifying the origins and agents of risk, as well as the 'natural' properties and behaviours of things by which the magnitude of such risks might be accurately measured, becomes impossible.

The evaporation of responsibility may be the final destination of these developments, but we are not there yet. Almost by definition, risk signals the limits of expert knowledge, the penumbra of uncertainty that surrounds the scientific project. 'Risk society' – which is today probably the most fashionable concept in the social sciences – is a society where science has lost its halo of neutrality. The controversies surrounding genetic engineering illustrate how increasingly risk-ridden technologies and practices undermine the power and influence of an expertocracy whose disinterested authority and condescending reassurances were in the past rarely questioned. For perhaps more than any other science, genetic biology cannot investigate it's object, cannot produce raw and insightful knowledge about nature, without simultaneously engaging in an uncertain and experimental alteration of that object. By inadvertently removing the veil of scientific objectivity, the development of biotechnology may thus have opened up a democratic space for the ethical evaluation and control of science. The problem, as we shall see in the next two chapters, is that there are strong economic interests seeking to close down that space.

Why Europe / U.S. so diff?

- EU valuing local food / growth
- US valuing spread of capitalism,
- Culture in EU is older more meaningful than modern capitalism
- Chain restaurants in U.S. vs local cuisine in EU

- Ethics of Genetically modifying humans before born

→ the market is older than the farm in the U.S., mindset is farm to make money, its the opposite in Europe

3

The Life Science Industry

Critical reflection on the full implications of biotechnology in agriculture requires more than a scientific appreciation of ecological risks. It also requires that we consider the social and economic context, the relations of power and prevailing distribution of influence and need, into which biotechnologies are inserted, and which determine the focus and goals of the conception, development and application of those technologies. As I shall argue more fully in the next chapter, the technology of genetic engineering lends itself to a particularly acute concentration of wealth and power, which has been consolidated in the western world by a legal and regulatory environment unsympathetic to the interests of the poor. In this chapter I want to focus on the most explicit manifestations of this power, and show how this has seriously undermined proper public scrutiny of the biotechnology industry.

A Century of Monsanto

As one of the leading biotechnology firms whose genetically altered seeds accounted for over ninety per cent of the global transgenic crop area in 2001,[1] Monsanto is rightly regarded as the most formidable commercial incarnation of the new genetic science. Founded as a chemical company in 1901 in

St. Louis, Missouri, where its headquarters still lie, the company gained fame for developing and producing Agent Orange – the dioxin-rich defoliant used to destroy native ground cover and crops during the Vietnam War – and has been implicated in several cases of employee and residential contamination, which in the Missouri town of Times Beach became so serious that the government ordered its evacuation in 1982. The company was also judged to have submitted fraudulent reports on its products to the US Environmental Protection Agency (EPA),[2] and in February 2002 embarrassingly lost a court case against lawyers representing 3,500 residents of a small town in Alabama, who accused Monsanto of covering up for several decades its contamination of local rivers and land with the widely-used industrial coolants 'PCBs'. Some of the plaintiffs were found to have levels of PCBs in their blood twenty-seven times higher than the national average, and many claim that cancer rates in the town are abnormally high. Although Monsanto closed the local factory in 1971 and ceased producing PCBs altogether in 1977 – shortly before they were banned for being, in the eyes of EPA and the World Health Organisation, 'probable carcinogens' – documents obtained by the *Washington Post* now suggest that by the mid-sixties the company was well aware of the toxicity of these chemical discharges, and as early as 1935 had acknowledged in an internal memo that PCBs 'cannot be considered non-toxic'.[3]

Monsanto's commercial portfolio includes the food and beverage sweetener, aspartame, thought by some neurologists to be a possible cause of brain cancer,[4] and recombinant bovine growth hormone (rbGH) – now better known as bovine somatotropin (BST). This hormone, which is produced from genetically engineered *E. coli*, is used to boost milk yields in cows in the US (and several Third World countries), but banned in Canada, Europe, and just about everywhere else, due to concerns over animal and human health.[5] By 1996 Monsanto was the fourth largest chemical firm in the US, and accordingly earned the dubious distinction of being one of America's top polluters, producing around five per cent of the total 5.7 billion pounds of toxic chemicals released into the environment in the US in 1992.[6]

Between 1996–98, having already declared its intention to reduce its toxic discharges, Monsanto used an $8 billion spending spree to switch its interests fully to the biotechnology field. Though it retained its agrochemical divisions – and hence its flagship herbicide Roundup – the company relinquished its $3 billion industrial chemical and synthetic fabrics operations (spun off in 1997 as a separate company called Solutia) in order to concentrate its investment in the 'life sciences'. Buying up a number of agricultural biotech businesses (including the California-based plant biotech company Calgene, and the W.R. Grace subsidiary and holder of a broad patent on genetically engineered soybeans, Agracetus), and making partners of others (Incyte Pharmaceuticals and GeneTrace), it also strengthened its strategic hand by acquiring several major seed firms. These included Asgrow Agronomics, DeKalb Genetics, the top corn-seed seller Holden's Foundation Seeds, Cargill's international seed division, Unilever's European wheat breeding business (Plant Breeding International – originally a public institution based at Cambridge University), and seed firms in India and Brazil (including Agroceres, the largest seed company in the Southern Hemisphere). Despite failing in its protracted efforts to acquire Delta and Pine Land, the company which partnered the US Department of Agriculture in developing and patenting the notorious 'terminator' seed sterilisation technology, by 1998 Monsanto had transformed itself into the second largest seed company in the world.

Monsanto's aggressive development and marketing of genetically modified crops has, however, not been kind to its public image, and over the last three years its fortunes have fluctuated. As consumer boycotts in Europe eventually provoked the suspicions of American consumers, its stock price fell by nearly two-thirds, and in March 2000 the company was bought up by the pharmaceutical giant Pharmacia and Upjohn (which was renamed Pharmacia Corporation). Pharmacia's principal interest was Monsanto's pharmaceutical unit, G.D. Searle, which developed the best-selling arthritis drug Celebrex, and in November 2001 the company announced it would be selling back to shareholders its majority stake in Monsanto's agro-biotechnology operations, claiming that the agriculture industry was too cyclical and, with genetic engineering thrown in, too contentious to reassure nervous investors.[7]

Monsanto remains a powerful player in the agro-biotech marketplace, however, not least because it continues to produce the world's best-selling herbicide. With sales of Roundup making up over fifteen per cent of Monsanto's total sales and half of its operating profit in the mid 1990s,[8] the company has a strong commercial interest in consolidating its lead in the production of herbicide-resistant seeds. The advantages of this technology became particularly apparent in 2000, when the company's patent for the Roundup herbicide expired. Roundup Ready cotton, rapeseed, corn and soybeans look like being the means by which the company will maintain its market dominance, as it prohibits buyers of those seeds from using any herbicide formulations other than its own, and uses intellectual property law to ensure that the technology involved in producing those seeds is not utilised by competitors.

In addition to paying a patenting or 'technology fee' charged by Monsanto at a 1998 rate of between $6.50–$32 per acre according to the particular crop,[9] farmers who buy Monsanto's herbicide-resistant products are legally contracted to refrain from using any glyphosate herbicide other than Roundup. The same 'technology user agreement' (TUA) also serves as an explicit contractual statement of Monsanto's patent rights, which forbids the buyer from re-sowing any seeds produced from the crop that is planted. An infringement of this contract will lead to penalties equal to '100 times the then applicable fee for the Roundup Ready gene, times the number of units of transferred seed, plus reasonable attorney's fees and expenses' – plus damages. Any signatory to the 1996 TUA also found themselves granting 'Monsanto, or its authorised agent, the right to inspect and test all of Grower's fields planted with soybeans and to monitor Grower's soybean fields for the following three years for compliance with the terms of the Agreement'.[10]

With around 25 per cent of the North American soybean crop currently grown from saved seed,[11] Monsanto had good reason to suspect that contravention of the terms of this agreement might be widespread, and duly dispatched detectives from the Pinkerton Agency, famous for its role in breaking early US trade unions, to investigate. By 1999 over 500 farmers in the US and Canada had been sued by Monsanto or were awaiting lawsuits, while over 100 had already avoided potentially ruinous legal action by destroying their own crops. In one extensively reported case, a Canadian farmer who found his rape fields contaminated with Roundup Ready plants which he said he had neither bought nor sown, and who was sued by Monsanto to boot, in turn launched a lawsuit of his own, accusing Monsanto of contaminating his seed stock, trespassing on his land, defaming his name and stealing his plants.[12] The tactics used by Monsanto to prevent what it sees as theft of its property included the setting up of a free telephone hotline inviting farmers to incriminate their neighbours, the placing of adverts in farming journals warning readers of its aggressive legal response to the planting of pirated seeds, and the use of radio advertisements to name farmers who had been caught.[13]

The Life Science Industry

Monsanto may have attracted the lion's share of public criticism, but the company is far from being the sole player in the agro-biotech industry, and its status and strategy are not unique. Indeed, the recent history of the company tells us some important things about the nature of developments in an industry which, despite Monsanto's public fall from grace, has seen biotech assets more than triple in their market value from $97 billion in 1998 to $350 billion in early 2000.[14] There is today an extraordinary concentration of economic power in the life-science field, with giant corporations spreading their investments over a range of related industries, including agriculture, food processing, and medicine. Propelled by mergers, strategic alliances and joint ventures between leading firms, the pattern of development is pointing towards the emergence of clusters of multinationals co-operating in achieving complete command of the food chain, from patent protection of transgenic germplasm, through chemical-assisted growing to the collection and distribution of harvests and their processing into food.

In contrast to the well-documented trend towards 'vertical disintegration' in post-Fordist industries, the goal here seems to be what Bill Heffernan describes as a 'fully vertically integrated food system from gene to shelf. Within this emerging system', he continues,

> there will be no markets and thus no "price discovery" from the gene, fertiliser processing and chemical production to the supermarket shelf . . . In a food chain cluster, the food product is passed along from stage to stage, but ownership never changes and neither does the location of the decision-making. Starting with the intellectual property rights that governments give to the biotechnology firms, the food product always remains the property of a firm or cluster of firms.[15]

The situation described in Heffernan's analysis conforms precisely to what Jeremy Rifkin calls the 'Age of Access' – a world in which 'markets give way to networks, sellers and buyers are replaced by suppliers and users, and virtually everything is accessed'. In the Age of Access, 'the farmer is granted only short-term access to someone else's intellectual property. The seeds are never technically sold or legally purchased, only rented'.[16]

Monsanto's competitors in the life science industry included Novartis AG, formed in 1996 with the merger of the Swiss agrochemical and pharmaceutical companies Ciba-Geigy and Sandoz. By 1999 Novartis was enjoying agribusiness sales of $4.7 billion – a quarter of its total sales – placing it second in this sector of the market. The new multinational subsequently signed a five-year $25 million contract financing genomics research at the University of California at Berkeley, a deal which gave the company first claim to licence a proportion of all the research products in the laboratories, regardless of whether or not Novartis funded them. A similar deal was struck with the

University of Maryland's Psychiatric Research Centre, allowing the Swiss multinational access to the Centre's brain-tissue bank and one of its labs in exchange for $24 million over six years. Novartis also forged an alliance with Archer Daniels Midland (ADM), an American company famous for its extensive grain collection and processing network (and, less honourably, for being levied a $100 million anti-trust fine in 1995).[17]

In Germany the chemical giant Hoechst, which in 1998 sold off its major American subsidiary, Celanese, merged with the French chemical manufacturer Rhône-Poulenc, creating the world's biggest multinational life-science company, Aventis SA, with a research and development budget of $3 billion and combined yearly sales of $20 billion. This record was soon broken by the 1999 merger of UK-based Zeneca Group Plc with Astra AB of Sweden, to form AstraZeneca, the world's third largest agrochemical company and a leading pharmaceutical firm. One year later third became first, as AstraZeneca and Novartis sloughed off and merged their agribusiness divisions in November 2000, in a move which some market analysts believed signalled the beginning of the end of agro-pharmaceutical conglomerates. The new company, Syngenta, is the first global corporation focused entirely on agribusiness, with an annual turnover of $7.34 billion, ranking first in global sales of agrochemicals, third for seeds, and holding forty-two per cent of biotech patents on 'genetic use restriction technologies' (techniques that make plants yield sterile seeds or respond only to the application of proprietary chemicals).[18]

To compete in the new marketplace the German corporation Bayer took similar action, selling off its photographic subsidiary Agfa in 1998 then spending $1.2 billion to acquire the diagnostic division of Chiron, one of the world's largest biotech companies, and investing $465 million in the leading genomics drug research company, Millennium Pharmaceuticals.[19] Bayer also holds a sixteen per cent stake in PPL Therapeutics, the Scottish biotech company that cloned Dolly the sheep, and is financing trials of the cystic fibrosis drug, AAT, extracted from the milk of genetically engineered cows.[20] In 2001 the German multinational announced it had agreed to pay $6.6 billion for Aventis's crop and agrochemical division (Aventis CropScience), a deal which will make Bayer, already the world's biggest insecticide producer, the second largest agrochemical business in the world.

In 1998, Monsanto's arch-rival DuPont relinquished its major petroleum subsidiary Conoco – which accounted for nearly half of DuPont's $45 billion sales and a quarter of its income – then paid $7.7 billion to acquire full control over Pioneer Hi-Bred International, the world's largest seed company and a leading supplier of agricultural genetics. It also spent $2.6 billion on a fifty per cent share in DuPont Merck Pharmaceutical, and bought a major processor of soya protein, Protein Technologies International, its company goal being to make the life sciences the source of thirty per cent of its income by 2002.[21]

As a result of these mergers and acquisitions, only four companies – Syngenta, Monsanto, Aventis and DuPont – today account for virtually the entire global market for transgenic seeds, while the addition of the German chemical giant BASF gives these top five companies a seventy per cent monopoly of the world's pesticide sales.[22] By the end of 1998, of the 1,370 agro-biotech patents granted by the US Patent and Trademark Office to the top thirty firms, three-quarters were held by six companies (Monsanto, DuPont, Syngenta, Dow, Aventis, and Grupo Pulsar).[23] The announcement in April 2002 that Monsanto and DuPont had reached an agreement to swap key patented technologies and drop outstanding patent lawsuits, means that the six dominant patent-holding companies are now effectively five, with the Monsanto-DuPont venture accounting for forty per cent of the 1,370 agro-biotech patents mentioned above.[24] Alliances like this one, which by stopping short of a full-blown merger strategically evades the grip of the main anti-trust regulations, account for a rising share of corporate revenue, with an estimated 20,000 of such 'non-merger monopolies' formed worldwide in 1996–98.[25]

The consolidation of power and concentration of ownership in the life science industry, and the concomitant growth of food chain monopolies, means that many farmers in the industrialised world are being forced into a position of what has been called 'bioserfdom', as control over farm-level management and decision-making – which crops to grow, which inputs to use, which buyers to sell to – is usurped by the global economic machine of the multinationals, who rapidly remove their profits from circulation in rural communities and transfer them to anonymous shareholders and investors. These kinds of economic conditions only favour the high-volume 'superfarms' – the top 1.2 per cent of farms in the US, for example, which pull in nearly forty per cent of the net farm income.[26] For the rest of the farming community, the situation is precarious, with economic independence destroyed by market oligopolies, and by the decline in flexibility and choice which this entails. Mergers and take-overs also function to eliminate the competitive pressure to satisfy smaller, diverse and less profitable markets, as illustrated in June 2000 when Seminis, a subsidiary of the Mexico-based conglomerate Savia, having bought itself into the position of being the world's largest vegetable seed company, announced that it would be removing a quarter of its marketed seed varieties as part of a 'global restructuring and optimisation plan'.[27]

Markets that are dominated by a limited number of giant companies also generate perfect conditions for so-called 'unfair practices', notably price-fixing arrangements between supposedly competing firms, and subsidised price-cutting used to drive out smaller competitors. Through the elimination of competitors, such companies become capable of reaping super-profits by exploiting not just the labour of their workers – between 1983 and 1999 the 14.4 per cent increase in the workforce of the world's biggest corporations was dwarfed by a 362.4 per cent growth in their profits[28] – but also the market dependence of those who must buy, rent or consume their products. In US agribusiness oligopoly is an indisputable fact, with forty per cent of vegetable seeds coming from a single source (Seminis). Three quarters of the US corn-seed market is controlled by two firms (DuPont and Monsanto), five multinationals command three quarters of the global vegetable-seed market, while Delta & Pine Land alone accounted for seventy per cent of the US cotton-seed market in 1998.[29] Similar trends are occurring in the livestock industry, with the consequence that domestic animal breeds are disappearing at an annual rate of five per cent a year worldwide.[30] We should not forget the danger this poses for food security: with more than ninety per cent of all commercially produced turkeys in the world deriving from three breeding flocks, for example, the damage wreaked by the emergence of a deadly new strain of avian flu would be unprecedented.[31]

The Privatisation of Science

The dominance of giant transnational corporations is of course not unique to the world of farming and agribusiness. For the global economy as a whole, an estimated two-thirds of international trade is controlled by multinationals – and around half of that, more strikingly, is 'trade' between the affiliated firms and subsidiaries of single corporations, thus giving maximum opportunity for the favourite tax avoidance strategy of the transnational firm, 'transfer pricing'.[32] Measured in sales terms, of the world's largest 100 economies, fifty-one are in fact multinational companies. It is this fearsome command of capital and its derivatives which gives the corporate giants the leverage to dictate to governments, electorates and public bodies the terms of social debate and policy.

This degree of economic muscle does not bode well for the poorer communities of the world, but the consequences of corporate consolidation in the biotech industry are cultural as well as economic. Indeed, it is the ability of multinational firms to shape the nature of scientific research and public debate on genetic engineering which is vital to preserving their monopoly of this technology and its profit-oriented application. One immediate consequence of the expansion of multinational capital relative to the tax-raising powers of the nation-state, is that academics and their employers have become increasingly dependent on the private sector for funding, support and advice. This is particularly evident in the US where, since 1992, industry's annual contribution to biomedical research has exceeded the funds allocated to the discipline by the federal government.[33] For university scientists unable to command satisfactory corporate funding, the Bayh-Dole Act of 1980 and the Federal Technology Transfer Act of 1986 compensated them with the right to apply for patents on federally-funded inventions, allowing researchers to keep up to $150,000 a year in royalties on top of the salaries they receive from the public purse.[34]

Research by Sheldon Krimsky into industry–university links suggests that scientists in the US are far more integrated into the values and goals of the commercial sector than the public are led to believe. Krimsky found that more than a third of biologists and biomedical scientists belonging to the prestigious National Academy of Sciences had clearly identifiable formal ties to private companies, either working as consultants, advisors, directors or managers for those companies, or possessing corporate equity large enough to be listed in a firm's prospectus.[35] In a subsequent study, which looked at 789 biomedical papers published by university scientists in Massachusetts in 1992, Krimsky and his colleagues found that a third were written by lead authors who had a financial interest – in every case *undisclosed* – in the results they were reporting.[36]

It is not unusual now for US universities to hold stock in companies formed by faculty members, and hundreds of millions of dollars are made every year by universities which have licensed their own inventions or established venture capital entities – such as Harvard Medical Science Partners – to invest in faculty research and develop commercial products. Sponsorship is also becoming an important source of income for university research, with over a tenth of all research grants received in the British higher education sector deriving from private sources.[37] Even a hallowed academic institution like Cambridge University has opened its arms to corporate sponsorship, with professorships and laboratories funded by Shell, BP, ICI, Glaxo, Price Waterhouse, Marks and

Spencer, Rolls-Royce, AT&T, Microsoft and Zeneca, while Oxford University has accepted sizeable donations from Rupert Murdoch and the alleged arms broker Wafic Said.[38] The acceptance by Nottingham University of £3.8 million from British American Tabacco (BAT), in order to fund a new 'international centre for corporate responsibility', was seen by most health professionals in the UK as such a shameless error of judgement that it prompted the editor of the *British Medical Journal*, Richard Smith, to resign in protest from his professorial post at the university in 2001.[39] BAT is not only the second biggest tobacco company in the world, but is currently the subject of a UK Department of Trade and Industry inquiry into accusations of its involvement in smuggling and money laundering in Asia and South America, and must contest an anti-racketeering writ filed in New York by a consortium of Colombian states.

When university biologists in the UK turn instead to taxpayers for financial support, they are likely to apply to the Biotechnology and Biological Sciences Research Council, which also pays for the secondment of academics into corporations. This public funding body is chaired by a former executive director of Zeneca, and that is not all:

> Among the members of the council are the Chief Executive of the pharmaceutical firm Chiroscience; the former Director of Research and Development at the controversial food company Nestlé; the President of the Food and Drink Federation; a consultant to the biochemical industry; and the general manager of Britain's biggest farming business. The BBSRC's strategy board contains executives from SmithKline Beecham, Merck Sharpe and Dohme and AgrEvo UK, the company (now owned by Aventis) hoping to be the first to commercialise genetically engineered crops in Britain. The research council has seven specialist committees, each overseeing the dispersal of money to different branches of biology. Employees of Zeneca, according to the council's Web site, sit on all of them.[40]

Government advisory panels also tend to be staffed by scientists with close economic ties to the industry. In 1998, for example, the thirteen-strong Advisory Committee on Releases to the Environment (Acre), which assesses applications for growing genetically engineered crops in Britain, had eight members with ties to companies or organisations involved in developing biotech agriculture.[41] In the US, after the National Academy of Sciences had released a report in April 2000 concluding that there was no strict distinction between the risks posed by genetically modified crops and those associated with conventionally bred foods, it was soon discovered that most of the twelve panel members had some professional link with the biotech industry. One of the study's directors left at short-notice to take up a job working for the biotech trade lobby, the Biotechnology Industry Organisation, four of the committee's members received research funds from companies including Monsanto, Novartis and Pioneer Hi-Bred, and two others had worked as, respectively, a lawyer and an advisor for the biotech industry.[42]

The Revolving Door

The interpenetration of politics and economics in the US biotech field is in fact so pervasive and well-established that a pro-industry attitude amongst those working for government agencies is a reliable guarantee of future employment in the private sector. In the late 1990s the Edmonds Institute, the American think-tank for environmental and public affairs, identified numerous instances of American senior civil servants, including many who worked for regulatory and safety assessment bodies like the Environmental Protection Agency (EPA) and the Food and Drug Administration (FDA), taking up jobs in biotech companies, as well as several corporate scientists who moved in the opposite direction.[43]

Among the latter was Margaret Miller, former chemical laboratory supervisor for Monsanto, who became deputy director of Human Food Safety and Consultative Services at the FDA. Miller's most significant accomplishment in her new job was to raise by a factor of a hundred the maximum permissible level of antibiotic treatment for cows, thus making what may have been a historic contribution to the growth of antibiotic-resistant bacteria. She apparently took this decision in order to make the use of Monsanto's recombinant bovine growth hormone – for which she herself had signed the human safety approval in 1992 – more attractive to farmers worried about the high incidence of mastitis amongst injected cows.[44]

The formation of George W. Bush's new cabinet at the beginning of 2001 strengthened the political influence of the biotech lobby even further. Included amongst the new executive is the new Attorney General, John Ashcroft, who led calls to the Clinton administration to promote genetically manipulated crops in the Third World, and who received $10,000 from Monsanto in the 2000 elections; the new secretary of health and human services, Tommy Thompson, who as the governor of Wisconsin formerly campaigned for the biotech industry with the help of Monsanto funds; Ann Veneman, the new agriculture secretary, who was a director at Calgene before it was taken over by Monsanto; the Defence Secretary, Donald Rumsfeld, who was president of Searle Pharmaceuticals when it was bought by Monsanto in 1985; and Larry Combest, a known supporter of genetically engineered food who received $2,000 from Monsanto during the elections and is now chair of the powerful House of Representatives agricultural committee.[45]

Conflicts of Interest in Medical Research

With politics and academia so thoroughly penetrated by biotechnology interests and funding, it would be surprising if public information, debate and key political decisions about the new biotechnologies did not fall victim to commercial bias. This is certainly an observable trend in the medical field,[46] where the growing dependence on corporate finance of researchers, non-profit healthcare organisations and advisory groups has led to a catalogue of scandals and controversies, some of which deserve a mention here.

Investigative journalist Jeanne Lenzer, for example, examined the decision in August 2000 of the American Heart Association (AHA) to upgrade its assessment of a drug called alteplase from 'optional' to 'definitely recommended'. Alteplase is an enzyme, produced from genetically engineered human cell lines, which is used as a thrombolytic agent to dissolve blood clots. Many health professionals were surprised at the AHA's decision, because the majority of controlled trials of the product have correlated the use of thrombolytics in cases of acute ischaemic stroke with increased mortality. This risk was all the more unacceptable, in Lenzer's view, in light of a study which found that a fifth of patients initially diagnosed with stroke by specialists were in fact misdiagnosed. Investigating further, Lenzer found that the manufacturers of alteplase, the US biotech firm Genentech, had contributed $11 million to the AHA over the previous decade, and that the commercial presence of Genentech also reached the nine-strong advisory panel responsible for drawing up the new guidelines. One of the panel opposed the revised recommendation, but his reasoning, submitted to the AHA in written form, was never published. Six of the remaining eight were discovered by Lenzer to have financial ties with Genentech, either as consultants, speakers at company-sponsored colloquia, or as recipients of research funding. None of these interests were publicly disclosed, and two of the six panellists initially denied them.[47]

Canadian researchers have also analysed the relationship between corporate affiliations and medical opinion over the use of another controversial treatment. Taking seventy articles that had commented on whether calcium-channel antagonists, a class of anti-hypertensive drugs used to treat cardiovascular disorders, were safe enough for clinical use, they divided the writers into 'supportive', 'neutral', and 'critical' categories, then sent out surveys questioning the authors' relationship with pharmaceutical companies. They found that ninety-six per cent of those who were supportive of the drugs had financial ties with at least one manufacturer of them (compared with sixty per cent of the neutral authors, and thirty-seven per cent of the critical ones), and this figure rose to 100 per cent when ties with any drug company were declared (compared with sixty-seven per cent of neutral authors and forty-three per cent of critical ones).[48]

In another study, American researchers analysed 106 articles published between 1980 and 1995 reviewing the scientific evidence on the health effects of passive smoking. Although only thirty-seven per cent of reviews concluded that passive smoking was not detrimental to health, three-quarters of these (seventy-four per cent) were written by authors known to have received funding from or participated in activities sponsored by the tobacco industry. With less than seven per cent of authors with industry affiliations agreeing with what is now an established medical fact, the researchers concluded that 'the only factor associated with concluding that passive smoking is not harmful was whether an author was affiliated with the tobacco industry'.[49]

Researchers have also looked at the way commercial interests influence the economic evaluation of medical treatments. One such study in the US examined the relationship between pharmaceutical industry sponsorship and published assessments of the cost-effectiveness of new cancer drugs. Whilst thirty-eight per cent of research papers funded by non-profit organisations reported unfavourable conclusions regarding the cost-effectiveness of the particular drug studied, only five per cent of pharmaceutical company-sponsored studies came to the same conclusion. Unsurprisingly, the percentage dropped to zero when the researchers considered the articles written by scientific teams which included a pharmaceutical or consulting firm employee.[50]

More specific research on the relationship between universities and the biotech industry itself has replicated the results of studies made in the medical domain. Researchers have found, for example, that company-sponsored academics in biotech faculties are four times as likely to report trade secrets from their work than those without such funding, and four times as likely to allow considerations of commercial applicability to influence their choice of research.[51] The same researchers also examined the prevalence and the effects of corporate gifts – discretionary funding, equipment, bio-materials, etc. – on life-science research. Of the 2,167 respondents, forty-three per cent said they had received a corporate gift independent of a grant or contract in the last three years, and two thirds of those said the gift was important to their research. Unsurprisingly, the gifts came with conditions attached: sixty per cent said the gift was not to be passed to a third party, and fifty-nine per cent said that the gift could only be used for the pre-arranged purpose. More worryingly, a third reported that the donor wanted pre-publication review of any research texts deriving from use of the gift, a similar proportion said the company expected the recipient to provide evaluation of the product supplied, and a fifth reported that the donor company wanted ownership of all patentable results from the research in which the gift was used.[52] In a separate study, the same team found that a fifth of life-science faculty had delayed publication of their research results for strategic, commercial or patent-related reasons.[53] Increasing secrecy and the hoarding of valuable knowledge now seems to be an inevitable consequence of the commercialisation of intellectual labour.

The Strong Arm of Capital

The economic interests of biotech companies are not just predetermining the pace and direction of academic study, but they may also result in the censorship and intimidation of scientists and researchers who express dissident opinions. Again some of the most alarming precedents have been set in the medical world. Nancy Olivieri, a senior scientist at the prestigious biomedical research institution, the Hospital for Sick Children (HSC) in Toronto, was one academic who experienced the power of the pharmaceutical industry to suppress research whose findings run counter to its commercial goals. Funded by the Canadian firm Apotex Pharmaceuticals, she was conducting clinical trials of a new oral treatment for the haemoglobin-deficiency disease, thalassemia. Although her preliminary findings in the early 1990s were favourable, in subsequent research she discovered that the drug could have serious side-effects after long-term use, including severe liver toxicity. When she reported her findings to Apotex, she was threatened with legal action if she made her discovery public, was sacked and then reinstated by HSC, and saw data from her study used by an Apotex-paid research fellow without her permission. Meanwhile the drug, Deferiprone, has been approved for sale in Europe.[54]

A similar case was that of Betty Dong, a pharmacy professor at the University of California, San Francisco, who was paid by the British pharmaceutical company, Boots, to compare its leading treatment for hypothyroidism with two generic drugs and a rival brand. Discovering that the cheaper drugs were no less effective or safe than the $600 million market leader, Dong submitted her study to the *Journal of the American Medical Association* and it was accepted and scheduled for publication in 1995. Ignoring the judgement of the scientific peers who had reviewed and accepted the article, Boots drew on a contractual clause allowing it to block publication of the article, and by threatening the university with a protracted lawsuit, persuaded Dong's employer to withdraw its legal support. Following a *Wall Street Journal* exposé of the affair, the drug company, which had by then been sold off to BASF of Germany and renamed Knoll Pharmaceuticals, agreed to allow the report's publication.[55] It subsequently paid $98 million in an out of court settlement to patients who had claimed compensation for the unnecessary premium they had spent on the drug during the time the study had been suppressed.[56]

Attempts by the media to expose the illegal or disreputable conduct of life-science companies have also encountered the powerful arm of commercial interests. In 1997, two award-winning broadcast journalists were sacked from a Florida television station owned by Rupert Murdoch's Fox Television. On the basis of three months of intensive investigation, the two journalists had produced a documentary report exposing how Florida grocers had secretly reneged on promises not to sell milk from BST-treated cows – now estimated to make up thirty per cent of all American dairy cows – until the genetically engineered hormone had gained wide acceptance amongst consumers. Despite expensive advertising and publicity for the scheduled four-part series, the programme was dramatically dropped following last-minute legal representations by Monsanto to the parent company, Fox Television (Monsanto also has a history of suing dairies who choose to label their non-BST milk as 'hormone-free'[57]). The journalists involved then submitted eighty-three different rewrites of the story to try to satisfy nervous managers, but were eventually sacked for refusing to accept a risk-free, Monsanto-friendly version of the broadcast. They launched a lawsuit claiming unfair dismissal for whistle-blowing, and in August 2000 a Florida state court jury, which had heard supporting testimony from presidential candidate Ralph Nader as an expert witness, found in favour of one of them. Although Jane Akre was thus awarded $425,000 in damages, Fox immediately initiated what will be a lengthy appeal process.[58]

Monsanto also failed to prevent *The Ecologist* from publishing a special issue devoted to the biotech industry. Although 12,000 copies of the September/October 1998 issue were suddenly shredded by the magazine's long-standing printers, who had acted out of fear of being sued by Monsanto for libel, the editors found another printer and saw 400,000 copies leave the press.[59] Similar defiance was expressed by the editors of the world's leading medical journals in September 2001, with the publication of a joint editorial that appeared in thirteen different publications, including the *Lancet*, the *New England Journal of Medicine*, and the *Journal of the American Medical Association*. In an unprecedented and much-publicised statement, the editorial attacked the way the dominant trend towards corporate sponsorship of pharmaceutical trials was undermining the autonomy of the participating scientists through 'contractual agreements that deny investigators the right to examine the data independently or to submit a manuscript for publication without first obtaining the consent of the sponsor'. Insisting that a 'submitted manuscript is the intellectual property of its authors, not the study sponsor', the editors declared their collective refusal to 'review or publish articles based on studies that are conducted under conditions that allow the sponsor to have sole control of the data or to withhold publication'.[60] Many hope this policy will also slow the growth in the practice of senior doctors and scientists accepting large sums of money from pharmaceutical companies in exchange for 'authoring' articles ghostwritten by the companies' own researchers.[61]

Unfortunately the wider scientific press seems less sympathetic to such initiatives, with *Science* magazine twice agreeing to waive the normal condition that genomics companies wanting to publish in the journal a resumé of their findings – full genetic sequences are too long and inappropriate for print – should deposit the complete set of data in one of three public online databases (all of which share their holdings). After it allowed Celera Genomics this exemption with the publication of the company's draft of the human genome in February 2001, the prestigious journal did the same fourteen months later when Syngenta was allowed to publish in its pages details of its sequencing of Japan's most popular rice variety, Japonica, whilst retaining the right to control access to the full data.[62] The spurious justification for this was that it was acceptable to publish results of research without disclosing all the data – spurious because in genetic sequencing the result *is* the data.

The Pusztai Case

In Britain the latent conflict between commercial and ethical interests in biotech research reared its head in August 1998, when a world-renowned expert on food safety, Arpad Pusztai, was sacked by the Rowett Institute, ostensibly for talking publicly about research he had performed which had yet to be published in a peer-reviewed journal. Having already published around forty scientific papers specifically evaluating the nutritional value of plant-based foodstuffs, and worried about the long time-lag before completed research is finally published, Pusztai decided to make public his preliminary findings that potatoes which had been genetically engineered to produce a pesticidal lectin damaged the immune system and intestines of rats.[63]

The Rowett Institute is a leading European centre for research into nutrition and food science. Its main funding comes from UK government departments, on the premise that it plays a critical role in assessing food safety and nutrition. It also has many research programmes that are funded by private companies, including Monsanto, and holds several patents for the results of its work. After suspending Pusztai, the Institute suggested that the genetically modified potatoes were not intended for human consumption – a claim still fiercely disputed by Pusztai, and which looks particularly dubious given that Novartis bought the patent for them, and Axis Genetics had at the time been given permission to grow them in field trials. The Institute then subsequently tried to discredit Pusztai's research on the grounds that the methodology he used was flawed.[64] This was despite the fact that Pusztai had used the same methodology in scores of previously published experiments, and that the design of this particular research had already been peer reviewed by experts of the Biotechnology and Biological Sciences Research Council and chosen ahead of over two dozen rival proposals.

Pusztai's research was also, at the time, the first and only independent scientific investigation of the safety of genetically engineered foodstuffs for mammalian consumption – even though people had been consuming genetically modified products for several years. After Pusztai's (now prematurely aborted) experimental work was endorsed by twenty-four international experts in the field, and the British Medical Association had published a report recommending an indefinite moratorium on genetically engineered crops, the biotech lobby hit back with the publication of no less than four favourable reports on biotechnology in the space of two days – two from government advisory bodies and two from medical associations, including a review from The Royal Society aimed at discrediting Pusztai's research altogether. The eagerness of the Rowett Institute to abandon what was a publicly-funded £1.6 million research project on the safety implications of genetically modified food, and the attempt by political and scientific figures to portray such food, in the absence of independent tests, as inherently safe, only makes sense when seen in the new commercial context.[65]

Privatising Knowledge

It is clear from the preceeding case studies that the concentration of economic power and ownership in the biotechnology and broader life science industries constitutes a serious threat to the independence and integrity of scientific research. While scientific authority in the biotechnology field is increasingly undermined by geneticists' risky and experimental modifications of nature, these risks and uncertainties are themselves subject to selective misrepresentation and censorship, as scientific impartiality gives way not to a new sense of civic responsibility and public accountability amongst scientists, but to a culture of corporate fealty, in which money is the ultimate determinant of truth.

Yet there is also an increasingly stark contradiction emerging from this burgeoning sector of the economy, which centres, as we shall see more closely in the next chapter, on the commercial treatment of scientific knowledge as private 'intellectual property'. The right to patent valuable information and ideas has long been seen as an indispensable precondition for economic growth and innovation, since it rewards inventiveness by preventing competitors from exploiting other people's ideas. Yet the historical evidence does not fully corroborate this orthodoxy. Indeed, Eric Schiff's 1971 study of the Dutch and Swiss economies during several decades when they lacked a patent system arrives at the opposite conclusion. From 1869 to 1912, during which time the patent law in the Netherlands was repealed, the absence of domestic patents in this country 'in all probability furthered rather than hampered expansion',[66] as two major indigenous industries – margarine and incandescent lamps – emerged as strong competitors in the international marketplace.

The newly discovered method for manufacturing margarine was by the 1870s patented in France, Britain and Prussia, but it was two Dutch butter merchants, Jurgens and Van den Bergh, who built the first margarine factories, improved the product's quality by exploiting technical talent and spontaneous initiative, and subsequently dominated the international market. It was these two companies, expanding in a patentless environment, which together formed Margarine Unie in 1927, then merged again in 1930 with the British soapmaker Lever Brothers to constitute the modern food giant – and today a fierce lobbyist for strict intellectual property regimes – Unilever. Another major European supporter and modern beneficiary of the patenting system, Philips, emerged in the absence of Dutch patent law in Eindhoven in 1891. Schiff notes that the smooth and steady expansion of the Philips brothers' company, which produced an improved version of Edison's carbon filament lamp, contrasted favourably with the lengthy litigation battles, heavy royalty payments and restrictive licensing conditions faced by its German competitor Deutsche Edison-Gesellschaft.[67]

Switzerland also lacked patent laws until 1888, and a comprehensive patent system was only introduced there in 1907. Yet Schiff points out that textile manufacturing, machine production, foodstuffs, and chemicals were all well-established industries in Switzerland by the last quarter of the nineteenth century. In the manufacture of dyes, in particular, Schiff notes that industry expansion was 'helped rather than hampered by the absence of a Swiss patent system'.[68] After all, the silk-dyeing company owned by Alexander Clavel in Basel, which in 1859 successfully adopted the method of manufacturing the synthetic dye aniline (discovered and patented three years earlier by the British chemist William Henry Perkin), is the same company that was bought in 1873 by Bindschedler and Busch and which subsequently grew to become the multinational chemical giant Ciba.[69]

Schiff's research strongly suggests that the absence of intellectual property rights does not hamper competitive innovation, entrepreneurialship and national economic growth. If we move forward a hundred years to today's 'information society', this argument can surely be stated more forcefully, for the free exchange of information and ideas is becoming increasingly inseparable both from productive work and from the activities by which workers become equipped to participate in the new knowledge economy. This is because knowledge, unlike conventional fixed capital, rarely functions productively when it is privatised, indeed in some cases the productiveness of knowledge – take the 'information' encoded in software programmes as an example – is directly proportional to the degree to which that knowledge is shared with and integrated into collective systems of information processing and understanding. Knowledge is in this respect like language: it is useless unless it is shared. The hoarding, either by secrecy or legal monopoly, of knowledge, is in fact inimical to the dispersal, refinement, and intelligent application of that knowledge, and stands in sharp contrast to the communal processes of communication, collaboration and intellectual exchange by which such knowledge is in the first place produced.

That the privatisation of information and ideas is now hindering scientific innovation and development is in fact a common observation made by biotech researchers themselves. In a recent survey of nearly 2000 university-based geneticists in the US, almost half reported that, in the previous three years, a request they had made to other researchers for further information or materials relating to their *already published* research had been denied, a common reason being the researchers' desire to protect their claims on future publications and on commercially exploitable material. A third of those surveyed agreed that the withholding of data was becoming more common in their field.[70]

In seeking to understand this trend towards the counterproductive privatisation of knowledge, we may justifiably return to the theoretical analysis and lexicon of Marx. Marx predicted that as capitalism matured, the social relations that define the private ownership and control of productive assets would become a fetter on the productive employment of those assets, and a dialectical antagonism would emerge between the potential creativity of labour and the needs of the powerful to monopolise its products.[71] The evidence indicates that intellectual property law and the marketisation of science are indeed becoming an increasingly serious constraint on productive collaboration and the free exchange of ideas, stifling innovation, consuming huge amounts of time and money in lengthy litigation disputes, delaying the dissemination of knowledge as scientists seek to protect planned patent submissions, and creating an atmosphere of secrecy rather than open debate and reflective inquiry. Under these conditions the values of economic growth, social prosperity and collective welfare cannot be invoked to justify the privatisation of common intellectual and biological resources, for as we shall see in the next chapter the purpose of such privatisation is inherently political – the appropriation and reproduction of power.

4

Manufacturing Scarcity

The Privatisation of Nature

The highly concentrated patterns of private ownership and investment in germplasm and plant breeding that lie at the heart of modern agricultural biotechnology, could only have begun to develop under appropriate economic, technological and legal conditions. Removing the obstacles to capital accumulation in the seed industry has historically been accomplished in three ways: biologically, by making plants produce sterile or agronomically deficient seeds (thus eliminating the farmer as competitor); politically, by persuading government-funded plant breeding research to withdraw from potentially profitable markets for finished plant varieties and limit itself to germplasm evaluation and maintenance (thus eliminating the government as competitor); and juridically, by extending proprietary rights to plants and seeds (thus eliminating smaller firms as competitors).[1]

Jack Kloppenburg traces the beginnings of the decline in the public sponsorship of US plant breeding to the 1930s, and the successful hybridisation of corn. Hybrids are produced by cross-breeding distinct, inbred (and therefore phenotypically deficient) lines, until a match is found which yields progeny that exhibit exceptional strength and vigour. Called 'heterosis', this tendency for the offspring of genetically diverse plants to perform much better than their parents is not fully understood by plant biologists, and there are different explanations for the phenomenon.[2]

When hybrid plants are reproduced, however, the superior agronomic traits are lost to the next generation of plants, which suffer an immediate drop in performance. Hybrid plants, which are also characterised by unusual uniformity (of structure, fruit size, yield volume, etc.), must therefore be renewed each year by farmers – particularly those who work for industrialised agricultural systems where the demands of mechanical harvesting, food processing, and choosy consumers mean high levels of product standardisation are required. Because the market for hybrid plant seed – the parental lines of which can in the US be protected as trade secrets – is thus inherently renewable, hybridisation inevitably attracts large capital investment. Before 1980, for example, estimated four-firm concentration in the US for the breeding of corn – a plant which is easily hybridised owing to its separate male and female flowers – was fifty-seven per cent, compared with fourteen per cent for the US seed industry as a whole.[3]

Amongst the other major grains, wheat is notoriously difficult to hybridise, because its flowers contain both male and female organs (which must be separated and emasculated to prevent self-pollination). This was still insufficient a deterrent for companies such as Pioneer Hi-Bred, Northrup-King and Monsanto, who in the 1980s began to experiment with costly genetic and chemical ways of producing hybrid wheat, aware that successful hybridisation of America's third largest crop would give those companies an unassailable economic grip over this sector of the farming industry.[4]

The efforts of these companies to produce hybrid wheat are telling, because they show explicitly what the corporate transformation of agriculture has in other respects been careful to conceal. The goal of this research was not, as it was for public sector plant breeding, to find more efficient or productive means of harnessing nature. Its aim, rather, was to sterilise nature's own prodigious and normally renewable productive and reproductive power so as to prevent it from creating for those who work it their own means of production: seeds.

'Hybridisation thus uncouples seed as "seed" from seed as "grain", Kloppenburg explains, 'and thereby facilitates the transformation of seed from a use-value to an exchange-value'.[5] Or as Lewontin and Berlan describe it, comparing the self-reproduction of seeds with the simple reproduction of digital information, hybridisation is an agricultural 'copy-protection device',[6] purposefully designed to maximise the user's dependency on the supplier.

Unsurprisingly, the state played a historic role in facilitating this development, particularly in the US. Before Henry A. Wallace – who went on to found what became Pioneer Hi-Bred – persuaded his father, as US Secretary of Agriculture, to replace traditional corn-breeding programmes with a centralised hybrid-breeding research programme and a tenfold increase in funds, all attempts by American agriculturalists to breed hybrid corn for commercial purposes had failed. Had the same volume of public resources been invested in improving open-pollinated corn varieties through traditional methods of mass selection, Berlan and Lewontin point out, the highly exaggerated but in fact quite modest yield gains delivered by hybridisation would have been exceeded: the only deficit would have been in the size of company profits.[7]

Although hybrid crops enabled the breeding companies to eliminate the farmer as competitor, monopoly profits were constrained by competition between different companies marketing the same or similar varieties. Since many of these varieties derived directly from government breeding programmes, pressure was put on public sector institutions in the US to stop work on hybrids. By the early 1950s, the public sector had withdrawn from the field of hybrid corn production, and state support for all hybrids steadily declined thereafter. By the 1980s, as pro-market administrations tightened their grip on power in Europe and the US, most public investment in plant breeding in the Western capitalist economies had all but dried up, with even the UK's prestigious Plant Breeding Institute and the National Seed Development Organisation sold off to Unilever in 1987.

With competition from the farmer and the government significantly reduced, the remaining obstacles to corporate monopoly in the seed industry required legal remedy. Critical here was the establishment of an international legal framework of Plant Breeders' Rights (PBRs), originating in 1961 with the International Union for the Protection of New Varieties of Plants (UPOV). This independent intergovernmental organisation encourages states to formulate, as a condition of their membership of the Union, PBRs that conform to the legal guidelines set out in a series of Acts.[8] By June 2001 there were 47 signatories to the Convention, a few of which were developing countries. An exemplary PBR that is consistent with the UPOV rulings is the Plant Variety Protection Act (PVPA) passed in the US in 1970 and amended in 1980. This gives breeders who produce distinct (i.e. novel), stable (i.e. which reproduce true-to-type), and uniform (i.e. which are stable within a generation) plant varieties, a degree of patent-like protection for their seeds (proper patents could not be granted because living things were not thought to meet the criterion of being a non-obvious and useful human 'invention'[9]).

Making a plant breed with homogeneity and stability of course contributes nothing directly to its long-term performance in the field (with its lack of genetic variation the reverse is more than likely), but this time and effort is crucial if the plant, its reproductive power, and any results thereof, are to be identified and protected as private commodities, as well as efficiently processed by large-scale monoculture harvesting regimes. Criteria such as uniformity and reproducibility systematically discriminate against traditional farmers, whose interests are often best served by maximising the adaptability of seed stocks to diverse conditions and climates. 'As a result', Naomi Roht-Arriaza explains, 'UPOV twice disadvantages traditional farmers by making it difficult for them to use the protected varieties of others and by making it difficult for them to use UPOV to protect their own innovations.'[10] This injustice is mirrored by the international collection and storage of Southern germplasm. Described by agencies in the wealthy world as a conservationist measure designed to preserve humanity's 'common heritage', the contents of the seed banks are widely used in the laboratories of the gene-poor Northern countries to create legally protected crops varieties for commercial sale. 'As a result, farmers from the areas where the germplasm was originally protected and selected may end up "paying for the end product of their own genius".'[11]

How does UPOV make it difficult for traditional farmers 'to use the protected varieties of others'? Legislation like the US PVPA makes it illegal to sell protected seed without permission from the certificate holder, but it originally allowed farmers to use protected plants both as a seed source for future planting and for exchange with neighbouring farmers (the so-called 'farmers' privilege'), and as a basis for developing new varieties or pursuing other plant research (the so-called 'research exemption'). The 1991 amendment of the UPOV Act then introduced a new principle of 'essential derivation', which removed farmers' right to use protected seed to breed new but similar varieties. This weakening of the breeders' exemption failed to satisfy an increasingly powerful agribusiness lobby, however, which had become emboldened by the first wave of merges between chemical, seed and biotechnology firms that began in the 1970s, tipping the balance of power between legislators and industry decisively in favour of multinational capital.[12] In the new economic climate of the 1990s, the major companies saw any property rights exemptions as a serious obstacle to maintaining their profit margins. Coinciding with the results of a 1994 Supreme Court case which ruled against a farmer accused of infringing the rights of Asgrow Seed Company, the American PVPA was amended to make it illegal, under any circumstances, to sell saved seed for planting purposes from varieties protected by PVPA after April 1994, as well as restricting the amount that could be saved, from all protected varieties, to a quantity equivalent to replanting the farmer's normal crop.

This amendment of the US Act was a considerable victory for corporate America, but its relevance to the global marketplace remained unclear. What was really desired by the agro-biotech industry was the extension of full and internationally recognised patent protection to humanly 'invented' plants, animals, seeds and bacteria, and, as a final guarantee, the creation of organisms with built-in obsolescence (that is, with the inability to reproduce). It was the rapidly evolving science of genetic engineering which made these desires a reality.

Enclosure of the Genetic Commons

The second major wave of mergers and acquisitions in agricultural biotechnology, which occurred in the late 1990s, had both technological and legal foundations.[13] Its technological basis lay in the rapid advancements in genome mapping, gene splicing and DNA insertion techniques, which made possible the engineering of organisms with previously unthinkable combinations of genetic material, and with it a degree of unprecedented control over the fertility, reproduction and flourishing of living things, including a dramatic compression in the otherwise lengthy processes of biological breeding. Its legal foundation was the extension of patent laws, primarily in the US, to cover both these new techniques and the 'products' they yielded, and then the commitment of GATT and subsequently the World Trade Organisation to harmonise, through the Trade-Related Intellectual Property rights (TRIPs) agreement, minimum levels of patent protection and breeders' rights in the global biotech marketplace. The ideology of genetic reductionism also helped legitimise this process, explaining biological functioning in terms of genetic 'information', and thus making it plausible to think of the gene an object of intellectual property. Once patents became a commercial asset, they too followed the well-worn path of oligopoly and concentration of ownership, with four-fifths of the 1,600 patents issued for genetically modified crops and related technologies by 1999 found to belong to just thirteen companies.[14]

Jeremy Rifkin describes the patenting of genetic material – material which represents 'the last remaining frontier of the natural world' – as the final step in a five-hundred-year programme to commercially privatise the Earth's ecosystems. Beginning with the land enclosure movements which, by forcibly dispossessing the European peasantry of their means of subsistence, created one of the main preconditions for capitalist industrialisation, this process soon included, in rapid succession, 'the commercial enclosure of parts of the oceanic commons, the atmospheric commons, and, more recently, the electromagnetic spectrum commons. Today,' Rifkin continues, 'large swaths of the ocean – near coastal waters – are commercially leased, as is the air which has been converted into commercial air corridors, and the electromagnetic frequencies which governments lease to private companies for radio, telephone, television, and computer transmission.'[15] Vandana Shiva refers to this same process as 'the second coming of Columbus'. Having fought over and colonised the distant outreaches of geographical space, capital is now searching for new sources of accumulation in 'the interior spaces of the bodies of women, plants, and animals. Resistance to biopiracy', she argues, 'is a resistance to the ultimate colonisation of life itself.'[16]

The beginning of the movement to enclose and privatise the genetic commons dates back to 1971, when Ananda Mohan Chakrabarty and his employer, General Electric, applied for a US patent on a genetically modified bacterium. By his own admission, Chakrabarty had 'simply shuffled genes', taking four strains of bacteria that contained plasmids capable of digesting different types of hydrocarbons in oil, and transferring three of those plasmids into the genome of the fourth. The result was a hybrid bacterium with an enhanced appetite for oil – an organism with potential ecological value as a treatment for industrial oil spills. The patent application was unsurprisingly rejected, however, since a genetically modified life form, living and reproducing by virtue of its own organic nature, could hardly be classified as an invention, and indeed living things were explicitly deemed 'unpatentable subject matter' by US patent law.

Appealing against the decision of the US Patent Office, however, Chakrabarty and GE won their case by a narrow majority, the court judges arguing that microorganisms such as this are 'more akin to inanimate chemical compositions such as reactants, reagents, and catalysts, than they are to horses and honeybees or raspberries and roses'.[17] Upon subsequent appeal by the Patent Office, the case was referred to the Supreme Court. By this time GE had abandoned its intention to market the product whose status was under dispute, having discovered that the bacterium was too fragile to function in the treacherous conditions of the open sea. The company was delighted, nonetheless, when in 1980 the Supreme Court finally ruled, by another slim margin, in favour of the applicants, emphasising that the bacterium should qualify for a patent because it was indeed a 'human-made invention'.

General Electric was not the only company to rejoice at this decision, for it was the business community as a whole which was first to recognise the full implications of this legal precedent. Within months of the Supreme Court ruling, the then fledgling biotech company Genentech, founded in 1976 by Herbert Boyer and Bob Swanson, was floated on the US stock exchange. Despite not having a single product on the market, the company raised $36 million within a day of frantic trading, enjoying the fastest price-per-share increase ever recorded on Wall Street.[18] In 1985, the US Patents and Trademarks Office (PTO) affirmed the legal precedent, ruling that genetically engineered plants, seeds and plant tissue could all be patented. Two years after that the prospects for investors in biotechnology were given another momentous lift. Despite the fact that legal support for the patenting of Chakrabarty's microorganism seemed to have been originally based on its evolutionary distance from higher organisms like 'horses, honeybees, raspberries and roses', the PTO made an astounding U-turn and ruled in 1987 that all multicellular organisms – including animals – were eligible for patent protection.

Although the first patented animal was an oyster genetically engineered for greater size and taste, it was the so-called 'onco-mouse' – engineered to carry and breed with human genes predisposing it to cancer – which gained the distinction in 1988 of being the first mammal to be patented. This patent, awarded to Harvard professor Philip Leder, but with the licensing rights surrendered to DuPont, is so broad that it covers virtually all non-human 'onco-animals' (and all subsequent generations of their doomed progeny). Since then fish, cows, pigs and sheep have all been successfully registered as patented inventions, as genetic engineering became an indispensable means of winning private ownership of the biochemical and reproductive functions of living things.

Power of the Patent

US patent law may not have widened sufficiently to include human beings – this being interpreted by the PTO to be an infringement of the anti-slavery principles of the Thirteenth Amendment – but it does permit the patenting of human embryos, foetuses, organs, tissues, cells and genes. Commercial ownership of body parts inevitably brings with it complicated legal disputes. One of the first of these concerned a Japanese researcher at the University of California who was involved in producing an antibody-yielding cell line with promising application in the treatment of cancer. The cell line was a 'hybridoma', formed by the fusion of an established human cell line developed by the university, with lymphocytes taken from the researcher's own mother, who was dying of cervical cancer. The researcher was accused by his American colleagues of stealing the hybrid cells, which were then developed further by a Japanese pharmaceutical company owned by the man's father, and eventually used to treat his mother. The legal dispute was eventually resolved with the assignment of patent rights to the University of California, while the Japanese medical and pharmaceutical conglomerate won exclusive licence for the product in Japan and other Asian countries. The researcher's original claim to familial ownership of his mother's body tissue was eventually dropped on agreement that the hybridoma be treated as a laboratory invention.[19]

In another famous legal case, a doctor and researcher at the UCLA Medical Centre received a patent in 1984 for a white blood cell line derived from the spleen of a leukaemia patient, John Moore. Unbeknownst to Moore, the physicians had suspected that the cancerous spleen contained cells with valuable anti-bacterial and cancer-fighting properties. Having filed a patent application, the senior doctor involved promptly signed contracts with several pharmaceutical firms, hoping to exploit the estimated $3 billion commercial potential of the cell line. Moore in turn launched his own legal attempt to claim a share of these profits, arguing in effect that he should have the right to make money from the commercial use of his own body parts.[20] After a series of decisions and appeals, the California Supreme Court finally ruled in 1990 that human tissues could not be sold or bartered by the person giving them up, but upheld the right of companies and individuals to patent cells, tissues and genes, and to use such patents for commercial gain.[21]

At the end of 1999, EU patenting legislation was relaxed, as the European Patent Office lifted its four-year moratorium on the patenting of life forms, leaving it with a backlog of some 15,000 biotechnology patent applications to process. A patent gives the holder the right, valid for a period of between 17 and 20 years, to prevent anyone from using the described 'invention' – even if the same item is developed by another person completely independently. In return for this monopoly the holder is obliged to disclose in public records the knowledge and expertise that made possible the invention.

Legislation in Europe and the US now allows scientists or their employers, once they have identified the structure and function of a particular protein, to claim exclusive ownership of that protein or gene, and to licence their 'product' to customers wanting to use it for medical or agronomic research or for commercial production. After isolating naturally occurring proteins by synthesising them in genetically engineered bacteria, and creating copy DNA (cDNA) versions of genes with their non-coding regions (introns) edited out, scientists have successfully argued that these substances are patentable inventions, not discoveries or products of nature.[22] Similar arguments are now being deployed in the nanotechnology industry to justify the patenting of purified natural elements.

By 1998, some 8,000 patents on human genes, or methods and techniques related to their isolation and manipulation, were thought to have been awarded by the US Patent Office, as private interests fought furiously to capitalise on the advances brought about by four decades of public funding in biomedical research.[23] Incyte Genomics was by 2000 the holder of 315 patents for genes, while Human Genome Sciences (HGS) held 90, including one for CCR5, the gene thought to allow the AIDS virus to enter human cells.[24] HGS knew that the gene produced a type of cell receptor, but assumed it played a role in inflammatory diseases like arthritis. After the company had won its patent, which allows it to claim royalties from any use of the genetic sequence, publicly-funded researchers at the Aaron Diamond Aids Centre in New York and at the National Institutes of Health discovered that the protein required for the AIDS virus to infect cells was the same CCR5 receptor. HGS's stock price soared as a result of this non-commercial research, with investors well aware that any attempts to develop a drug to block the production of the CCR5 protein would bring massive returns to the company.

Among many other controversial patents is one for the so-called human 'psychosis gene' held by Wellcome / Smithkline Beecham,[25] and a patent on the gene for the human hormone 'relaxin', which is expressed in the ovarian tissue of women in the late stages of pregnancy. Having seen opposition to this patent defeated by a ruling from the European Patent Office in 1995,[26] the holder – the Howard Florey Institute of Experimental Physiology and Medicine at the University of Melbourne – has licensed the patent to Genentech, who plan to synthesise the hormone using genetically engineered bacteria, then market the product as a means of softening the cervix and loosening restrictive ligaments either during childbirth or for women undergoing assisted conception and embryo implantation.

By October 2000, over 160,000 patent applications had been filed on human DNA sequences by firms based in the US, Western Europe and Japan, with the top ten companies accounting for seventy per cent of them, and the French firm Genset leading its competitors with 36,000 applications alone.[27] This privatisation of human biology seems a particularly flagrant violation of the patenting criteria of 'novelty' and 'non-obviousness', for even if it is the act of *discovering* the already existing genetic sequence which is deemed to be an inventive step, this step is today accomplished not by human ingenuity but by increasingly powerful gene-sequencing computers. As James Watson complained in 1991 of the decision by the US government's National Institutes of Health (NIH) to file patents on hundreds (and subsequently thousands) of gene fragments known as 'expressed sequence tags' (ESTs), the identification of these sequences was work that could be done by 'virtually any monkey'.[28]

In 1996, the US company Biocyte was awarded a patent by the European Patent Office covering the use of cryopreserved stem cells from umbilical-cord blood (granted because the company was the first to isolate and freeze them). This effectively made it illegal for any individual or institution wanting to use such cells for research or medicine (they offer a particularly attractive alternative to bone marrow transplants for patients with systemic blood diseases) to do so without gaining (i.e. buying) permission from Biocyte. Though a legal appeal against the EU patent succeeded in rescinding it in 1999, the challenge was upheld on technical rather than ethical grounds, with the product deemed to 'lack novelty or an inventive step'.[29] The original US patent remains in force.

Opponents are also challenging the patent awarded in the US to the California firm Systemix in 1991 which covers the process of isolating human bone marrow stem cells,[30] as well as two broad patent licences issued to the US company Geron Corporation by the UK patent office which not only cover the use of nuclear transfer technology (the cloning of human and non-human cells as pioneered by state-funded researchers at the Roslin Institute), but also any 'embryos, animals, and cell lines made using the technology'.[31] A consortium of medical groups has also launched a legal challenge to US Patent No. 4,874,693. This was awarded to researcher Mark Bogart in 1989 for the discovery that the raised concentration of certain proteins in the maternal blood indicates a significant risk that the foetus the woman is carrying has a genetic disorder such as Down syndrome or spina bifida. Bogart is currently receiving over $1 million a year in royalty payments from laboratories which administer the test for these chemicals, no matter how the tests themselves are carried out.[32]

Where patents apply to particular bioengineering techniques or processes – such as the use of *Bacillus thuringiensis* genes as a built-in pesticide, the creation of glyphosate-resistance plants, specific gene-insertion techniques or vectors, or cloning techniques (and its products) – the degree of legal protection conferred by them can be extremely broad. The Wisconsin-based biotechnology firm Agracetus, for example, won a US patent in 1988 on 'particle bombardment technology' (the so-called 'gene gun') as a means to transfer foreign genes into plant cells. In 1994, the same company (by then owned by W.R. Grace), having successfully used this technique to manipulate the genome of soybeans, won a European patent essentially covering the 'idea' of genetically manipulating soybean varieties, thus giving it exclusive rights to *all* such products no matter what method is used to create them. While an equivalent US patent is still pending, the US Patent Office did award Agracetus a 1992 patent giving it the same broad ownership rights over genetically engineered *cotton* (legally challenged by the US Department of Agriculture, the patent was eventually rescinded and is now under appeal).

Interestingly, Monsanto became an unlikely ally of the Rural Advancement Foundation International (RAFI) in mounting a legal challenge to Agracetus's European soybean patent, arguing that the latter should be 'revoked in its entirety' because the techniques employed were common knowledge in the industry at the time. Having recalculated the full value of the patent, however, the biotech giant soon changed its stance, withdrew its 292-page statement of opposition, and instead paid W.R. Grace $150 million to make Agracetus's intellectual property its own.[33] Monsanto is well aware that expansive patents like these are not only a source of royalty revenue, but also function as a highly effective deterrent to new entrants to the industry, threatening researchers and competitors with extremely expensive and time-consuming litigation suits.[34] And despite the legal contests and public controversies, the trend is continuing, with US Patent No. 6,174,724, awarded in January 2001, giving Monsanto exclusive rights to all plants genetically engineered using antiobiotic resistance markers.[35]

The livelihoods of farmers are also being jeopardised by the patenting of life forms. Probably the most important benefit to the biotech industry deriving from the extension of patent law to living organisms is that ownership of an organism includes rights over its natural capacity to reproduce itself, and with it the natural products – the offspring – of that reproductive function. This effectively outlaws the saving and re-sowing of seed from fully patented plants – thus deepening farmers' dependence on seed suppliers, and destroying their right to develop their own genetically diverse and adaptive seed stocks. It will also almost certainly mean that farmers who breed transgenic livestock will have to negotiate with patent-holders the right to own each new generation of animals.[36]

The contradictions exposed by these developments are plain to see. While the biotech industry boldly insists that there is nothing new in genetically modified organisms nor in the 'natural' sequences of DNA that are extracted to create them, it is at the same time reaping enormous financial rewards by patenting those organisms, and the genes they have ingeniously isolated, as novel inventions.[37] Moreover, whereas patents fiercely penalise those who want to use the patented object to develop derivative products and innovations, the 'inventions' that are patented are themselves typically derivative both of indigenous knowledge systems and of life forms which have been discovered, cultivated, selectively bred and husbanded by generations of non-Western peasants and farmers who have worked the land over millennia. This blatant contradiction between the power of the Western patent-holder and the powerlessness of the non-Western innovator is legally enshrined in Section 102 of the US Patent Act of 1952, which refuses patents for things that are already in use in the US (the so-called 'prior art' principle), but permits them for inventions that are in use in another country unless they have been previously described in a publication.[38] This was precisely the loophole exploited by two American medical researchers who won a patent in 1995 for the use of turmeric for accelerating the healing of wounds.[39] How else did the US – a country with only one native food crop (the sunflower) – become the breadbasket of the world, if not by expropriating the genetic resources of other continents?[40]

Biopiracy and the Third World

Biopiracy – the capture and patenting of natural resources and their derivatives – represents a particular threat to the countries of the South, whose tropical ecosystems house most of the planet's biodiversity, but whose citizens are largely spectators in the scramble for lucrative patents. According to the UN Development Programme, OECD countries accounted for eighty-six per cent of the 836,000 patents filed in 1998.[41] Of all utility patents on plants, four-fifths are held by multinationals while most of the remainder are controlled by research institutions and universities in the Northern hemisphere.[42]

The fact that one of the biggest and most profitable industries in the world – the pharmaceutical industry – derives nearly three-quarters of its plant-based prescription drugs from material used in indigenous medicine,[43] with drugs derived from plants in the Southern hemisphere earning the pharmaceutical industry an estimated $32 billion a year,[44] shows just how easy it has been for Northern capital to exploit the natural resources of the South, and of course the cultural knowledge systems of indigenous farming and medicine. The pharmaceutical industry is today one of the mainstays of the 'knowledge economy', spending millions of dollars on the research, development and testing of the average drug. But its outlays would be significantly greater if it did not have free access to the healing practices and medical lore of indigenous communities. According to Roht-Arriaza, 'by consulting indigenous peoples, bioprospectors can increase the success ratio in trials for useful substances from one in 10,000 samples to one in two'.[45] Moreover, the cost of actually manufacturing drugs – as in the commercial reproduction of music and software – is almost negligible by comparison, while economic demand for medicine is wonderfully inelastic, with sick people prepared to forgo numerous other goods in order to afford the high cost of a patented drug if it promises to relieve their suffering.

New global patent laws have accelerated the phenomenon of biopiracy. There is, for example, Phytopharm's patent for P57, the appetite-suppressing agent from the Hoodia cactus that grows in the Kalahari desert. The cactus has been used by bushmen in southern Africa for thousands of years as a means of staving off hunger and thirst during long hunting trips. Phytopharm, a UK bioprospecting company, bought the original biochemical information from South Africa's Council for Scientific and Industrial Research (CSIR), and has now sold the rights to licence P57 to Pfizer, the makers of Viagra, for $21 million. With the drug working successfully in phase-two clinical trials, Pfizer hope this deal will give them a priceless stake in the £6 billion market for slimming aids.[46] (In an unusual post-script to this tale, the bushmen lodged a legal claim for compensation from the South African government, and in March 2002 were granted the right to receive a share of the CSIR's royalties.[47])

Other cases of biopiracy are notable not just for the exploitation of Third World resources, but also for the way that biotech patents threaten to damage the indigenous industries of the developing world. In 1997, for example, the

American company RiceTec, Inc., filed for several US patents covering rice varieties hybridised from Indian and Pakistani basmati strains and American long-grain rice. Basmati is the undisputed queen of rice, deriving its unique aroma, flavour and unusually delicate texture from the sub-Himalayan climate of the Punjab region, and from the long selection process of generations of farmers and breeders. Had RiceTec been completely successful in its patent application, it would have been able to market home-grown grains as authentic basmati, thus jeopardising an annual basmati export market worth $277 million to Indian and Pakistani farmers. In the final decision by the US Patent and Trademark Office, the company was allowed to register its products with hybrid names – Texmati, Jasmati and Kasmati – and to market the rice varieties as 'superior to basmati'.[48]

Enterprising plant scientist and genetics expert Chris Deren at the University of Florida has also succeeded in creating strains of indigenous Thai jasmine rice which may be cultivated in North America. Thailand, which is the world's leading rice exporter, produces over a million tonnes of jasmine rice every year, and exports a third of it to the US where it is rightly cherished as a gourmet product. Because jasmine rice is photoperiod sensitive, the plant flowers late in the season when daylight hours are fewer but the weather is still warm. This means that in temperate, sub-tropical climates, falling night temperatures late in the season normally halt the growth of the plant before it can seed. By bombarding the plants with gamma rays, Deren was able to induce a genetic mutation which made the rice insensitive to day length. He has also manipulated the plant to grow twelve inches shorter than the original Thai jasmine, which means that it can be harvested by machine (tall plants bend and become too tangled for the machinery), and will be more resistant to hurricane damage. If all goes to plan, Florida farmers will be growing commercially viable jasmine rice in about ten years' time. Though Deren has evidently pledged not to patent the new strain of rice – an action which would in any case violate the terms of the Material Transfer Agreement (MTA) which should have been signed on acquisition of the original rice germplasm from the International Rice Research Institute – the commercial development of the crop will inevitably lead to deep cuts in Thailand's $300 million per annum export market to the US, as well as stiff competition to meet the demands of European consumers.[49]

A similar story pertains to a raft of patents taken out by Japanese and US firms (including Monsanto) on formulas for stable solutions and emulsions deriving from the native Indian neem tree (*Azadirachta indica*). The tree has been used for centuries by Indian villagers mindful of its medicinal, anti-bacterial, and pesticidal properties, yet this collective knowledge did not stop W.R. Grace and the US Department of Agriculture from winning in 1994 a European patent covering the use of antifungal agents extracted from the tree.[50] Or there is the case of the small American seed company which in 1999 won a US patent on a bean, called 'Enola', directly derived from the popular Mexican 'azufrado' landrace. The company's owner had planted commercial beans bought in Mexico, then selected them for a yellow colouring until it bred true. Granted a patent on the grounds that the beans were a distinctive colour and not grown before in the US, the firm filed a lawsuit in November 2001 against sixteen small bean seed companies and farmers in Colorado who were known to be selling a similar yellow bean, alleging infringement of its intellectual property rights. The controversial patent is now facing a legal challenge, supported by the UN's Food and Agriculture Organisation (FAO), from the Colombia-based international plant breeding institute, the International Centre for Tropical Agriculture (CIAT).[51]

Further controversy surrounds the patenting in 2000 of the Andean nuna bean by the US firm Appropriate Engineering and Manufacturing. The patent for the nutritious South American bean, which is toasted rather than boiled and therefore regarded as a resource-friendly staple in fuel-scarce environments, covers crosses involving at least thirty-three varieties of the bean, the novel crosses being patentable because they enable the bean to grow outside the Andes. If the patent is recognised by the South American authorities themselves – the 'inventors' have already received a World Intellectual Property Organisation patent, and are planning to apply for patents in as many as 121 countries – further breeding development of this crop in these countries may become illegal, whilst the efforts of indigenous farmers to sell the crop to fuel-scarce economies in other parts of the developing world are likely to face strong competition from North American exporters.[52]

Manufacturing Diversity

Agricultural genetic engineering threatens to consolidate this trend because it permits, with an unprecedented depth of intrusion, the isolation and removal of value-producing and essentially multipliable genes, and thus also the disposal of the plants, environments and communities in which those genes were previously embedded. Just as the bioprospectors want to 'save the genes' of threatened populations – witness the intentions of the progenitors of the Human Genome Diversity Project to 'build a repository of genetic diversity by collecting and preserving DNA samples from vanishing indigenous peoples'[53], or the ambitious attempts to produce clones from the cells of extinct species, or the Northern policy of 'banking' the diverse germplasm created and conserved by now-doomed Southern agricultural communities – so the same interests aim to liberate the production of potentially lucrative cash crops from the ecological conditions 'monopolised' by the Third World. Having aggressively restructured the economies of the South, transforming them from self-sufficient food producers into export economies dependent on the changing whims of rich Northern consumers, the dominant powers in the developed world now have the technology at their disposal to produce their own cash crops. Using cloning technology and *in vitro* micropropagation techniques, modern biotech companies will soon be able to mass produce – in the laboratories and temperate climes of the North, using multipurpose technology, and in ways which will avoid the limitations and uncertainties of geopolitical instability, weather, seasons, labour, transportation, and the long-term storage of perishable goods – high-value food crops, sweeteners, drugs, flavourings, fragrances and dyes which were previously only found in the tropics.

Efforts to modify the oil composition of Northern seed crops like rape and soybean so as to yield speciality oils or more stable oils which do not need chemical hydrogenation – as has already been accomplished through the insertion of laurel and coconut tree genes into rapeseed to boost its fatty acid content[54] – could, if successful, have a catastrophic effect on Southern economies dependent on the export of tropical oils (a quarter of the population of the Philippines, for instance, is at least partially dependent on the coconut oil industry[55]). Research is also in progress to produce coffee, tea, rubber, citrus pulp vesicles and cocoa *in vitro*.[56] Mars UK has already been awarded two patents in the US on genes from a West African plant thought to be responsible for the distinctive flavour of the region's cocoa,[57] while two US companies have claimed they are ready to mass produce natural vanilla flavour from plant cell cultures. Promising that it will sell at a fraction of the price of natural vanilla extract, the creators of this laboratory-produced vanilla could displace in Madagascar alone over 70,000 small farmers who grow vanilla orchids.[58]

In a more recent development, several multinational companies have licensed from University of Wisconsin scientists the genetic resources necessary to produce a super-sweet protein normally produced in the berries of the West African 'j'oublie' plant (*Pentadiplandra brazzeana*). The scientists are thought to have discovered the berries in Gabon, where they are a traditional part of the indigenous human and animal diet. The researchers were then awarded several US and European patents covering the isolation, sequencing and use of the DNA which codes for the protein 'brazzein'. Because brazzein, unlike other non-sugar sweeteners, does not lose its taste when heated, and is reported to be between 500 and 2,000 times sweeter than sugar, it has obvious commercial potential as a non-calorific sugar substitute. One company which has licensed the technology, the Texas-based ProdiGene (a spin-off of Pioneer Hi-Bred International), is currently developing a transgenic maize plant which expresses the protein in its seeds. Following in the path of Unilever's microbial synthesis of the sweetener thaumatin (a protein discovered in the West African 'katempfe' bush, and which Tate and Lyle have extracted and marketed since the late 1970s), other scientists are attempting to use *E. coli* to synthesise brazzein *in vitro*.[59] The threat this is likely to pose to the sugar cane industries of the Caribbean, the Philippines, Kenya and elsewhere – industries which have already been hit hard by the enormous growth in high-fructose corn sweetener (isoglucose) produced mainly in the US and Japan using immobilised enzyme technology[60] – may be perilous indeed.

Revisiting the Green Revolution

'Worrying about starving future generations won't feed them', Monsanto admonished conscience-stricken readers through an advert taken out in the UK *Observer* in August 1998, 'Food biotechnology will'. The same message was repeated to the public by various representatives of science and industry, who solemnly warned that the rejection of biotechnology was a rich person's luxury which the developing world could ill-afford to share. This advertising offensive was a direct response to diminishing consumer support for genetically engineered products in Europe, and to the dawning realisation that the chief attraction of the new biotechnology lay in its capacity to yield unprecedented concentrations of power and wealth.

Every year hunger and its related diseases kills around twelve million children under the age of five, which is alone more than the total annual death toll from World War II.[61] Yet anyone familiar with the West's milk, cheese and butter mountains, with the growth of obesity in the wealthy world, and with the extraordinary wastefulness of Western producers and consumers, has good reason to question whether population growth and technological capacity are the most important factors in the war against hunger.

In April 2000, for example, only a month after the Worldwatch Institute in Washington announced that the number of overweight people in the world – 1.2 billion – had for the first time equalled the number who were malnourished, the *Guardian* reported how up to half-a-million tons of edible food was being thrown away each year in Britain.[62] According to figures gleaned from the UN's Food and Agriculture Organisation (FAO) by Frances Moore Lappé and her colleagues at the Institute for Food and Development Policy, during the past thirty-five years the increase in global food production has outstripped the world's population growth by around sixteen per cent, with gains in food production staying ahead of population growth in every region except Africa.[63]

Does food scarcity explain why more than three-quarters of all malnourished children under five in the developing world live in countries with food surpluses? Is underproduction the reason why India exported nearly $2 billion worth of rice, wheat and flour in 1995 while more than a fifth of its population went hungry? Does it explain why 100,000 children die from hunger every year in Brazil despite it being the world's third largest food exporter, or why Ethiopia was growing animal feed for Western livestock farmers during the height of the devastating 1984 famine?[64] The root cause of hunger, as these facts bear out, is not insufficient food production but the unequal distribution of the means of consuming it.

With both critics and supporters of agricultural biotechnology likening it to a 'second' Green Revolution, understanding the precise impact of the Green Revolution on the developing countries is central to the debate over the value of the new biotechnologies to the poor. The term was coined in the 1960s to draw attention to the creation of new seed varieties which produced unusually high yields. The revolution was termed 'green', incidentally, not because it was a proto-ecological movement, but because it was the West's ideological attempt to limit the influence of Mao's 'red revolution' on the rest of the Third World. The new seed varieties were developed in a dozen or more International Agricultural Research Centres (IARCs), funded by a range of public and quasi-public organisations – including the Rockefeller and Ford Foundations – in collaboration with the World Bank, a number of chemical companies, and experts from public agricultural research institutions in the developed world. It would certainly be wrong to characterise the Green

Revolution as a mainly corporate agricultural programme, for such a description would, as has been pointed out elsewhere,[65] obscure what is probably the most distinguishing feature of the *new* biotech revolution. That the first Green Revolution successfully accelerated the capitalist penetration and restructuring of world agriculture is, nonetheless, beyond doubt.

By the 1970s the new varieties of wheat, rice and corn had spread widely in Latin America and Asia (Africa, on the other hand, was not a major target), and by the 1990s it was estimated that around forty per cent of Third World farmers were using Green Revolution seeds.[66] The secret of the so-called 'miracle seeds' lay in their improved responsiveness to controlled irrigation and petrochemical fertilisers. This fact alone already begins to explain why the food increases delivered by the Green Revolution did not find their way to the mouths of the poor. Despite an eight per cent growth in per capita food supplies in South America during 1970–90, for example, the number of hungry people in the region rose by nineteen per cent over the same period, while in South Asia a nine per cent increase in per capita food production was matched by a nine per cent growth in hungry people.[67] In one extensive study of research reports on the effects of the Green Revolution, it was found that four-fifths of reports which addressed the distribution effects of the new technology had concluded that inequality had increased.[68]

There is no mystery behind these figures. The induced need for complementary irrigation systems and expensive chemical inputs manufactured by Western multinationals immediately favoured the larger and more wealthy farmers, but offered little for poorer people working on more diverse and marginal lands. The new seeds were in any case almost exclusively designed to produce exportable cash crops, with research and credit rarely directed towards supporting traditional peasant-produced staple foods such as millet, sorghum, pulses, and tuber crops like cassava and sweet potato. As outputs increased following the take-up of the new seeds, grain prices fell, and more and more smaller farmers were forced out of business. Subsistence or near-subsistence farmers, many of whom previously supplemented their income by working for wealthier landowners, saw their land bought up and their jobs replaced by herbicides and machines, as the larger growers grasped the opportunity provided by standardised, monoculture crops to eliminate the potential threat posed by trade unions and obstinate tenant farmers. In regions where land and labour is abundant but capital scarce, and where the infrastructure and incentives for service sector growth are lacking, labour-saving technology had a profoundly dislocating effect on the distribution of wealth and well-being.

For those who resisted migration from the land, the Green Revolution brought further difficulties. Diminishing soil fertility, caused by the abandonment of intercropping and excessive dependence on chemical inputs, and the inevitable emergence of pesticide-resistant weeds and insects, forced down net returns and compelled farmers to use more and more fertiliser and pesticides to maintain the same level of production. Lappé and her colleagues point out that 'over the past thirty years the annual growth of fertiliser use on Asian rice has been from three to forty times faster than the growth of rice yields', while 'the quantity of agricultural production per ton of fertiliser used in India dropped by two-thirds during the Green Revolution years'.[69]

Increased used of pesticides became unavoidable once the traditional methods of mixed farming had been replaced by genetically uniform crops vulnerable to predators and disease, and once landraces – indigenous seeds selected and exchanged by peasant farmers in order to maximise the adaptability of plants to specific conditions – were displaced by single-generation hybrids which do not breed true. Demand for herbicides also grew due to the spread of semi-dwarf hybrid wheat and rice varieties which, though strong enough to carry larger loads of grain and withstand heavy applications of fertiliser, were disadvantaged against higher-growing, sunlight-craving weeds. Although most farmers who converted to dwarf hybrids enjoyed an increase in grain yield, part of the cost of this increase was a reduction in the volume of straw produced and with it supplies of animal fodder and organic fertiliser. A similar diseconomy is faced by the smallholding farmers who decide to slaughter their cattle young to satisfy the wants of the meat-eating minority, and then export with the carcasses an important source of renewable energy, fuel and fertiliser.[70]

Rich Solutions

Today the growth in yield from Green Revolution crops has largely flattened out, and in a number of areas begun to fall, as soil degradation has worsened.[71] Statistical evidence of improved output has in other cases concealed a more sobering reality. Third World farmers who produce directly for their own nutritional needs, then trade their surplus to buy industrial commodities and non-subsistence goods, have from the perspective of economists often enjoyed significant increases in output and earnings when they have transformed their diverse harvests into monoculture export crops. Whilst the economic value of their tradable output may have increased, however, what goes unaccounted for in this reckoning is the destruction of farmers' nutritional self-sufficiency and the creation of new economic needs. Because economic rationality recognises exchange-value only, these diseconomies become visible only when they are translated into a form which simultaneously masquerades as wealth: economic demand. This deception is apparent even in commentaries on the supposedly improving outlook for farmers in the wealthy world, which routinely confuse farmers' incomes with incomes from farming. As Lewontin points out, in the US in 1997, '60 percent of farm operators were also employed off the farm and 40 percent worked at alternative employment for more than two hundred days a year'.[72] It is the forced proletarianisation of rural communities which is the real underlying trend here, and there can be no satisfactory economic representation of this qualitative transformation of social relations.

Meanwhile, the largest ever survey of worldwide sustainable agriculture has shown that remarkable productivity gains can be made by the adoption of farming techniques based on minimal chemical applications and maximum use of natural regenerative processes combined with local knowledge and skills. In the study by Jules Pretty and colleagues at the University of Essex, which examined the results of 208 projects covering nearly thirty million hectares spread over fifty-two countries, sustainable agriculture was found to have brought average yield increases of 50–100 per cent for rainfed crops, with small-to-medium sized root crop farmers reporting average increases of 150 per cent.[73]

Following one innovative strategy now spreading across east Africa, thousands of maize farmers are benefiting from the discovery that yields can be increased by up to seventy per cent by the judicious planting of weeds which reduce pest damage. The smell of the common napier grass (*Pennisetum purpureum*) is so attractive to the rapacious stem borer that it is now grown on the margins of cereal fields to lure the insects away from the precious crops. Instead of attacking the insect with genetically engineered *Bacillus thuringiensis*, the grass – which is subsequently recycled as animal fodder – produces a natural sticky substance that traps the larvae and prevents them from pupating. The other serious threat to maize yields in Africa is the potentially devastating root parasite witchweed (*Striga*), which Kenyan researchers have found will not grow in soil already occupied by the ground-covering plant silverleaf desmodium (*Desmodium uncinatum*). Since the smell of desmodium, like the companion weed known as molasses grass (*Melinis minutiflora*), is also repellent to the stem borer, African farmers are now using weeds to implement an effective 'push and pull' system as biological protection again the insect.[74] Advances like these are complemented by the widespread success of Cuba's conversion from capital-intensive monoculture farming to low-input and organic sustainable agriculture, a dramatic structural transformation necessitated by the collapse of the socialist bloc in 1989 and the evaporation of Eastern European exports which previously met over half of Cuba's calorific needs.[75]

Stuff the Rich

Jules Pretty's research into alternative sustainable agriculture indicated less impressive results for irrigated crops, however. This is significant, because huge pressures have built up since the 1980s to bind Third World farmers firmly to the global economy by converting production for household and local markets into the cultivation of thirsty export crops. Not only has this entrenched farmers' dependence on unsustainable irrigation systems, it has also exposed millions to the pernicious effects of fluctuating commodity prices, interest rates and currency values. The pressures that have induced these dependencies include the need for Third World governments to earn hard currency to pay interest on foreign bank loans, the channelling by foreign aid donors of income to high-profile, export-oriented projects, the dumping of surplus Western food and tied aid on indigenous markets (which not only cheapens the price of locally produced food, but also changes the tastes and preferences of Third World consumers to suit the needs of Western exporters) and, finally, the enormous government subsidies enjoyed by Northern farmers, while the governments of the South are enjoined by the World Bank, IMF, and WTO rules to deregulate their economies, privatise national industries and promote free trade.

In Kenya, for example, which was self-sufficient in food until the 1980s and which remains an agricultural exporting economy, eighty per cent of food consumed in the country is now imported. In explaining this transformation one cannot ignore the fact that in 1993 EU wheat was sold in Kenya at half the price paid to European farmers, leading to the collapse of Kenyan wheat prices (and thus wheat producers) in 1995.[76] Haiti, similarly, was almost self-sufficient in rice at the beginning of 1990, but by the end of the decade, as the effects of a World Bank-sponsored trade liberalisation programme began to bite, imports of subsidised US rice were swallowing up more than half of local spending on the product, pushing eighty per cent of the rural population below the poverty line and – despite rice prices falling by a quarter – leaving at least half of the country's children undernourished.[77] Meanwhile in India, where soil nutrients are disappearing, the price of imported farm inputs are increasing, and fertiliser subsidies are being phased out under the instruction of the IMF, thousands of small- and medium-sized farmers were recently pushed into bankruptcy, including 500 cotton farmers who committed suicide to enable their families to escape legal responsibility for their debts.[78]

Because the Green Revolution was a 'rich' (i.e. a technologically- and commodity-intensive) solution to what is essentially a social and economic problem, it inevitably worked in favour of the wealthy and against the interests of the poor. It seems quite apparent that the 'second' Green Revolution – the biotech revolution – is destined for the same outcome. With three-quarters of all transgenic crops currently modified to tolerate the herbicide products of the same companies, widespread recognition amongst scientists that the genetic foundations of the most agronomically valuable traits are a long way from being understood, and the ferocious scramble for patent applications by the leading life-science companies (an estimated eighty per cent of all crops planted in the developing world, it should be noted, are currently from saved seeds[79]), the benefits which biotechnology may deliver to the malnourished millions look meagre indeed.

To this grim scenario we should also add the fact that the vast majority of transgenic crops grown worldwide are destined for the feeding troughs of animals. In 1998 soybean and maize accounted for fifty-two per cent and thirty per cent respectively of global transgenic acreage.[80] Over ninety per cent of soybean harvests and around sixty per cent of traded maize are consumed by livestock.[81] Yet meat from grain-fed livestock is not only an unaffordable luxury for the majority of the world's population, two-thirds of whom have a primarily vegetarian diet – with US cattle consuming six times more cereal, legume and vegetable protein suitable for human consumption than the animal protein they provide as meat, it is also an extremely inefficient and ecologically irrational means of producing food.[82]

A Golden Future?

I have attempted to show how agricultural biotechnology is a technology shaped by the interests of its progenitors, and that its primary purpose is to transform the practices and products of agricultural life into forms compatible with the highly concentrated system of ownership and control that characterises capitalist industry. The traits it is most concerned with – uniformity, transportability, extended shelf-life, visual appeal, suitability for the food processing industry, herbicide tolerance – are the means by which the products and needs of farming communities and environments are adapted to the requirements of the highly mechanised, vertically integrated, chemical-industrial complex, as well as to the legal preconditions for the isolation and privatisation of living things. 'Machines are not made to harvest crops', claimed two forward-looking agricultural researchers in 1968, 'in reality, crops must be designed to be harvested by machine'.[83]

As for those needs which, because of their stubborn simplicity, and their rootedness in time, place and specific social relations and environments, cannot be converted into markets for profitable commodities – these needs do not capture the interest of the biotech giants. This is why, instead of facilitating the kind of transformation of lifestyles and redistribution of resources which defines the only truly 'ethical' response to the related diseases of scarcity and affluence, the life science industry and its scientists have their eyes most firmly set on the emerging 'neutraceutical' market – food products designed to offer the overfed a way of consuming their way out of the afflictions of overconsumption.[84] As with the highly profitable dieting industry – which in the US alone is worth $33 billion annually[85] – those who need to consume less are, by virtue of their excess income, a far more lucrative market for commercial producers than those who cannot afford to meet their basic needs.[86]

The other set of needs which agricultural biotechnology shows an interest in addressing includes those which are a result of the chemical-industrial transformation of farming itself. Just as the world's third largest tobacco company, Japan Tobacco, has established exclusive licensing and partnership deals with US biotech companies with the aim of developing vaccines for the same disease – lung cancer – which the company's own product is responsible for,[87] so many agricultural biotech companies and researchers are looking to produce high-tech treatments for the scarcities and blights engendered by the last Green Revolution. This is the case, for example, with the transgenic rice engineered by scientists at the International Rice Research Institute (IRRI) to resist bacterial blast (*Xanthomonas oryzae*) – a disease which is almost exclusive to Green Revolution varieties – as well as with the much-vaunted 'golden rice'. The latter was produced by Swiss-based plant scientists Ingo Potrykus and Peter Beyer, using two genes from a narcissus plant and an enzyme-coding bacterial gene to produce beta-carotene (converted by the body into vitamin A) in the rice endosperm.

It has been estimated that around 400 million people worldwide are at risk from Vitamin A deficiency (VAD). It is responsible for up to thirty per cent of deaths among pre-school children, causes vision damage to millions, and affects seventy per cent of children under five in South and Southeast Asia.[88] When questioned about the prospects of golden rice before a session of the Group of Eight Summit in Okinawa in July 2000, President Clinton promised 'it could save 40,000 lives a day', a claim reiterated by numerous biotech companies, including Syngenta who warned that a single month's delay in bringing the product to market would leave 50,000 children blind.[89] Yet the main sponsors of the golden rice project, the Rockefeller Foundation, later conceded that a normal diet of the genetically modified rice would provide only eight per cent of the required daily intake of vitamin A, and that a pregnant woman, in the absence of other sources of the vitamin, would have to eat around eighteen kilos of the cooked rice to meet her daily needs.[90]

It is well known, moreover, that the main cause of VAD in the developing world is the loss of dietary diversity, and that this loss is also the cause of deficiencies in other essential vitamins and minerals (iron deficiency, for example, is estimated to affect around a quarter of the world's population). Not only is VAD normally accompanied by other nutritional deficits, but the proper absorption of a variety of vitamins and minerals depends on the balanced consumption of others. Vitamin A is fat-soluble, for example, which means it must be consumed with fats or oils to be properly absorbed. In other cases artificially enriched foods or pharmaceutical supplements may present serious risks of over-absorption of vitamins or minerals. Nutritional imbalances, in other words, are best prevented by a balanced but varied diet. Yet the concentration of VAD in major rice-consuming countries is itself partly a reflection of the way high-yielding but low-nutrient Green Revolution crops have displaced locally-grown fruit and green leafy vegetables from people's diets. These include a range of nutritious wild greens – bathua, amaranth and mustard leaves in India, for example – which in monoculture, chemical-intensive farming become redefined as 'weeds' and, because they compete with the cash crops, subjected to chemical liquidation.[91] Nutrition surveys in regions that have been converted from subsistence to commercial agriculture routinely find that increased yields and incomes are not accompanied by improved nutrition, as self-sufficiency in food is replaced by dependence on refined and processed commodities that are typically high in fat, sugar and salt, but low in vitamins.[92] And this is to say nothing of the ecological dangers of transgenic 'golden rice',[93] nor of the economic implications of a project which has cost several hundred million dollars in (mostly public-funded) research, but commercial control over which, because of the worries which Potrykus and Beyer had about committing costly patent infringements, has now been transferred to the more capable hands of AstraZeneca (now Syngenta).[94]

Frankenstein Farming

Of course, were rice grains left unpolished – as they were before the Green Revolution's aggressive marketing of white polished rice stigmatised the consumption of brown grains – then the rice husks would themselves provide a valuable source of important vitamins, including a small amount of vitamin A. To make sense of this process, of the way scarcity and deficiency are continually manufactured as the inseparable partner of global wealth, requires an analysis of the political and economic interests which agricultural biotechnology is designed to serve. For if there is a logic at work in removing from nature its value to human beings, then reintroducing that value by technological means, then one prime candidate is the logic of domination – a logic which resides in using advanced scientific technologies like genetic engineering, under the auspices of furthering economic growth, to convert the productive and reproductive capacities of both people and nature into a source of corporate power and profit. Marx conceptualised this well: the forms of domination characteristic of capitalist social relations, though originally a means of increasing the collective production of wealth, eventually become a fetter on the development and efficient marshalling of society's resources, as capitalism preserves itself only by being increasingly destructive of people's capacity to contribute to their own well-being.

This logic of domination achieved its most perfect expression when in March 1998 the US Department of Agriculture (USDA) and Delta & Pine Land were jointly awarded the first of many patents issued in the US for genetic seed sterilisation, or what RAFI dubbed 'terminator technology'. Falling under the category of what the biotech industry calls 'genetic use restriction technologies' (GURTs), the original terminator patent (US Patent 5,723,765) describes the creation of transgenic plants which yield pollen or seeds made sterile by the release of a toxin, such as that expressed by a gene from the soil bacterium, *Bacillus amyloliquefaciens*. When triggered by a promoter specific to a developmental stage of the plant, such as the formation of the sexual organs or the drying out of the mature seed, the released toxin typically works by breaking down RNA and making impossible the synthesis of the proteins necessary for the maturation of viable gametes or embryos.

Because crop varieties rendered sterile in this way cannot produce fertile seeds, researchers also had to create a means – ingenious in its conception – of turning off the terminator gene, for otherwise seed suppliers would be sabotaging their own business. The expression of the toxin is therefore blocked by a sequence of DNA inserted between the terminator gene and its promoter. This DNA sequence is flanked by nucleotides which are recognised by a 'site-specific' recombinase enzyme, called 'Cre', the gene for which is also added to the plant. This latter gene, which originally derives from a bacteriophage, is given a promoter which is deliberately inactivated by the presence of a 'repressor protein' synthesised by another foreign gene. The effects of the repressor protein can, however, be overridden by the application of a chemical, such as the antibiotic tetracycline. The company can hence reserve a proportion of the seed harvest for future seed production, then treat the remainder with the antibiotic. The latter, by neutralising the repressor protein, precipitates the activation of the recombinase promoter, and the resulting enzyme splices out the blocking sequence and catalyses the recombination of the terminator gene with its promoter. The seeds are sold, and when the right developmental stage in the plant's lifecycle is reached the promoter will activate the synthesis of the toxin and no further seeds will set (or those that do will fail to germinate).[95]

The original purpose of this technology, which is also being studied with a view to creating seedless fruit, is to provide a biological means of policing patents on life forms, and to prevent seed saving from crops – especially soybean, rice and wheat – where hybridisation has not been commercially successful. It does this by introducing planned obsolescence into the organism itself. As the president and CEO of Delta & Pine Land, Murray Robinson, told RAFI, 'We expect [the new technology] to have global implications, especially in markets or countries where patent laws are weak or non-existent'. With around 1.4 billion people thought to be dependent on farmer-saved seed, terminator technology promises to reward Delta & Pine Land by 'opening significant worldwide seed markets to the sale of transgenic technology for crops in which seed currently is saved and used in subsequent plantings'. As for the USDA, the goal, according to spokesman Willard Phelps, is 'to increase the value of proprietary seed owned by US seed companies and to open up new markets in Second and Third World countries'.[96] Rice and wheat farmers in China, India and Pakistan are apparently the primary markets here. Murray Robinson told a US seed-trade journal in 1998 that the company's seed sterilisation technology could be used on over 405 million hectares of land worldwide – an area almost equal to the size of South Asia – and that this could generate revenues for his company in excess of a billion dollars per annum.[97]

In October 1999 Monsanto and AstraZeneca responded to international pressure by publicly vowing not to commercialise terminator technology. But by 2000 all the major biotech companies had been issued with patents for terminator-type systems, and Delta & Pine Land, the world's largest cotton company, was aggressively pressing ahead with its development.[98] According to Mae-Wan Ho, the technology has already been field-tested in Europe since 1990, with crops of oilseed rape engineered by AgrEvo, a subsidiary of Aventis, to be sterile, currently undergoing trials in the UK, while the US has already apparently licensed the commercial release of four different crops with the barnase terminator gene (a corn and a canola by AgrEvo, a chicory by Bejo, and a corn by Plant Genetic Systems).[99]

At least forty-three patents have also been issued for 'trait-specific' genetic use restriction technologies (T-GURTs). Also known as 'inducible gene control systems', these involve engineering crops to express desirable traits only when activated by the application of a proprietary chemical,[100] such as the potatoes field-tested in the UK in 1999 and 2000 containing a promoter which triggers sprouting only after application of an alcohol solution.[101] Included in this category are several patents originally issued to Novartis, which cover a method for making the immune system of plants responsive to chemical treatment.[102] When disease threatens the transgenic plants, the chemically-equipped farmer will be able to activate the hyper-expression of their immune systems. But the patent covers the use of the same technology to *disable* plants' immune systems, raising the prospect that farmers will soon be buying 'invalid crops' which, like the culturally manufactured diseases that are the cherished objects of the new medical industry (the diseases of ageing, ugliness, hyperactivity and 'maladaptive' behaviour), can be brought to full and immaculate health only through consumption of a prescribed commodity.

The existence of these kinds of technologies are a source of outrage to many, but they also serve a political purpose in demonstrating both the ultimate stakes in the biotech transformation of agriculture, and indeed the increasingly contradictory dynamics of modern capitalism. Armed with a reductionist philosophy of genetic science, which denies the fluidity and adaptability of natural organisms in favour of a computer model of reprogrammable machines, the biotech industry, while promoting itself as the creator of life, seems irreversibly committed to the death of living things, sterilising and dismantling people's organic environments in order to sell nature back to them as a corporate invention. The doctrine of DNA serves this political function with deadly effect: it divorces life, harvests, health and well-being from the self-sustaining, self-renewing networks of genes, cells, organisms and their environments, and reduces the flourishing of plants, animals and human bodies to the patentable and commodifiable form of the gene.

Yet the commodification of nature goes against everything ecologists have taught the world since the 1960s. To treat organisms as assembled inventions presupposes the destruction of those organisms as self-sustaining things, and hence the destruction of sustainable forms of life. For all the novelty and invention displayed in the marketplace of consumer societies, the accomplishments of modern capitalism are dwarfed by the monumental diversity of natural forms, by the phenomena of adaptive growth and development, of self-replenishment, reproduction and renewal, on the basis of which all human industry ultimately rests.

A culture obsessed with the technical mastery and manufacturing of life also risks abandoning its respect for the mysteries of organic things. It risks becoming intolerant of nature's remaining independence, insensitive to forms of flourishing which cannot be controlled, packaged and purchased as private goods, and impatient to correct the 'inefficiencies' and exploit the 'redundancies' of all living organisms, including the human. Think – as do many of today's reproductive scientists – of the millions of 'unused' eggs which are produced in a woman's ovaries, a tiny fraction of which are released for possible fertilisation, and which could therefore be harvested and utilised in fertility treatment and research into so-called therapeutic cloning. By committing ourselves to the complete domination and control of nature, might we relinquish our ability to tolerate the different and the unexpected, our acceptance of the imperfect, our wonder for that which defies functional explanation, or our sympathy for or interest in the accidental and unintended deviations from what we believe is nature's correct course? We shall explore the implications of these questions for our understanding and treatment of humans' capacities, limitations and sensibilities, and our sense of health and illness, in the following chapters.

5

Animal Biotechnology and Ethics

The Power to Create Life

The corporate strategy, exemplified in the development of terminator technology, of manufacturing scarcity in order to convert into markets needs which were previously unrecognised or satisfied in non-commodified forms, is not confined to the world of plant genetic engineering. It also has an analogue in the animal biotechnology industry. In January 2001 a domestic American cow, attended by scientists from the Massachusetts company Advanced Cell Technology (ACT), gave birth to a gaur, an endangered species from South-east Asia, related to the bison.[1] The animal had been cloned from skin cells preserved from a gaur which had died in San Diego zoo in 1993. The nucleus of a cell was successfully fused with an enucleated cow egg in a process made famous by the Roslin Institute in cloning Dolly the sheep. Collaborating with Spanish conservationists, ACT is also working on cells extracted from the now extinct Pyrenean mountain goat, the bucardo, a clone of which they hope to be able to gestate in the womb of an Ibex, which is the bucardo's nearest living relative. The same Massachusetts-based company is also known to be looking into efforts to clone the panda. Chinese scientists have already brought such an effort to the embryo stage, having fused the nuclei of panda cells with enucleated rabbit eggs (the plan is to use black bears as surrogate mothers, though this faces problems of immunological incompatibility). In Australia, meanwhile, researchers have extracted genetic material from a nineteenth-century Tasmanian tiger pup preserved in alcohol in the Australian Museum in Sidney, the aim being to use nuclear transfer techniques to resurrect an animal that has been extinct since the 1930s and in doing so, as the director of the museum explained, to 'lift the burden of guilt' carried by Australians.[2]

The possibility of using modern biotechnology to 'conservationist' ends is also endorsed by the animals rights philosopher Bernard Rollin. Rollin believes that, once the natural habitats of wild species have been regrettably degraded by human encroachment, then it is better to use genetic engineering 'to modify the animals to cope with this new situation so that they can be happy and thrive rather than allow them to sicken, suffer, starve and die'.[3] The question, of course, is whether the technology and ambition to modify animals in order to adapt them to inhospitable environments will serve to legitimise, by virtue of its apparent success, humans' increasing unwillingness and inability to protect and care for those environments and the innumerable species of animals, insects and plants that comprise them. The disturbing image engendered by these developments is thus of a scientific community indifferent to the natural patterns, features and divisions of organic life, and which is content to address the latter instead as an assemblage of inherently disposable artefacts to be manipulated and reconstructed according to human whim.

How indeed would the cloned product of an extinct species generate the genetic diversity to enable that species to survive uncaged, or acquire the character traits and practical know-how normally passed on by adult animals to their progeny?[4] Wouldn't the hostile environment that endangered that species in the first place require that genetic scientists take over the reproductive, social and evolutionary mechanisms of natural selection of 'wild' animals, that they control fertility and sex-selection, regulate numbers, introduce adaptive traits through the genetic manipulation of cells and embryos, determine the patterns of social interaction and membership, and thus assume responsibility not just for the reproduction but also *for the very definition of species themselves*?[5]

The possibility of resurrecting extinct species, or of modifying and 'improving' endangered ones, has been welcomed by the biotech industry as an opportunity to draw attention to the way the genetic engineering of animals – the creation of 'transgenic' animals – may have 'ecological' as well as medical and industrial applications. Yet all such applications appear to rest on a biological reductionism which abstracts organisms from their species-specific habitats and forms of life, divides them into component parts, and then justifies the reassembling of those organisms so as to suit the technology-intensive environments and the mechanistic prejudices of humans. The functioning of this reductionism, the implications of which I shall return to at the end of the chapter, is well captured by Vandana Shiva:

> Primarily, the ontological and epistemological assumptions of reductionism are based on uniformity, perceiving all systems as comprising the same basic constituents, discrete, and atomistic, and assuming all basic processes to be mechanical . . . The epistemological assumptions of reductionism are related to its ontological assumptions: uniformity permits knowledge of parts of a system to stand for knowledge of the whole. Divisibility permits context-free abstraction of knowledge, and creates criteria of validity based on alienation and non-participation, which is then projected as 'objectivity'.[6]

Biotechnology and Animals

Aside from the putative ecological potential of the genetic manipulation of mammals, what are the medical and scientific applications of animal biotechnology available to us today? Developments in this field have yielded technologies which range from artificial insemination and *in vitro* fertilisation to the cryopreservation and splitting of embryos, pre-implantation genetic diagnosis, the microinjection of foreign DNA into early animal embryos, and cloning by nuclear transfer. According to Michael Banner,[7] the application of new genetic technologies to animal breeding and research falls into three general categories. These are 'indirect' genetic engineering, which uses genetic analysis to make more informed selection of desirable breeding stock; pre-natal selection, which uses both *in vitro* and *in vivo* genetic diagnosis of embryos to screen out and eliminate undesirable genotypes; and 'direct' genetic engineering, which uses recombinant DNA techniques, including cloning, to create animals with new characteristics.[8] These applications are of course not mutually exclusive, and are invariably combined.

There are, in turn, four major uses to which the direct genetic engineering of animals is put. First, there is the manipulation of the genetic material of animals to turn them into so-called 'pharmaceutical factories' or 'bioreactors'. By the late 1980s scientists had succeeded in inducing the expression of various human proteins in the milk of lactating mice, and it was on the basis of these achievements that experiments began on larger farm animals, focusing on milk, blood, and even urine and semen, as carriers of medically important proteins.[9]

Today at least twenty-nine different human proteins have been produced in transgenic animals.[10] One of the first of these was a goat that was genetically altered to express in its mammary glands, and hence secrete into its milk, a variant of human tissue-type plasminogen activator, which is used to break down blood clots.[11] Transgenic sheep have since become a more popular source of medically valuable proteins. They have been engineered to produce human clotting factor IX,[12] which aids blood coagulation and is deficient in people who suffer from haemophilia B (Christmas disease), as well as alpha-1-antitrypsin for emphysema sufferers.[13] Cows have been successfully engineered to express in their milk human lactoferrin, a milk protein crucial to the development of infants' immune systems.[14] Pigs have been genetically manipulated to produce human haemoglobin in their blood,[15] and the genome of poultry has been altered – normally using retroviral vectors, because the 60,000 or so cells of the laid egg cannot be adequately targeted by microinjection – so that the yolk or albumin of their eggs contains pharmaceutically useful human proteins such as immunoglobulins (antibodies), alpha interferon (produced by the body's cells in response to viral infections, and used in the treatment of certain forms of cancer), and insulin.[16]

Second, there is the creation of transgenic animals, normally mice, both for toxicology testing and as 'models' for human genetic disorders. In the early days of this research, mice were bred with their siblings for consecutive generations (a minimum of twenty, and in many cases over a hundred), then selected for medically interesting traits deriving from spontaneous genetic mutations which the mice, being inbred, carried on both chromosomes. In later research genetic mutations were deliberately induced by stressing mice with radiation and chemical toxins, then selecting them for unusual phenotypes and breeding them as pure lines. This technique yielded mice biologically programmed to develop, amongst other ailments, obesity, diabetes, muscular dystrophy, cataracts, deafness, immune deficiency and baldness (in the last case produced solely through repeated inbreeding).

By the early 1980s the techniques of genetic engineering were sufficiently developed to allow scientists to transfer foreign DNA into the genomes of mice, either by injecting naked DNA into fertilised eggs, or by manipulating embryonic stem cells then injecting these cells into the early mouse embryo.

Mice could now be bred with specific mutations which predisposed them to a range of genetic diseases found in humans, ranging from cancer and Huntington's disease to cystic fibrosis, asthma and Alzheimer's. A review published in 1997 listed mouse disease models for 110 different human conditions, including a form of brittle bone disease which causes the animals to suffer extensive bone fracturing.[17] Strains of mutant mice with deleted, inactivated or supplementary genes have been bred and used to investigate 'neuropsychiatric, cardiovascular, pulmonary, oncological, inflammatory and immunological diseases, as well as for studying mechanisms and disorders of human metabolism, reproduction and early development'.[18]

Third, there is the creation of transgenic animals for use as a source of replacement organs (so-called 'spare parts factories') for humans. Though pig heart valves have been used in human cardiac surgery for some years, the future prospects for xenotransplantation using living organs like hearts and kidneys are naturally limited by the aggressive rejection and often destruction of foreign matter by the human immune system (even in successful cases the recipients must take immunosuppressant drugs for the rest of their lives). For this reason progress is being made in adding human DNA sequences to pig embryos in order to produce animals which express human proteins that prevent or reduce hyperacute rejection of the organ by the recipient. A second strategy being developed involves deleting the porcine gene that codes for an enzyme (called 'alpha-1,3-galactosyltransferase') which makes the sugar that, recognised as a foreign antigen by the primate immune system, causes hyperacute rejection in humans.[19] This was achieved, with much public fanfare, by a US subsidiary of PPL Therapeutics, which oversaw the birth of a litter of cloned transgenic piglets on Christmas Day 2001. The female piglets, each of which has one of the two copies of the critical gene deleted, will eventually be mated with males cloned in the same way. One in four of the resulting offspring should then have both their copies of the gene deleted, in a project which PPL, along with competing companies such as Immerge BioTherapeutics and Advanced Cell Technology, both of Massachusetts, hope will yield clinical human trials in four to five years.[20]

Fourth, genetic engineering is being used or proposed as a method to increase the productivity of farmed animals. Examples of this application include the genetic manipulation of salmon to grow faster and bigger; several unsuccessful attempts to introduce disease resistance traits into farm animals; proposals to make cows produce milk containing the anti-bacterial properties of human breast milk, reduced lactose to make it more palatable to Asian consumers, and increased protein content for cheese producers; and sheep manipulated with the aim of increasing their production of wool.[21]

In all four categories mentioned above, cloning by nuclear transfer will probably play an important future role in maintaining the purity of genetic lines, particularly when only one sex is required (cloning by inducing the splitting of embryos, it should be pointed out, is already established practice in the cattle industry). A case in point here is the breeding of dairy cows. A prize dairy cow, for example, can produce nearly three times the volume of milk as its ordinary counterparts, but passing that genctic prowess onto subsequent generations through sexual reproduction is an unreliable process because of the lottery of meiosis (the formation of genetically unique haploid gametes). Though intergenerational stability is to some extent achieved today using animal sperm banks, the most desirable option from a productivity point of view would be the cloning of the mature cow's full complement of chromosomes – a practice which could also take place before the donor animal has reached sexual maturity.

Risks to Animal Welfare

Because of the imprecise and experimental nature of the new biotechnologies, animal mortality rates in this field are extraordinarily high, and so is the volume of animals required for experimentation and research. Figures released by the Home Office showed that in 1999 alone there were 2.66 million experiments on animals in the UK – most of which were rats and mice. With transgenic animals comprising over half-a-million of that total, this was a fourteen per cent rise over the previous year, but also crowned a *ten-fold increase* from the 48,000 experiments conducted in 1990.[22]

The genetic engineering of animals to produce experimental models of human diseases is perhaps the most disturbing use of biotechnology on animals, since the suffering of those animals is a necessary feature and goal of the research. In one early research project, for example, transgenic mice were produced with malignant tumours in the lenses of their eyes – a region where tumours have never been observed to occur naturally in vertebrates. The researchers described the technique of inducing tumours in the eyes of mice as 'a potentially useful system for testing tumorigenicity of oncogenes', essentially because the eyes 'are not an essential tissue for viability, and tumours can be removed for study without losing valuable breeding animals'.[23] Indeed, removing the mice's eyes would have been a necessity for the propagation of further generations of the valuable creature, since the cancer would have killed its host before the latter became mature enough to reproduce.

The production of transgenic animals is also an inefficient and still experimental procedure which can be extremely wasteful of animal lives. As already noted, the dominant method for the genetic engineering of animals is the microinjection of foreign DNA into the nuclei of fertilised eggs or 'zygotes'. Because of the trauma of microinjection, combined with the random insertion of often multiple copies of the foreign gene (hundreds of copies are normally injected at once), this typically results in forms of genetic and chromosomal disruption, leading to a high incidence of lost embryos, developmental defects and perinatal deaths. Of the estimated ten per cent of microinjected zygotes which survive to birth, only five per cent of cattle, seven per cent of goats, and at best thirty per cent of mice, show take up of the foreign gene.[24] The remainder, as well as those transgenic animals which fail to express the desired phenotype, are destroyed.

Herman, the world's first transgenic dairy calf, was one of only two transgenic animals resulting from an experiment which began with the injection of over a thousand fertilised bovine oocytes (immature eggs). Nineteen calves were born in all, but seventeen of these failed to integrate the foreign DNA.[25] In the recent procedure that produced the four transgenic piglets cloned by Immerge BioTherapeutics and University of Missouri researchers as part of the project to develop xenotransplant-friendly pigs, over 3,000 embryos were implanted in twenty-eight sows, resulting in seven live piglets (a 0.2 per cent success rate).[26]

In the cloning experiment which created Dolly, 277 eggs were fused with nuclei from cells taken from the mammary gland tissue of a six-year-old ewe, resulting in the development of twenty-nine blastula-stage embryos which were transferred to the wombs of thirteen surrogate ewes.[27] Dolly was the only clone carried successfully to term, and developed normally until her sixth year, when she was discovered to have arthritis in a hip and knee joint (arthritis is extremely rare in sheep of that age, and normally only affects the elbow joint).[28] It is known that Dolly, along with other cloned animals, was born with shorter than normal telomeres – stretches of non-coding DNA which protect the tips of each chromosome but which shorten over time until the cell is unable to divide and so dies – and this may be one explanation for the animal's symptoms of premature ageing (since the nucleus of the somatic cell from which she was born was six years old, some argue that Dolly is much older than her post-natal years).

It has also been found that the normal reprogramming of methylation patterns that occurs during sexual reproduction in mammals – where the highly methylated genomes of spermatozoa are progressively demethylated after fertilisation and during the first couple of rounds of cell division, then reacquire characteristic methylation patterns in subsequent stages of embryonic development – is not the same in cloned embryos. Researchers have found that demethylation of the genomes of early cloned embryos does not occur to the same extent, and is often proceeded by a 'precocious' reorganisation of methylation patterns so as to resemble those of more mature, differentiated cells. These aberrant patterns of epigenetic reprogramming are strong candidates in the search to explain the substantial losses of cloned embryos during pre-implantation and early post-implantation development.[29] According to one review of the success rates in mammalian cloning, the percentage of animals reaching adulthood per manipulated egg is as low as 0.3 per cent for cows and less than one per cent for sheep.[30]

The production of eggs for both microinjection and cloning procedures is also problematic from an animal welfare point of view because it requires the use of hormones to induce female animals to superovulate, followed by surgical insemination. The temporary surrogates whose ligated oviducts (tied up to prevent the embryos passing to and implanting in the sheep's womb) are used to culture the developing embryos, are often slaughtered when the blastocysts are removed and implanted in final-stage surrogates which have also been made receptive with hormones (in mice this state of 'pseudo-pregnancy' is achieved by mating the surrogates with sterile males). Both eggs and embryos are normally removed by surgical incision (laparotomy), and any males that are born, notably when pharmaceutical milk production is the goal, are usually destroyed because, being genetically engineered animals, they cannot be used for food. Sheep and cattle developed from *in vitro* embryos are also associated with longer gestation periods, increased birth weight ('large offspring syndrome'), and greater likelihood of Caesarean delivery.

Animals successfully engineered to express endogenous or foreign proteins can also suffer a range of pathological side effects, which may derive from the overproduction of the desired protein, its leakage into or expression in non-target cells, or the unpredictable effects of the promoter. The notorious 'Beltsville' pigs, for example, were produced from single-cell embryos that had been injected with a human growth hormone gene. The pigs showed improved weight gain, greater feed efficiency, and reduced subcutaneous fat, but at the cost of a wide range of pathological side-effects, including 'gastric ulceration, severe synovitis [joint inflammation], degenerative joint disease, percarditis and endocarditis [inflammation of the outer and inner lining of the heart], cardiomegaly [enlargement of the heart], parakeratosis [cracking of the skin], nephritis [inflammation of the kidney], and pneumonia. In addition', the researchers disclose, 'gilts were anestrus [infertility due to quiescence or involution of the reproductive tract] and boars lacked libido.'[31]

The expression of human growth hormone in mice has in one experiment caused severe damage to the liver, kidneys and heart of the animals, while an attempt to produce human erythropoietin in the milk of rabbits led to low level expression of the protein in non-mammary tissues, as a result of which most animals died prematurely and were infertile. In another experiment the use of the mouse promoter, Whey Acidic Protein (WAP), in transgenic pigs interfered with the development of the mammary tissue and prevented the sows from expressing sufficient milk for their offspring.[32]

The Consequentialist Paradigm

How have philosophers and ethicists responded to the moral concerns raised by the growing use of animals in biotechnology research? For most of the last century, reflection on the human exploitation of animals, both critical and supportive, has been guided by the dominant doctrine of 'consequentialism'. Consequentialism has its historical roots in the rise of bourgeois rationalism, the waning of religious authority, and the post-Kantian critique of transcendent moral norms. The simplest form of consequentialism is classical utilitarianism, exemplified in the philosophy of Jeremy Bentham, and to some extent John Stuart Mill, which measures the good of an action in terms of whether it maximises the net balance of pleasure over pain. Consequentialism is in this sense a 'teleological' theory because it defines as virtuous actions which deliver good results. The motives, convictions, beliefs and goals of the agent are at best morally neutral, if not irrelevant.

Since the early development of utilitarian moral philosophy, thinkers from this tradition have strived to modify this seemingly dry and mathematical doctrine in an attempt to accommodate more refined definitions and ingredients of the good life – not just 'pleasure', but also life, health, freedom to choose, aesthetic appreciation, truth, knowledge, friendship, justice, and so on.

A new variant of the doctrine, 'rule-consequentialism', also emerged to make the perspective more workable as a guide for public policy. Instead of proposing that each individual act be judged in terms of its immediate consequences (so-called 'act-consequentialism'), rule-consequentialism holds that rules (policies, laws, maxims) are necessary to harmonise individuals' actions over space and time, and hence to make possible moral judgement over actions which may have limited immediate or individual significance. Theft, for example, while condonable in some circumstances, may thus be condemned because it would, if generalised, lead to the breakdown of property relations; telling lies, similarly, if done consistently and by many, would generate a dysfunctional degree of distrust; and animal experiments, though the value of any particular one may be negligible on its own, may thus be defended for the ultimate breakthroughs to which they collectively lead.

For all their differences, however, most consequentialists agree on one thing: moral judgement in the domain of public affairs must be the result of a pragmatic and disinterested deliberation which removes from consideration potentially divisive and irreconcilable categorical moral and religious convictions. As a method of ethical assessment consequentialism is thus supremely suited to highly professionalised, technologically complex, market-dominated societies, where the prediction and calculation of the benefits, risks, consequences and costs associated with an action are typically entrusted to experts and accountants.

In the sphere of public debate about the rights and welfare of animals relative to the needs of the human community, consequentialist arguments have largely held sway. In Peter Singer's utilitarian reasoning, for example, the pain and discomfort suffered by farm animals outweighs the benefits enjoyed by meat-eating humans, thus making the industrial rearing of animals for food morally unjustifiable.[33] Singer's position, which represents a version of act-consequentialism, is not a defence of animal 'rights', and does of course imply that the exploitation of animals – including the human animal – may be acceptable if the resulting benefits are great enough. In this respect his perspective was in fact prefigured a century earlier in the musings of Bentham and Mill. Bentham, for example, famously anticipated a future time 'when the rest of the animal creation may acquire those rights which never could have been withholden from them but by the hand of tyranny'. In Bentham's view the question to be asked when evaluating the moral status of animals 'is not, Can they *reason*? Nor, Can they *talk*? But, Can they *suffer*?'[34] And Mill, despite differing from Bentham in his attempt to reconcile utilitarianism with an appreciation of the importance of 'higher' values, sentiments and pleasures, also used utilitarian reasoning to challenge the prevailing disregard for the well-being of animals:

> We are perfectly willing to stake the whole question on this one issue. Granted that any practice causes more pain to animals than it gives pleasure to man; is that practice moral or immoral? And if, exactly in proportion as human beings raise their heads out of the slough of selfishness, they do not with one voice answer "immoral", let the morality of the principle of utility be forever condemned.[35]

In the first few decades of the twentieth century, consequentialist arguments were widely employed by the eugenics movement, which sought to limit the proliferation of defective genes, and to some extent they still inform modern public health policy in matters, for example, of child immunisation and holiday inoculations, where the liberties of individuals may be curtailed by the need to prevent the spread of infectious diseases.

Consequentialism is also the approach which the medical and scientific communities believe best justifies their affirmative stance towards the use of animals in genetic and biomedical research. The experimental manipulation, suffering and destruction of animals is here seen as a price worth paying – a cost outweighed by the benefits – for an increase in the prosperity and health of society's human members. Most advisory reports in the UK and the European Union on the 'ethical implications' of biotechnology have thus taken the utilitarian view that there is nothing inherently wrong with the genetic manipulation of animals, but scientists involved in such practices must demonstrate that there are tangible human benefits to be gained from their work. This reasoning is well illustrated in the report by a working party set up by the British Medical Association (BMA) in the early 1990s, which claimed that 'biotechnology and genetic modification are in themselves morally neutral. It is the uses to which they are put which create dilemmas. The challenge which faces us is to try to achieve an optimal future: one which maximises the benefits of genetic modification and minimises the harms'.[36]

The Limitations of Animal Models

The consequentialist paradigm is certainly not without its problems, and there are strong arguments, which I shall consider in a moment, in favour of a different ethical framework for guiding our relations with animals. Even on its own terms, however, consequentialism may not provide the justification for animal experiments which most medical ethicists believe it does. One reason for this is the lack of factual evidence to demonstrate the reliability of animals as models for research into human disease and its treatment. In their book, *Sacred Cows and Golden Geese*, the medical scientists Ray and Jean Greek bring together considerable evidence in support of their claim that the genetic, physiological, biochemical and metabolic features of non-human animals are insufficiently similar to humans to justify their use in medical experiments.

Advances in medical knowledge and treatment, they point out, owe much more to clinical observation, test-tube research, epidemiological studies, post-mortems and serendipity than they do to the practice of animal experimentation. The lack of parity between humans and non-human animals is evident from the fact that numerous medicines considered indispensable to modern health systems – aspirin, ibuprofen, penicillin and other antibiotics, fluoride, isoniazid, anti-hypertensive drugs – are ineffective or pathogenic when given to animals, and were discovered and approved despite rather than because of the results of experiments on animals. It also explains why tens of thousands of people die every year in the US due to adverse reactions from drugs which passed successfully through animal-experiment protocols, and why, despite extensive animal testing, more than half of the 198 new medications approved by the US Food and Drug Administration (FDA) between 1976 and 1985 had to be withdrawn or relabelled due to severe unpredicted side effects.[37]

The use of genetically engineered animals in drug testing is the latest stage in a long and increasingly aggressive attempt by the pharmaceutical industry to persuade the public and the regulators that chemico-pharmacology offers the best and most comprehensive medical paradigm for the treatment of human illness. Yet the evidence justifying the genetic manipulation of animals to further this paradigm is not strong. After World War II, animal testing for proposed cancer treatments, for example, was initially conducted on mice bred to exhibit high rates of spontaneous tumours. When the seemingly effective anti-cancer compounds subsequently failed to work on human patients, scientists began, in the mid-1970s, to implant human tumours into the bodies of mice with faulty immune systems. Here again biological reductionism led to flawed expectations, as the xenograft tumours, once implanted into mice, behaved more like endogenous mouse tumours than human ones, refusing to spread to other tissues, for example. Not only did the anti-cancer treatments, though effective in human tumours implanted in mice, often fail to work in humans, but the reverse also occurred. When a US National Cancer Institute team, investigating twelve anti-cancer agents already used successfully to treat human patients, tested the agents on forty-eight human cancer cell lines that had been transplanted into mice, they found that thirty of the tumours failed to respond significantly to the otherwise effective drugs.[38]

The creation of animals – mainly mice – genetically modified with tumour-promoting human oncogenes or deleted tumour-suppressor genes,[39] emerged in the 1980s as the most likely answer to the search for an adequate animal-model of human cancer. Yet once again genetic reductionism, by obscuring the wholistic nature of genomic functioning, led to faulty expectations of human–animal parity. In a mouse engineered with an inactivated tumour-suppressor gene, for example, cancer of the colon develops, as it often does when an homologous gene[40] is disturbed in humans. In the later stages of the disease, however, the development of the cancer begins to diverge from the human phenotype (it spares the mouse's liver, for instance). The loss or mutation of another tumour-suppressor gene is strongly associated, in humans, with cancer of the retina (retinoblastoma), but when the gene is disrupted in mice the rodents remain free from the disease while some develop pituitary gland tumours instead.[41] And when geneticists in 1996 deleted in mice the BRCA1 gene – mutations of which in humans are associated with a higher incidence of breast cancer – rodents with one remaining copy of the gene stayed fit and healthy, while embryos which had both copies deleted died before birth.[42]

Similar surprises awaited genetic scientists studying other diseases, including the discovery that when the gene for the 'transmembrane conductor regulator' – a protein which is deficient in people suffering from the lethal lung disease cystic fibrosis – is knocked out in mice, this has limited effect on the animals' lungs, but leads instead to lethal intestinal obstruction. Attempts to produce a mouse model of the progressive neurological disorder Lesch-Nyhan syndrome by introducing a defective version of the relevant gene in mice, also failed to produce animals with the distinctive behavioural abnormalities exhibited by human sufferers of the disease.[43] And when a transgenic rabbit was produced to model the hormonal disorder acromegaly, the animal was unexpectedly sterile.[44]

Deontological Ethics

The utilitarian argument that the suffering of animals is a cost outweighed by the benefits which animal experiments bring to human well-being is thus contestable on the grounds of factual evidence. It may also be challenged from the perspective of a different ethical paradigm: what philosophers call a 'deontological' approach. This rests on the belief that there are qualities of experience, action and judgement which have inherent value, regardless of whether or not they make an instrumental or incremental contribution to the sum total of human welfare. Protecting these inherent values requires moral obligations, such as the duty to avoid inflicting harm, which cannot be neutralised or traded off against equivalent goods or compensations. In its strongest form this moral perspective confers intrinsic value – and usually, by implication, 'rights' – on all conscious and sentient beings.[45]

Modern consequentialism, its critics often point out, is in fact already tainted by a deontological ethics in so far as it recognises that individuals' right to liberty, security, and freedom from pain, indignity and violence, is an inviolable good which, at least formally, cannot be sacrificed to increase aggregate well-being. Indeed, since the happiness of individuals depends in part on their *own* preferred goods and values (the things that really, intrinsically, matter to them) being respected, it is apparent that a strict utilitarian perspective, in which happiness and suffering are objective weights of which humans are the passive carriers, is an incoherent philosophy. By disregarding the subjective motives, commitments and ideals of people the consequences of whose actions the utilitarian seeks to judge, consequentialism disqualifies its own claim to moral authority. It does so because it thereby forfeits any claim to explain why the quest for moral certainty and a concern for the good actually *matters* to people at all.

Animal rights thinkers have often exploited the anomalies of the consequentialist orthodoxy, arguing that acknowledgement of intrinsic or absolute goods is an inescapable facet of modern morality, and that this necessarily brings into the frame the intrinsic value of non-human beings. Of course, a strict interpretation of the deontological position would invalidate this claim, since it derives from Kant the belief that it is the individual's capacity for rational choice and autonomy, in defiance of their animal nature, which is the fundamental human right to be protected. But a strict interpretation is rarely adhered to, and in fact most deontological theorists implicitly acknowledge that moral freedom is both grounded in humans' natural faculties and is given substance, order and meaning through the emotional experience of needs and drives – for interaction, kinship, social recognition, physical flourishing – many of which humans share with non-human animals. Reason alone, from this perspective, cannot account for the fact that reason *matters*, and the capacity to *care about one's life* – whether a rational capacity or otherwise – is certainly not a privilege of humanity.

Though he accepts in principle that there may be situations in which the rights and needs of human beings take precedence over those of animals, Tom Regan, who rejects the utilitarian characterisation of individual beings as merely receptacles for qualities whose value exists independently of those beings, is probably the best known exponent of this revised 'deontological' position. Instead of regarding animals as mere carriers or contributors of values that can be formalised, quantified and aggregated, Regan maintains that animals, like human beings, have inherent value as living beings whose well-being matters to them, and who therefore have a corresponding right to be free from pain, suffering and premature death.[46]

One possible challenge to the view that modern animal biotechnologies are unethical because they violate animals' intrinsic right to be free from suffering, is the argument that many methods of animal husbandry that do not involve genetic engineering – ranging from selective breeding to battery farming – cause as much, if not more, suffering to animals, which are largely treated as disposable processing factories employed for the conversion of fodder into food. Modern strains of pigs, for example, have been bred to manifest larger muscle blocks, more anaerobic fibres and smaller hearts than their ancestral relatives, with the result that they are more likely to die of heart failure and suffer distress during activity. Chickens, similarly, have been selected for speed of growth and muscle composition – a modern broiler chicken reaches its slaughter weight in half the time it took thirty years ago. But these animals lack the skeletal and cardiovascular system to cope with these changes, leading to leg problems – exacerbated by lack of exercise for battery hens – and heart dysfunctions. Turkeys are also selected for maximum growth rates, with one consequence being that male birds are too heavy to mount their mates, who must instead be artificially inseminated. Meanwhile dairy cows bred to produce a volume of milk ten times greater than they would need to suckle their calves routinely suffer gait problems and hind foot damage, as well as high rates of mastitis, while some breeds of beef cattle have become so muscular they suffer calving problems.[47]

For many proponents of genetic engineering, these facts show both that animal suffering is an inherent feature of modern farming, and that human manipulation of animals' genomes is nothing new. In Andrew George's words, which echo the sentiments of proponents of agricultural biotechnology, 'humans have, albeit unconsciously, been carrying out highly effective genetic engineering of animals for several millennia by selective breeding programmes'.[48] And since transgenic animals could, in principle, be created without causing suffering, and for that matter engineered – with stronger hearts, disease resistance, improved immune systems, or even diminished sentience, for example – to suffer *less*, there is in the eyes of most scientists nothing inherent in genetic engineering which makes it an unethical tool.

The fact that genetic engineering may be interpreted as the latest stage in a historical process of 'domesticating' animals, however, does not, for some animal rights thinkers, make this latest technology immune to ethical critique. Some critics agree, for example, that the creation of transgenic animals may well be only a novel manifestation of a brutalising attitude to sentient beings which has survived for centuries, but argue that this constitutes weak grounds for justifying it. If genetically engineered cows provoke outrage from previously complacent observers, then their outrage, instead of being required to explain its choice of object, should in fact be invited to embrace a wider range of practices. As Stephen Clark explains:

> We ought not to manipulate the germ-line to produce creatures more to our taste (or to our medical need) for just the same reason that we ought not to have been domesticating, imprisoning, mutilating, breeding, killing them before. It is perfectly normal for people not to realise that they are doing wrong until their imagination is caught by some particular example of wrongdoing. It may well be that the wrong that awakens them is not really much more serious than many they have winked at or not seen . . . So the first objection to genetic engineering, even when there are no particular adverse effects to the animals involved (which of course there already have been), is that only someone unreasonably or immorally convinced that they had a right to interfere in the lives of others would even consider doing it.[49]

David Cooper presses this point further, arguing that what is novel about genetic engineering may simply be that it represents the last stage in a continuous process which is deemed to have reached its moral limit. 'The straw that broke the camel's back was just like the previous ones in the bale', Cooper points out, 'yet from the camel's point of view it was a very special straw. Sometimes, indeed, we only appreciate something as distinctive and novel by seeing it as the culminating stage – one that reaches a limit – of a continuous process.'[50]

In the view of Dutch biologist Henk Verhoog, modern animal biotechnology, though certainly part of a long historical process of domestication, is both quantitatively and qualitatively different to the techniques that preceded it, and therefore does indeed represent the breaking of a limit. It is quantitatively different in the sense that it increases the biological distance separating domesticated species from their natural (wild) ancestors. And it is qualitatively different because it entails processes of objectification and reductionism which depress and debase relationships amongst animals – particularly when novel transgenic mammals are created outside the ancestral bonds of animal communities – as well as between animals and humans. 'To accept certain stages in a process (such as domestication) as morally acceptable', Verhoog asserts, 'does not necessarily imply that all of the latter stages in that same process are also acceptable, especially when there are good reasons to believe that there are morally relevant differences between these stages.'[51]

The Preservation of Animals' Telos

While I shall return to the issue of reductionism in a moment, Verhoog's argument offers a useful illustration of one side of the philosophical debate over whether humans are justified in manipulating and interfering with the 'telos' of other species. The concept of telos derives from Aristotle's belief that underlying the chaotic and changing forms of the world there are structural categories of being, distinct types of things – the paradigm for which is the differentiated categories of biological species – each of which has its own form of flourishing, of being true to itself. This respectful regard for the 'norms' of the natural world was, as Jonas points out, the main casualty of the scientific revolution. By divesting nature of any will, purpose or essence of its own, the aspirations of science were transformed from the humble imitation of nature to its domination by the powers of human invention. 'In a nature that is its own perpetual accident, each thing can as well be other than it is without being any the less natural.'[52]

It has been argued that the orderly development and maturation of living things, the tendency towards wholeness that is expressed through the organic capacity to self-heal, and the exhibiting of what even Darwin recognised as 'that perfection of structure and co-adaptation which most justly excited our admiration', demonstrate that natural organisms can indeed be thought of as having an 'end' or 'purpose' which must be respected if their existence is to be in any sense their own.[53] In discussions of animal welfare issues, the concept of telos has consequently been adapted to describe 'the set of needs and interests which are genetically based, and environmentally expressed, and which collectively constitute or define the "form of life" or way of living exhibited by that animal, and whose fulfilment or thwarting matter to the animal'.[54]

For many of those who adopt a deontological approach to judging the ethical treatment of animals, the telos of animals has intrinsic value and should therefore be respected. This means that nocturnal and burrowing animals should not be kept in hard-based and illuminated cages, herbivores should not be fed meat, animals should not be slaughtered before they have reached adulthood, birds should not be prevented from nesting, flapping their wings, dust-bathing and so on, and that essential species-capacities – the capacity for locomotion, for social interaction and family membership, for exploration, courtship, grooming and play – should be allowed expression. Most traditional animal husbandry practices, it is argued, involved rearing animals in environments congenial to their telos, enhancing their most desirable or useful features by caring for their general well-being, and thus working with those animals much as a sculptor works with the grain of the wood.

Modern biotechnology, on the other hand, allows human beings to disregard animals' natural form of life, and even to create for them a new telos – which means, by implication, a new type of being altogether. The distinction between traditional animal farming and the modern biotech approach is, as Alan Holland puts it, a distinction 'between using another creature's ends as your own – which is acceptable – and disregarding that other creature's ends entirely – which is not'.[55] 'Old-style pastoralists identified the animals' ends, and worked out how to profit from them', Clark elaborates. 'New-style artificers work out where the profit lies, and mould the ends to suit them'.[56]

The argument that the creation of transgenic animals is an inherent violation of their telos must certainly meet the challenge posed by the fact that the closed gene-pools we call 'species' are themselves only temporary moments in the slow but continually evolving processes of biological mutation, selective adaptation, differentiation and hybridisation. As mentioned in chapter two, some degree of horizontal gene transfer is in any case inevitable between organisms which do not normally interbreed (hundreds of human genes, for example, are now thought to have been imported into the human genome from bacteria). Yet the argument against the notion of inviolable species boundaries cannot refute the fact that living things *do* exist in discrete units of interbreeding populations, and that the precise divisions by which we define these units are employed almost universally, even by tribes with no knowledge of Western taxonomy.[57] The rejection of the concept of species is thus only comprehensible if we abstract living things from their temporal contexts and bounds – the contexts of the life-span, of birth and death, of a living or remembered lineage – and treat those organisms as merely transitory moments in the grander, boundless historical narrative that is Evolution.

It must be recognised, nonetheless, that the alteration and violation of animals' telos is already an established part of animal farming. Through selective breeding and the intensive use of vaccines and antibiotics, farmers have proved capable of divorcing the productive growth and behaviour of livestock from the natural constraints of their species-telos, adjusting them to human needs while maintaining their general health and productive functioning. And since the telos of domesticated animals is already to some extent a product of human manipulation, changing their telos using a different technology cannot be legitimately represented as an unwarranted intrusion into a pristine nature. Hence some ecological thinkers have even argued that, though the genomes of 'wild' species deserve recognition and protection of their intrinsic value, once an animal has been adapted to the requirements of the human community there is 'no ethical justification for any bar on genetic alteration of domesticates, by whatever technical means'.[58]

It has also been argued that genetic engineering offers greater opportunity to change animals' telos the better to suit the environments they are reared in, hence to create, in effect, new varieties of animals for whom modern farming environments are not a violation of their 'nature', and hence not a source of frustration and suffering. In this respect, as Bernard Rollin believes, genetic engineering may have an important contribution to make to animal welfare. Acknowledging the strength of people's 'aesthetic revulsion' to the alteration of animals' genomes, Rollin suggests that 'if we could genetically engineer essentially decerebrate food animals, animals that have merely a vegetative life but no experiences, I believe it would be better to do this than to put conscious beings into environments in which they are miserable'.[59] The same logic applies to research animals, which Rollin thinks may be created with no brains so that they are 'incapable of perceiving their hideousness'.[60] In support of this Rollin proposes that the genetic engineering of animals be regulated by the principle of 'conservation of welfare':

> Any animals that are genetically engineered for human use or even for environmental benefit should be no worse off, in terms of suffering, after the new traits are introduced into the genome than the parent stock was prior to the insertion of the new genetic material. We may call this the principle of conservation of welfare. In other words, genetic engineering should at least be neutral regarding the well-being of animals. Ideally, it should improve the well-being of the animals engineered over that of the parent stock. Introducing genes for additional disease resistance that do not otherwise harm the organism is a clear example of a genetic modification that meets this test.[61]

In theory one could, therefore, produce chickens which lack a desire to nest, calves which are adapted to solitary confinement, high-volume dairy cows which are immune to mastitis, disease-resistant pigs or poultry which can remain free of infection in cramped and unsanitary conditions, cattle which can digest protein-rich cereals and hay without suffering from bloat, diarrhoea, acidosis – conditions which damage the immune system and increase the need for antibiotics – and other genetically engineered livestock which can tolerate extreme temperatures, drafts, or excessive humidity. Israeli scientists have already created, by means of low-tech crossbreeding, a featherless broiler chicken which can be raised in hot climates (overfed broilers generate so much internal heat that a feather coat plus high external temperatures are often lethal to them), and which promises to save the poultry industry significant outlays on ventilation systems and plucking labour.[62] Since animals are not naturally at home in laboratory environments, genetic engineering may also enable scientists to conquer the age-old problem of distinguishing between the pathological effects of stressful environments and the effects of experimental interventions, essentially by creating new laboratory species. Where animals are used to model particularly destructive human diseases, Rollin believes genetic engineering may permit the creation of such animals – but ones lacking in cerebral functioning, and therefore free of suffering.

Rollin also considers the transfer of the gene for hornlessness to cattle which must otherwise suffer from surgical dehorning, the use of gene therapy to treat the genetic diseases of purebread dogs, and the genetic alteration of endangered species to make them reproduce faster or adapt to hostile environments. If current developments run their course, the first commercial application of this form of genetic engineering will be the breeding of turkeys which, instead of being discouraged from brooding, as they are today, by rearing them in uncomfortable settings, are genetically manipulated to prevent the production of the hormone that triggers the desire to incubate their eggs – thus increasing egg-laying by up to twenty per cent.[63]

Taken to its logical extreme, genetic engineering could be used to manufacture biological factories of meat, pharmaceutical proteins or replacement organs – living things which lack any noticeable resemblance to the conscious, sentient, sociable animals from which they derive. These utility organisms may, like the 'deaf, blind, legless, microcephalic lumps' reluctantly envisaged by Stephen Clark,[64] be incapable of suffering, and their production and use may therefore make a significant contribution to the total sum of human well-being while respecting the principle of animal welfare. But are we then wrong to feel offended by such oddities, and the techno-scientific attitude which makes their creation possible?

One critical response to this possible application of genetic engineering is Alan Holland's dictum that 'there should be an objection to any practice which involves taking a form of life with a given level of capacity to exercise options and reducing that capacity significantly'.[65] Does this principle constitute an objection to the *improvement* of animals' capacities and options by genetic engineering, to scientific interventions designed to accelerate the evolutionary processes of genetic adaptation, mutation and survival? Presumably not. In truth it seems to me that there is a more important argument which needs making here, which centres not so much on the violation of animals' natural integrity or the diminishing of their species-capacities, but rather on the degradation and devaluation of *humans'* species-being.

The Enchantment of the World

Of all the virtues that make for a flourishing human life, humility is, for David Cooper at least, one which occupies a special place.[66] Humility entails a respect for the world as it discloses itself to us, and as something more than the knowledge we may have and the uses that we may make of it. Humility means taking the world as it is, and wondering at its strangeness. The value of humility is also recognised by a particular philosophical perspective. This is a phenomenological approach which, instead of claiming a transparent understanding of the objects of human inquiry, admits the ambiguity of our encounters with the world and the inescapable disunity of experience and knowledge. Central to this perspective is a recognition of the need to continually reflect upon and account for the way we come to understand the world, and to ensure that the accounts we give of the production and appropriation of knowledge do not prevent this understanding of the world from understanding itself.

All plant and animal husbandry is a form of intervention in the natural world, a shaping and harnessing of nature for the benefit of the human species. There is no doubt, moreover, that the 'naturalisation of nature', the attribution of unmediated innocence, purity and untarnished vitality to the organic world, is a project with suspect legitimacy. The valorisation of Nature is, for one, only meaningful to a species which has irreversibly fallen from Nature, and whose prospects of permanently returning to it are closed. It is also a project which collides with the all too ubiquitous imprint of that species' existence on the nature it reveres, thus making the 'preservation of nature' a misguided ambition in a world where few pockets of untouched wilderness remain. And it is a project which, by associating the value of nature with that which is stubborn and enduring, has too often served as an ideological prop for forms of domination.[67]

While there are, therefore, obvious dangers involved in romanticising traditional agricultural practices for being 'closer to nature', it may still be argued that the genetic approach to farming, and the reductionist science which underpins it, marks a watershed in the development of humans' relationship to nature in so far as it involves a decisive degradation in their subjective qualities of feeling and perception. To begin with, conventional selective breeding, however much it involves the subjugation and domestication of wild species, is a practice which must in turn be adapted – often with great patience and waiting – to the specific rhythms, processes, ancestral relations and regenerative capacities of the plants and animals concerned. One reason for this is that breeder and farmer are, in traditional agriculture, the same person, which means that the selection and production of desirable phenotypes does not arise out of a technical or laboratory process cut off from the cultivation and husbanding of plants and animals, but is merely one stage in the rearing and reproduction of life forms. Even where selective breeding or hybridisation forms a major part of the agricultural enterprise, this is ordinarily an attempt, limited by the constraints and preconditions of sexual reproduction, to combine embodied species-attributes and not, as it is for the genetic engineer, a manipulation and exchange of raw genetic material whose relationship to the organism in which it resides is wholly contingent.

The exploitation of nature is, in this respect, bounded and constrained by a recognition and respect for organisms *as organisms*, and an implicit contract between the farmer and the organism, the flourishing of the latter being the means for profiting the former. Instead of dividing and abstracting the organism to suit the controlled environment of the geneticist's laboratory, the farmer's understanding of living things, of their strengths and weaknesses, their characters and limitations, how they grow, develop, and interact with their social and biological surroundings, comes directly from the experience of handling, feeding, nurturing and harvesting them. This understanding also necessarily spreads outwards to grasp the environmental conditions which are implicated in the organism's evolutionary identity. Perhaps most importantly, to see organisms as living things, developing through unique interactions with the endless vagaries of their environments, is always to recognise that the ultimate truth of those things, the essence of their life-processes and behaviour, will never be fully disclosed to us. The parallel here between a non-reductive biology and the intrinsic limitations of atomic analysis as recognised by quantum physics was eloquently made by Niels Bohr.

> In every experiment on living organisms there must remain some uncertainty as regards the physical conditions to which they are subjected, and the idea suggests itself that the minimal freedom we must allow the organism will be just large enough to permit it, so to say, to hide its ultimate secrets from us. On this view, the very existence of life must in biology be considered as an elementary fact, just as in atomic physics the existence of the quantum of action has to be taken as a basic fact that cannot be derived from ordinary mechanical physics. Indeed, the essential non-analyzability of atomic stability in mechanical terms presents a close analogy to the impossibility of a physical or chemical explanation of the peculiar functions characteristic of life.[68]

It may be said, of course, that there is no value inherent in the opacity of nature's existence, in its independence or resistance to complete comprehension and control.[69] Without the value we attribute to that otherness, nature is merely brute, indifferent, 'being-in-itself', beyond both meaning and meaninglessness. But the fact that we *do* attribute value to the independent existence of living organisms, and have done since time immemorial, suggests that such organisms lend themselves, with varying degrees of hospitality, to the expression of a fundamental human species-capacity. This is the capacity to make the objects of our interventions in nature, and the manner in which we appropriate them, a source not just of physical sustenance, but also of aesthetic, affective and cognitive nourishment.[70]

On a purely affective level, animals both wild and domesticated naturally invite, through observation of their physical sensitivity and social instincts, a heightening of people's awareness of their own embodiment and social relatedness. In a broader context, all living things can function as sources of cognitive and aesthetic stimulation because their existence harbours potentialities, dynamic complexities and idiosyncrasies which cannot be mechanically appropriated, cannot be exhaustively understood, owned and controlled, but which make strong appeals to our sense of imagination, curiosity and sense of wonder. It is the diversity, fluidity and unpredictability of the natural world, in other words, which invites the sublimation of human's desire to possess nature and the conversion of this desire into cultural forms of nourishment. John O'Neill, who is actually keen to challenge those ecologists who see humans' duties to their environment as based on a continuity of self and nature, seems to concur with this observation when he describes an awareness of the independent existence of natural things not as a form of disenchantment of the world, not as the disillusionment of human's desire for unity with or mastery over their environment, but rather as 'the basis for a proper enchantment with it'.[71]

But what, exactly, does it mean to be 'properly enchanted' with nature? 'Appreciation of the strangeness of nature', O'Neill continues, 'is a component of a proper valuation of it'.[72] But the sustaining of our enchantment with nature also entails making sense of its strangeness, finding meaning in its indifference, imputing character, feeling or beauty to its behaviour and appearance. This is, curiously enough, precisely what we do when we anthropomorphise non-human life forms and attribute personality, emotion, desire and sensibility to animals, plants, and even elementary forces and forms which, like hurricanes or floods, remind us of the fragility of human life and which have always been objects of religious veneration and belief. In this way we accommodate and make sense of the temperamental and unpredictable character of the natural world, and in doing so restrain and mediate our desire for full mastery over it. Imputing character or feeling to nature thus represents and sustains our valorisation of its independent existence – an existence we may admire, fear, love, marvel at or feel repulsed by, but which we nonetheless recognise as something more than naked matter to be bent to our own intentions. The fact that even scientists, when removed from the pressures and expectations of their academic culture, are prone to anthropomorphise animal behaviour, suggests that there is something stubbornly human about this practice.[73]

The Religion of Technology

It is doubtless a misconception that scientists do not share a wonder for and appreciation of the objects that they study. As O'Neill stresses, science can and often does play an important role in developing and refining the skills and capacities that make possible a perception of the natural world's independent qualities, its complexity and beauty – often where non-scientists see only uniformity, ugliness or danger.

Yet the genetic engineering of living things is a scientific endeavour which is difficult to reconcile with a respect for the integrity and independence of natural organisms. The terms that have been widely used by genetic scientists to describe to the public their discoveries, accomplishments and ambitions – particularly those borrowed from the world of information technology and the science of inanimate things – make this endeavour especially suspect since they convey the conviction amongst those scientists that the capacity of organisms to withhold the essence of their being has been finally overcome, and the 'secret of life' has indeed been laid bare. This conviction reflects the belief that 'life' should occupy no special place in the scientific worldview – that organisms, like inanimate things, are merely the sum of their parts, and that dismantling the organism, and exposing those parts to the forensic eye of the molecular biologist, can extinguish the last vestiges of opacity and unintelligibility which otherwise make the organism a source of imagination, meditation and wonder.

At the very least, the representation of life in terms of transferable units of information, in terms of chemically encoded programmes commanding the development and behaviour of organisms, provides the language necessary to legitimise the treatment of these organisms as artefacts defined not by the essence they may fleetingly disclose to us, but by the function we would have them perform. By dismantling organisms into components that can be exhaustively known, owned, patented and controlled, the reductionism of genetic science seems to present us with the possibility of gaining absolute dominion over nature, just as Francis Bacon envisaged would be accomplished through mechanical techniques which, by torturing nature to reveal its secrets, would in turn be employed to create the natural world anew.[74]

Here the momentous hopes invested in the biotech industry reflect a religious enchantment with technology that, as David Noble shows, dates back to the early Middle Ages, flourished during the Enlightenment, and continues through to the apocalyptic evangelism of the atomic engineers, the biblical visions of rocket scientists and astronauts, and the rapturous enthusiasm for Artificial Intelligence displayed by today's leading researchers in robotics and information technology. In Noble's account, genetic engineering is the latest in a long series of technologies which have been venerated for promising transcendence and redemption to a fallen world by realising humanity's divinely ordained role to be God's companion in creation. 'DNA spelled God' – Noble recalls the highly religious pursuit of biology's Holy Grail – 'and the scientists' knowledge of DNA was a mark of their divinity'.[75]

Of course, this promise of omnipotence is a delusion and, as ecologists are keen to point out, a dangerous one at that. But genetic engineers' favoured language of 'bioreactors', 'spare-parts factories', 'nutraceutical fruits', 'livestock pharming', 'biofacture' and so on, are not just commercially sanitised descriptions of what remain stubbornly recalcitrant natural phenomena, and thus a potential source of future ecological disaster: this mechanistic language and philosophy is also disturbing for the way it seems to express the degradation of scientists' own physical and moral sensibilities, and a diminishing care for the integrity of life which has troubling ramifications for human relations themselves.

Is the hardening of the human heart and imagination an inevitable consequence of the biotech revolution? The danger, certainly, is that a science determined to uncover abstract and presuppositionless mechanisms that explain the natural world, and which defines its purpose as the unhindered accumulation and possession of objective knowledge, may soon find itself unable to explain *itself*, its original subjectivity and its rootedness not in the *absence* of knowledge, not in the experience of deficiency and ignorance, but in the fullness of mystery, ambiguity and wonder. We have to consider the cost – which cannot be measured and exchanged for a commensurable benefit – which the instrumental analysis and manipulation of sentient life, and the utilitarian logic which justifies it, imposes on our capacity to find meaning in the world, and to find ourselves as sources of meaning.

After all, the future world envisaged by the protagonists of biotechnology is not shaped in the image of the creative dissident, the anguished artist or the troubled moral thinker. It reflects, instead, the efforts of a meticulous assembler, a diligent conformist who produces a world according to rules which, by their indisputable objectivity and predictability, may well end up producing their subjects in turn. This process may begin with the experimental objects of molecular biology, the immaculate cells of a long-dead organism, the lifeless DNA, once extracted from the processes of its intercellular environment, cloned by the action of bacteria in a petri dish. It starts with material degraded to a form which is impervious to aesthetic or affective apprehension, which is emptied of all mystery, contingency, and potentiality, and which reveals its manner of functioning only to mathematical computation or the prosthetic appendages of laboratory scientists. Its goal is to transform the natural world into a universe of functional bio-machines, like the finely tuned animal-instruments which register the progress of human diseases, or the plant-factories which generate medicinal proteins. Yet these machines will, in turn, require operators, wardens and maintenance engineers, people devoid of indulgent sentiments, of humility and empathy for life, who can adapt themselves to this cold, functional universe and its predetermined and morally incontestable techniques. They will also require the consent of a citizenry schooled in the virtues of utilitarian reckoning, detached enough to measure means and ends, unswervingly committed to the principles of scientific and economic progress.

The ultimate danger is that the technocratic spirit which underlies the genetic paradigm will, by building a world in its own image, become a universal social requirement and goal. Transforming nature into an insensate, decerebrate, wholly objectified product, devoid of independent well-being, the biotech programme may thus deliver its golden promise by *relieving us of care*. The mechanisation of nature will lead to the mechanisation of ourselves, our sentiments, judgements, fears and dreams. This process is already foreshadowed by the cultural fetishisation of the cyborg, and of course in the assertion that human nature itself should now become the object of the biotech enterprise. I shall return to this subject in the final chapter.

6

Health and Disease: The Limitations of Genetic Determinism

Reductionism and Medical Science

The purpose of the preceding chapters was to explore the science of genetic engineering, as it has been developed and applied in the world of agricultural and animal biotechnology, and to highlight some of the ecological, political and ethical problems associated with these technologies and practices. My intention in this chapter is to throw some critical light on the Human Genome Project, and on the promised benefits to human health and well-being which its supporters believe it will bring. My principal concern here is with genetic reductionism in medical science. For understandable reasons, sociological critiques of genetic determinism have invariably focused on contesting the ambitious claims of sociobiologists, for whom people's tastes, behaviours, social accomplishments and personal characters can and should all be explained by reference to inherited biological determinants. While such critiques are not without merit, a potentially more fruitful challenge to genetic determinism is to face genetic science on its home ground: that is, critically to examine the far less controversial area of human health and disease.

Evidence of biological reductionism in the medical field is apparent, first of all, in the way 'health' is everywhere conceived as an intrinsic and exclusive property of the organism, rather than a relationship of equilibrium between human bodies and their environments. In homogenised and unresponsive social environments, natural genetic diversity throws up problems of incompatibility or maladjustment which biological reductionism treats exclusively as symptoms of organic disease, of 'defects' rather than socially unsuitable variations in the body's functioning. In medical genetics, this biological reductionism runs even deeper, with an increasing obsession with diseases 'caused' by single-gene defects. Though there is no doubt a humanitarian impulse to this project, the focus on monogenic diseases clearly serves a more pressing commercial agenda, which is to create markets for new and more profitable medical interventions. Such interventions, as I shall show in the next chapter, are likely to involve a level of intrusion into the human body which, given the complexity of the organism's functioning, carry considerable risks. This is one reason why the linear account of single-gene disorders has been popularised, since by simplifying the aetiology of disease it enhances the prospects of genetic correction.

Single-gene disorders represent, in any case, a tiny fraction of the total disease load. As I shall endeavour to show, in numerous cases they also prove to be much more complicated and variable than the phrase 'single-gene disorders' implies. With a huge variety of genetic mutations associated with identical phenotypes, and a wide variety of phenotypes associated with identical genetic mutations, some medical scientists have gone so far as to reject the notion of 'single-gene diseases' altogether. Where geneticists recognise the significance of the more numerous cases of so-called 'multifactorial' diseases – diseases, such as cancer, which result from the interaction of numerous internal and external factors – the danger of genetic reductionism remains. The problem is that genetic 'predispositions' to diseases triggered by external pathogens are themselves being reduced, in the medical and popular imagination, to inherited 'defects' amenable to corrective or preventative intervention. As with the focus on single-gene disorders, this is particularly worrying in the way it draws attention away from the collective social causes of illness and impairment. It also implies a medico-genetic programme whose goal I find deeply troubling: not the humanisation of the world, but the adaptation of humans – of human nature – to environments to which they are unsuited.

The Human Genome Project

The completion of the first 'rough draft' of the human genome was announced in June 2000. It was a joint declaration made by the public group of research laboratories that is the International Human Genome Sequencing Consortium, and its chief private competitor, the American company Celera Genomics. Senior figures in the scientific and political communities arose in unison to celebrate the magnitude of the achievement. The mapping of human genome 'will, in the fullness of time, rank as a cultural achievement on a par with the works of Shakespeare, or the pictures of Rembrandt, or the music of Wagner', proclaimed Colin Blakemore, president of the Forum of European Neuroscience, convinced that the explanatory power of art, literature and philosophy had been eclipsed. 'In the end,' he continued, 'it will inform us more reliably and fully about the human condition.'[1] While Tony Blair spoke of a 'breakthrough that opens the way for massive advances', President Clinton was less restrained, declaring, with the kind of mixture of scientific and biblical imagery that seems to characterise the religion of the new technology, that 'we are learning the language that allowed God to create life'.[2]

With a public budget of $3 billion, the Human Genome Project (HGP) officially began in the US in 1990, though dedicated human genome research centres at Los Alamos and elsewhere had already been set up and funded by the US Department of Energy since 1987. The British contribution was financed by the world's leading medical charity, the Wellcome Trust, as well as by the Medical Research Council, and complementary research in other countries was funded by the governments of France, Japan, Germany, Russia, China and elsewhere. Almost from its inception, the international effort to identify the complete sequence of nucleotide bases in the human genome courted huge interest, criticism and controversy.

To justify both the huge sums of public money involved and the exploitation of its findings by patent-hungry biotech companies and pharmaceutical firms,[3] sponsors of and participants in the project have made bold claims about the medical and scientific advances the project will yield. Francis Collins, who in 1993 succeeded James Watson as the director of the US arm of the HGP, claims that by 2010 there could be predictive tests for at least a dozen hereditary diseases, the elimination of hereditary conditions like haemophilia and porphyria, and by 2020 successful treatments for repairing mutated genes. The viability of these advances – most notably those developed in the field of genetic testing, in the manufacturing of drugs tailored to specific genetic profiles and in the use of gene therapy to correct or compensate for mutated DNA – is clearly founded on the promise that the mapping of the human genome will yield groundbreaking insights into the aetiology of disease, finally uncovering the 'hereditary contributions to manic depressive illness, to schizophrenia, to obsessive compulsive disorder, to autism'.[4] And this is to say nothing of the ambitious expectations of sociobiologists like Charles Murray, who claims that 'we are on the edge of understanding how human nature in individuals produces social and political institutions'.[5]

In one sense, the auspicious promises of the massive sequencing effort are well-founded. Evelyn Fox Keller, who has only recently been convinced of the merits of the HGP, argued that the most valuable achievement of this research programme is likely to be the laying bare of the incoherencies of biological determinism and the quasi-religious veneration of the gene. By February 2001 her prediction looked uncannily accurate. Scientists revealed that their analysis of the human genetic map, whose completion was announced eight months earlier, had exposed striking flaws in the prevailing wisdom. Some geneticists, struck by the fact that the biological complexity of different organisms seemed to bear no consistent relationship to the volume of DNA and chromosomes in their genomes (the cells of frogs, for example, contain a lot more DNA than those of human beings), had of course been challenging the presumed linear relationship between DNA and phenotypic traits for many years. Yet what the scientists discovered when they examined the human genome was quite unexpected. Instead of the anticipated 100,000 to 150,000 genes, the public consortium now estimated that there may be as few as 31,000 genes in the human genome, with their protein-coding sequences making up as little as 1.1 per cent of total DNA. These 31,000 genes are around twice the number in the cells of a fruitfly, only 10,000 more than in the simple roundworm, and merely 5,000 more genes than in the cells of the weed thale cress *Arabidopsis thaliana*.[6]

Since the human body creates eight times the number of proteins as there are genes, the Central Dogma that treats genes as the determining cause of observable traits now looks impossible to sustain. Human DNA is eighty-five per cent identical to the DNA of a dog, and there are only 300 genes in the human genome which are not shared by the mouse.[7] If genes alone cause the formation and behaviour of organisms, the difference between the laboratory scientist and the rat is trivial, and we may as well be experimenting on ourselves.[8]

Even more surprising, perhaps, is that it was Craig Venter, director of the private company which raced the public consortium to complete the sequencing project, who spoke most candidly about the real implications of this discovery. 'We simply do not have enough genes for this idea of biological determinism to be right', he conceded. 'The wonderful diversity of the human species is not hard-wired in our genetic code. Our environments are critical'.[9]

Health, Disease and Diversity

Venter may have abandoned a belief in genetic explanations for complex human behaviours, but he is unlikely to renounce the view that genetic science will be the source of radical breakthroughs in clinical medicine, since this was certainly on the minds of the investors who helped push Celera's stock market value up 600 per cent from its 1999 level (its share price later plummeted when Clinton and Blair expressed support for the HGP's free information policy). The genetic approach to understanding disease appears most pertinent to the 1,600 or so disorders thought to be caused by mutations to a single gene. Let us be clear at this point that when scientists use the term gene they are referring to a specific stretch of DNA on a specific chromosome which the cell uses as a template for the synthesis of a functional protein. However, because the exact sequence of bases in this stretch is subject to enormous variation across members of a species, the whole notion of a generic 'human genome' has been challenged by many as an extravagant and misleading reduction. According to current estimates, for example, there are in the human genome as many as ten million 'single nucleotide polymorphisms' (SNPs) – specific DNA sites where one per cent or more of the population exhibit variations in a single base – with tens, and possibly hundreds, of thousands of these sites located within protein-coding sequences. 'For the first time,' Aravinda Chakravarti explains, 'nearly every human gene and genomic region is marked by a sequence variation'.[10] Given this amount of genetic variation between individuals, how is it that scientists claim to be sequencing *the* human genome, as if all our DNA were identical? The answer is that they are, in fact, producing a 'composite' genome assembled from chromosomes taken from a number of different human cell lines held in laboratories in various parts of the world.[11] This attempt to make a fictitious, generic genome the standard by which disease, deviance and abnormality will be identified and measured, has not surprisingly provoked ethical criticism. It has also encountered the scientific complaint that the identification of a 'normal' human genome presupposes a full understanding of all the genes' 'proper' functions, an understanding which in turn requires that each sequence and its effects has been studied in a multitude of different individuals and environments.[12]

The launch of the Human Genome Diversity Project (HGDP) was partly an attempt to quell these concerns.[13] Conceived in 1991, ostensibly in order to document the true diversity of human populations, this project has also been handicapped by the prejudices of the genetic mindset, making a fetish of the biological distinctiveness of genetically 'pure' population groups, treating the targeted communities (including poor minority groups) as biological facts rather than social and cultural artefacts, and raising the widespread suspicion that capturing and preserving the DNA of 'unique, historically vital populations that are in danger of dying out'[14] is more important to the scientific establishment than the protection of the people themselves. Not only has the search for genetically isolated enclaves or races ignored the historical evidence of widespread ethnic intermarriage due to migration and military conquest in ancient societies; it has also disregarded the simple fact, originally identified by Richard Lewontin in his study of protein variants and blood-group data, and more recently corroborated by DNA analysis, that eight-five per cent of overall human genetic diversity is represented by differences between members of the same population.[15]

The hunt for the normal, healthy human genome and, correspondingly, for the unmutated ('wild type') allele, is also informed by, and seems to perpetuate, the fallacy that biological health is an intrinsic property of the organism. It is in hope of exploiting this fallacy that a number of biotech companies have launched their own private versions of the HGDP, prospecting for rare but precious human genes that could allow them to synthesise new pharmaceutical and health products.[16] Yet even from a strictly genetic perspective, it is plain to see that health is pre-eminently a relationship of equilibrium or concordance between bodies and their environments, and that changed environments can turn 'normal' genes into maladapted ones, can eliminate the deleterious effects of pathological mutations (by medical treatment, for example), and can even positively select for mutations with adaptive effects.

We should bear this last point in mind when considering the medical prospect of manipulating the human genome to therapeutic effect, a prospect that has been crudely defended on the grounds that it 'merely aims at restoring an order of things that obtained previously, but was disturbed by genetic mutation'.[17] The inability to digest lactose, for example, affects over a quarter of Western Europeans and up to three quarters of people from parts of Africa, eastern and south-eastern Asia and Oceania. Although it is commonly thought of as a recently acquired biological deficiency, it is in fact the capacity to *digest* lactose, and thus to take advantage of the successful domestication of cattle, which is the result of a genetic mutation to the gene which naturally switches off the expression of lactase – the enzyme that breaks down milk sugar – at the end of infancy.[18] This mutation was positively selected for by the development of pastoral cultures: by changing their way of life and culture, human beings created evolutionary pressures which led to a change in biological capacity. This is not Lamarckianism, but it does illustrate both the creativity and the malleability of human nature. As Kenan Malik puts it in a slightly different context: 'Because humans possess language, and because we have created institutions which allow us to possess knowledge not simply as individuals but collectively as a society, so acquired habits and knowledge can be passed from generation to generation, transforming human life – and human nature – in the process.'[19]

Another area where genetic abnormalities may be difficult to define is in many recessive genetic conditions where two mutated alleles are necessary for the disease phenotype, but where the carriers of single copies of mutated genes find themselves with improved resistance to certain environmental threats. For example, the mutation to the haemoglobin gene which, when both alleles are affected, causes the often fatal disease sickle-cell anaemia, confers increased resistance to the effects of the malaria parasite on those who have only one copy of it (hence the relative prevalence of the mutation among people native to or descending from equatorial Africa). The same genetic advantage seems to be conferred by thalassaemia mutations.[20] A single copy of the cystic fibrosis allele has now also been found to protect the body against typhoid, while heterozygous carriers of the Tay-Sachs mutation are thought to possess increased resistance to tuberculosis.[21]

The consecration of the faultless genome implies that correcting these mutations, especially amongst populations where the diseases they offer protection against are no longer a viable threat, will be one of the long-term goals for genetic science. But from an ecological perspective, genetic diversity is a resource as well as a burden, and one which may become increasingly valuable if future social and environmental changes – consider the threat of malaria-carrying (and pesticide-resistant) mosquitoes spreading northwards with global warming, or the current worldwide resurgence in multi-drug-resistant tuberculosis, or the growth of antibiotic-resistant bacteria – bring with them diseases which human beings are no longer equipped to withstand.

Mendelian Diseases

Sickle-cell anaemia, cystic fibrosis, and Tay-Sachs disease are examples of so-called 'single-gene diseases'. Single-gene diseases are inherited in classic Mendelian fashion, and are thought to account for around two per cent of all known human diseases.[22] They generally fall into three main categories: autosomal dominant disorders, which require a mutated allele on only one chromosome to be expressed; autosomal recessive disorders, which require the mutated allele to be present on both chromosomes (one inherited from each parent) to be expressed; and X-linked conditions, most of which are recessive, but which tend to affect males more than females because the X and Y sex chromosomes (males are XY and females XX) have little homologous DNA, with the consequence that the effects of a recessive allele on the X chromosome of a male has little chance of being masked by the expression of a normal (dominant) allele of the same gene on the other sex chromosome, as it often is for the female.

There are around 200 hereditary conditions recorded as autosomal dominant disorders, occurring at an average rate of seven in 1,000 live births, or 0.7 per cent. These include hypercholesterolemia (1:500), polycystic kidney disease (1:1,000), retinitis pigmentosa (1:4,000), tuberous sclerosis (1:5,800), myotonic dystrophy (1:8,000) and Huntingdon disease (1:10,000). Autosomal recessive disorders are more numerous – around 900 have been identified – but they occur (as homozygotes[23]) even less frequently at an average rate of around 2.5 in 1,000 live births, or 0.25 per cent. They include haemochromatosis (1:200), sickle-cell anaemia (1:500 among African-Americans), cystic fibrosis (1:2,500), Tay-Sachs disease (1:2,500 among Ashkenazi Jews), alpha-1-antitrypsin deficiency (1:3,500), phenylketonuria (1:10,000), beta-thalassaemia (1:20,000). The range of recorded X-linked disorders is much narrower and their incidence less common than autosomal disorders because the X chromosome makes up only five per cent of the total genome. They include colour vision defects (8:100), fragile X syndrome (1:1,500 in males, 1:2,500 in females), Duchenne muscular dystrophy (1:3,500 in males), haemophilia A (1:5,000 in males), Lesch-Nyhan syndrome (1:10,000 in males), adrenoleukodystrophy (1:20,000 in males).

There is also a fourth category of genetic disorders which result from chromosomal abnormalities. These are normally caused by errors in the formation of the germ cells during meiosis, and include structural changes involving the loss or relocation of chromosome segments (often precipitated by environmental agents such as chemicals or radiation), as well as the production of gametes with an excess or lost chromosome. An extra copy of chromosome twenty-one, for example, results in the condition known as Down's syndrome, which occurs at a frequency of one in 800 live births, while an extra X chromosome among males results in Klinefelter syndrome, which occurs in one of out 1,000 males and results in infertility.

Many of the conditions listed above are painful and distressing. They can involve serious disabilities and impairments, require frequent medical care, and lead to shortened and sometimes terribly brief lives. Yet the standard medical account of these conditions as 'single-gene diseases' which are, due to the simplicity of their aetiology, prime targets for corrective genetic intervention, is not without its flaws. To begin with, many of the symptoms that are diagnosed and categorised as the manifestation of a single disease are, in many cases, actually a group of related but different conditions. A good case in point is the classic monogenic disease, cystic fibrosis, which is the most common lethal autosomal recessive disease amongst Caucasians. Cystic fibrosis is caused by a mutation to a gene on chromosome seven which in healthy individuals is associated with the expression of something called a 'transmembrane conductance regulator', a protein which allows mucus-producing cells to secrete chloride ions, and with them water. In the absence of the protein, water is retained in the cells and the mucus becomes abnormally thick and sticky, accumulating in the sweat glands, airways and intestines, and causing, among other symptoms, pulmonary obstruction, infertility, chest infections, progressive lung-tissue damage, digestive problems and the blocking of the pancreatic duct.

Around two-thirds of cystic fibrosis sufferers exhibit a three base-pair deletion in the cystic fibrosis transmembrane conductance regulator gene (CFTR), which results in the failure to express a single amino acid, phenylalanine. However, the frequency of this mutation among people with cystic fibrosis varies considerably between different ethnic groups and populations, ranging from around a third of Ashkenazi Jewish and Hutterite cystic fibrosis sufferers, to eighty-seven per cent of sufferers in Denmark.[24] More remarkable is that 877 *other* CFTR mutations had by 1999 been identified worldwide, as recorded by the Cystic Fibrosis Genetic Analysis Consortium.[25] Many of these other mutations are rare, though in some countries a small number (three to six) account for between sixty and ninety per cent of identified cystic fibrosis cases.[26] In addition, around a hundred sequence alterations in the CFTR gene had by 1995 been identified and classified as 'benign' variations or polymorphisms. The picture is further complicated by the discovery that while over eighty per cent of males with cystic fibrosis also suffer from congenital bilateral absence of vas deferens (CBAVD) – meaning they are born without the tubes which transport sperm from each testis to the urethra – around sixty per cent of men with CBAVD exhibit at least one known CFTR mutation *but lack the symptoms of bona fide cystic fibrosis*.

In other cases CFTR mutations have been associated with mild pulmonary presentation, diffuse bronchiectasis, obstructive azoospermia, and other partial or atypical symptoms of cystic fibrosis, while a number of studies have reported different CBAVD and CF phenotypes exhibited by siblings with identical mutations.[27] Wide variation in symptoms has also been correlated with socio-economic status, diet, and exposure to infectious bacteria, as well as with the presence or absence of particular alleles of 'modifier genes' which may, for example, play an important role degrading the mutant proteins produced by defective genes.[28] All of this suggests that the idea of a 'cystic fibrosis gene' – if not the diagnostic category of the illness itself – is a rather extravagant reduction from what is a much more complex and puzzling phenomenon.

Cystic fibrosis is by no means the only inherited condition which resists simple genetic reduction. The symptoms of haemophilia B, for example, have been associated with 200 different nucleotide variations, while many DNA variations are thought to cause the potentially serious haemoglobin-deficiency disease thalassaemia. Today thalassaemia experts recognise that the disease is not in fact a single disorder, but a group of clinical symptoms which, according to the type and number of genes affected as well as their degree of disruption, range from mild to potentially fatal anaemia. Around 180 different mutations have been identified in the beta-globin gene of patients with beta-thalassaemia, for example, and divergent phenotypes are often observed in siblings who have inherited the same mutation. 'What at first sight seemed to be a relatively simply monogenic disorder', David Weatherall explains, 'is, in effect, an extremely complex syndrome in which many different genes are involved together with equally numerous ill understood effects of the environment.'[29]

Another haemoglobinopathy, the autosomal recessive condition sickle-cell anaemia, is in contrast caused by the precise mutation of a single base (a 'single nucleotide polymorphism') in the sequence that codes for a sub-unit of the oxygen-transport protein, haemoglobin, with the result that one constituent amino acid – valine – is replaced by another – glutamic acid. But despite the apparent simplicity of cause, the symptoms produced by this mutation are subject to wide variation, both between people (some sufferers are seriously ill from childhood, while others show only mild symptoms which may occur later in life), and according to circumstances (depending on physiological and biochemical conditions such as the degree of concentration of haemoglobin in the blood corpuscles, the degree of hydration, the level of constriction of the capillaries).[30]

Hereditary haemochromatosis is another disease popularly related to specific mutations in a single genetic sequence, in this case the gene HFE. The condition involves a failure to metabolise iron, which builds up in the body with potentially lethal effects. The precise role of the HFE protein is still unknown, but screening has shown that more than eighty per cent of people with hereditary haemochromatosis are homozygous for one specific HFE mutation, and such homozygotes are relatively common – comprising as many as five people in every thousand. Many will be surprised to hear, therefore, that in one recent study of 152 people found to be homozygous for this common mutation, only *one* was described as possessing the clinical symptoms of the disease. 'Our best estimate', the researchers concluded, 'is that less than 1% of homozygotes develop frank clinical haemochromatosis', thus casting considerable doubt on the wisdom of genetic testing for this disease.[31]

Another example of a 'single-gene disease' with varying symptoms is the rare recessive metabolic disease phenylketonuria (PKU). This has been traced to the mutation of a gene on chromosome twelve which normally expresses phenylalanine hydroxylase, an enzyme responsible for converting the amino acid phenylalanine into tyrosine. When levels of the enzyme are deficient, phenylalanine accumulates in the blood stream and becomes toxic to the central nervous system, causing the disease's characteristic symptoms of neurological dysfunction and mental retardation. More than a hundred different alleles of the gene have been found, however, with corresponding enzyme activities ranging from zero to normality. Perhaps most importantly, by monitoring blood phenylalanine levels and controlling, usually by means of a specialist diet, the amount of phenylalanine-rich protein that is consumed as food, sufferers from PKU can today expect healthy development and a normal life-span.[32]

Familial hypercholesterolemia is an autosomal dominant disease deriving from a mutation to the receptor gene for low-density lipoprotein, a type of cholesterol. This gene normally prevents the build up of cholesterol in the blood by enabling the liver cells to absorb, metabolise and disperse any excess. Defects to this gene lead to the clogging of the arteries with fatty deposits and an increased risk of heart attacks and strokes. Yet several genealogical studies – of several centuries of Dutch, Finnish, and American pedigrees – have shown that, in the absence of today's main environmental contributors to heart disease, familial hypercholesterolemia is compatible with a normal life-span, and that raised low-density lipoprotein concentrations may even have protected people from infectious diseases that were more common in earlier centuries.[33]

The X-linked recessive diseases, Becker and Duchenne muscular dystrophies, are serious conditions characterised by progressive degeneration and atrophy of skeletal muscle. Both diseases are associated with mutations in the gene that codes for dystrophin, an extremely large membrane-associated protein, with disease severity varying according to the degree of gene disruption and the resulting decrease in dystrophin synthesis (in the milder Becker muscular dystrophy the protein is produced in truncated form). Even these monogenic diseases show some puzzling features, however, which suggests that the gene alone cannot tell us the whole story. For example, in mice bred with a natural mutation to an homologous gene on the X chromosome, which results in the same absence of the dystrophin protein, the well-documented process of degeneration and death only affects the first generation of the mice's muscle fibres, which are subsequently replaced by healthy ones. Why is it that, despite the loss of dystrophin, the muscle tissue of mice does not exhibit the continuous rounds of fibre death that characterises the bodies of human sufferers?[34] Although a transgenic mouse has subsequently been bred to mimic all the main symptoms of Duchenne muscular dystrophy, this was achieved by deleting the gene for a similar muscle protein called utrophin, which is suspected to have played a compensatory function in the dystrophin-deficient mice (but which does not perform the same corrective role in humans).

Other anomalies have now been observed in human patients, such as the identification of dystrophin-positive muscles fibres in Duchenne sufferers, or the discovery of a still ambulant sixty-one-year-old with very mild Becker muscular dystrophy, in whose genome almost half the coding region of the dystrophin gene had been deleted.[35] Researchers who have studied this phenomenon in mice suggest the explanation lies in the cells' capacity to instigate alternative splicing arrangements, where enzymes prepare the primary nRNA transcript for protein synthesis not only by excising the redundant introns, but also by a corrective skipping of the exons containing the mutated sequences.[36]

Alzheimer's disease is another serious condition the genetic basis to which has been routinely simplified and exaggerated. The symptoms of Alzheimer's, which include the progressive and irreversible deterioration in memory, reasoning and behavioural control, are related to the degeneration of nerve cells and the build up of plaques of an insoluble protein in the brain. Mutations in three genes have been identified as highly penetrant sources of familial (inherited) early-onset Alzheimer's, yet this form of Alzheimer's is thought only to make up between one and five per cent of all Alzheimer's cases. Three quarters of medical cases are characterised as 'sporadic', with no clear inheritable cause, although a number of genetic mutations – such as a particular allele of the Apo E gene on chromosome nineteen – are now thought to increase susceptibility to the disease. Meanwhile, scientists are digesting the discovery that African-Americans are more than twice as likely to suffer from Alzheimer's as people of similar age in Nigeria,[37] a finding which corroborates previous evidence showing higher prevalence of Alzheimer's amongst older Japanese-American men in Hawaii than amongst the same age group in Japan.[38] With the Nigerians studied also exhibiting lower cholesterol levels and less incidence of high blood pressure and diabetes, one conclusion suggested is that the mainly vegetarian diet of the Nigerians offers protection against Alzheimer's and dementia as well as cardiovascular disease.[39]

Let me give one final example of a 'monogenic' disorder which does not live up to the ambitious connotations of the term. This is the degenerative autosomal dominant eye disease, retinitis pigmentosa, the cause of which is thought to be mutations which change the amino acids in the protein rhodopsin, the light-sensitive visual pigment of the retinal rods. Yet in one family with two sisters carrying the same mutation, the younger sister was found to be blind while the older one saw well enough to drive a truck at night. Small base changes are also thought to affect the rate and scope of retinal degeneration.[40]

Multifactorial Diseases

From this detailed analysis of monogenic diseases we can see why many medical geneticists actually argue that 'there is no such thing as a single gene disease'.[41] As Dipple and McCabe put it, reflecting their belief that no genes can exert their influence alone: 'genetic diseases represent a continuum with diminishing influence from a single primary gene influenced by modifier genes, to increasingly shared influence by multiple genes'.[42] This observation seems even more valid when we move away from monogenic diseases and consider the vast majority of medical conditions which do not, even in the eyes of mainstream geneticists, conform to the deterministic gene-phenotype model. Though most attempts to make firm associations between genes and conditions like asthma, obesity, hypertension, schizophrenia, and even more nebulous traits like aggression, addiction, sexual preference, intelligence and creativity, have repeatedly failed,[43] some medical conditions have been successfully correlated with genotypes that are thought to raise people's 'susceptibility'. Otherwise known as 'incompletely penetrant' genetic diseases, these include breast cancer, colon cancer, rare early-onset forms of type-two diabetes, and Alzheimer's disease. Yet in each of these medical conditions the identified mutations account for less than three per cent of all cases,[44] and even these cases are thought to be 'polygenic' (caused by the combined actions or interactions of more than one allele) or 'multifactorial' (caused by the interaction of inherited genotypes with physiological and environmental factors).

One of the biggest concerns raised by the extension of genetic categories to diseases with external origins is that it effectively medicalises what are often essentially social, economic and environmental problems. This ideological strategy is certainly not novel. In both the popular and the professional scientific imagination, medicine is seen as almost exclusively responsible for producing the remarkable decline over the last 150 years in deaths from infectious diseases (tuberculosis, scarlet fever, diphtheria, measles, whooping cough), and thus lengthening people's life-spans, despite the fact that mortality rates for most infectious diseases had been steadily decreasing in the industrialised world for decades before vaccinations and antibiotics were introduced.[45] The real cause of increasing longevity and reduced mortality rates was improvements in public hygiene, nutrition, sanitation systems, housing conditions, sewage disposal and clean water.[46] Just as clinical medicine claimed the credit for delivering individuals from the lethal grip of infectious diseases, so medical genetics now purports to be able to identify and eventually eliminate inherited predispositions for diseases which are disproportionately environmental in origin. This is nowhere more apparent than in the prevailing discourse on cancer.

Cancer

As is well known, cancerous growth of cells is associated with mutations in the tumour cells' DNA. One of the most interesting finds in the genetic fight against cancer was the discovery in 1979 of a gene on human chromosome seventeen called TP53. This was believed to belong to a family of 'tumour suppressor genes' or 'anti-oncogenes', the function of which appears to be to prevent uncontrolled cell growth. TP53 provides the template for a protein, called p53, which responds to broken or miscopied DNA by activating other genes which either halt DNA replication until the mutations have been corrected, or else kill the cell outright. Indeed, one theoretical explanation for the variable effectiveness of chemotherapy and radiation therapy is that they work in the same way as vaccinations: they exacerbate the rate of transcription errors in order to alert TP53 that it is needed; but when TP53 is already damaged, their effects can only be destructive.

Mutation to the TP53 gene is implicated, at least according to Ridley, in over half of all human cancers.[47] The other genetic basis of cancer is said to be 'oncogenes' – genes which directly cause cells to divide and proliferate in an abnormally aggressive and uncontrolled way. It is unclear whether any gene can become, by mutation, an oncogene, or whether only specific genes – sometimes called 'proto-oncogenes' – are susceptible to the copying errors which will turn them into oncogenes (it is also thought that some genes may be turned into oncogenes when they are mistakenly transposed onto another chromosome).

The identification in 1994 of another gene on chromosome seventeen, mutations in which appear to predispose people to breast and ovarian cancer, supported the tumour suppressor theory. When functioning properly the gene, called BRCA1, is used to synthesise a large protein which appears to repair damaged DNA, though curiously not just in breast and ovarian cells. The discovery of BRCA1, and subsequently BRCA2 on chromosome thirteen, seemed ample justification for the expansion of public investment in medical genetics in the US. When Myriad Genetics won patents for the genes in the US in 1999, and subsequently in Europe as well, then proceeded to demand royalty fees from university researchers developing and performing breast cancer testing programmes, the wisdom of this arrangement began to look less certain.[48] With private capital dictating the agenda, moreover, the geneticisation of cancer predictably began to obscure the social and environmental causes of the disease.

The observation that cancer is intrinsically a 'genetic disease', a disease based on mutated DNA, does not of course mean that cancer is *inherited*. Almost all forms of cancer in fact occur due to mutations in *somatic* cells which are damaged by environmental stresses largely deriving from mutagenic industrial pollutants (chemicals and radiation), pharmaceutical products, dietary habits, and other lifestyle activities like smoking and sunbathing. Genetic damage associated with cancerous cell growth is thus normally restricted to the cancerous cells themselves, and is rarely transferred to the cells of the reproductive organs and hence the next generation. There is, of course, evidence to suggest that there are many inherited mutant alleles or polymorphisms which influence the susceptibility of individuals to cancer, but these are generally thought to affect how well the human body deals with carcinogenic environments (in detoxifying chemicals, for example, or in resisting the mutagenic effects of radiation). This is why the link between so-called 'cancer genes' and the disease itself remains one of 'increased risk' rather than linear cause, and why the term 'cancer susceptibility genes' is becoming the preferred expression.

For example, in the case of breast cancer, which is the most commonly diagnosed cancer after non-melanoma skin cancer, and the second leading cause of cancer deaths after lung cancer, there are certainly well-documented patterns of family history to the disease. Inherited mutations to BRCA1 and BRCA2 (over 600 and 450 of which, respectively, have been recorded) are now thought to be the cause of this pattern, though contrary evidence – such as one study which found that only seven per cent of women from families with a history of breast cancer but not ovarian cancer had BRCA1 mutations[49] – is also accumulating. Moreover, even with an estimated one in 800 individuals carrying a pathogenic autosomal dominant mutation in BRCA1, the degree of penetration is far from complete, with the risk of a carrier developing cancer predicted at roughly sixty-three per cent.[50] And although it is estimated that carriers of inherited mutations in BRCA1 and BRCA2, who also come from families with a history of cancer, have an eighty-five per cent risk of developing breast cancer, when 120 Ashkenazi Jewish women and men, selected because they carried one of three specific BRCA gene mutations, were studied, the breast cancer risk was found to be significantly lower – fifty-six per cent – than the estimates based on subjects from high-risk families.[51]

Quantifying risk in these terms of course leads to further misconceptions, since the risk is calculated by averaging the statistical evidence of past trends, and by ignoring the possible impact of future changes in lifestyle, work environment, and so on. And while the tendency for some cancer specialists to express the same risk as '40–85 per cent probability' reveals just how imprecise the science of genetic prediction is in this area, the obsession with risk quantification still colludes in sustaining the myth that cancer is a risk that resides within us *as individuals* – much in the same way that shortness of height, to take one example from the sociological field, is seen to put pupils 'at risk' from bullying, and may even be referred to as 'the cause of their victimisation'.[52] In this way a person's increased vulnerability to the effects of an external danger (bullies, carcinogens) becomes subtly redefined as an abnormality or a disease – the disease of vulnerability or predisposition – regardless of whether or not the hazard to which the person is susceptible has actually exerted its effects, and regardless of the fact that the real affliction (oppression, cancer) is ultimately encountered or acquired rather than inherited.

Even inherited *susceptibility* to cancer, one should point out, is only associated with between six and ten per cent of all breast cancer cases.[53] Commenting on the general incidence of cancer as whole, the UK Government's Public Health Genetics Unit advises that only '1 per cent of cancers arise in families that show a very strong, hereditary susceptibility to a particular type of cancer',[54] a statistic reiterated by the genetic scientist Eric Fearon in *Science*.[55] Even an editorial for the *New England Journal of Medicine* expressed agreement with 'the widely accepted estimate that 80 to 90 per cent of human cancer is due to environmental factors'.[56] This approximation of the limited role played by germ-line mutations in contributing to cancer has been corroborated by numerous studies, including an exhaustive analysis of data on nearly 45,000 pairs of Scandinavian twins which concluded that '[i]nherited genetic factors make a minor contribution to susceptibility to most types of neoplasms',[57] as well as research that demonstrates how, when Chinese, Japanese, or Filipino women migrate to the US, their risk of developing breast cancer increases eighty per cent after ten years or more of Western living, and over several generations rises six-fold to approach the incidence rate among white Americans.[58] It is also substantiated by the figures recording a fifty-four per cent increase in cancers in the white American population since 1950[59] – a real growth rate of one per cent per year which cannot be accounted for by spontaneous genetic mutations, but which occurred during a period in which global production of synthetic organic chemicals rose 600-fold.[60]

The genetic approach to the study of cancer thus seems increasingly out of step with epidemiological knowledge of the disease. Not surprisingly, the 'war on cancer', famously launched in the US by President Nixon in 1971, has from the beginning been waged without drawing swords with the chemical and pharmaceutical industries. Because a preventative approach would require socially organised changes in people's living and working environments, if not a completely new model of social and economic development from which private companies and industries would be ill-placed to profit, national strategies for tackling cancer have remained centred on medical detection, therapy and treatment.

This has led to advertising campaigns like the Breast Cancer Awareness Month, which was co-founded in 1985 by the British chemical conglomerate Imperial Chemical Industries (ICI), together with a support group called Cancer Care Inc. and the American Academy of Family Physicians.[61] The annual campaign originally used the slogan 'early detection is your best prevention', recommending regular self-examination and medical screening. Since a detected cancer cannot obviously be prevented, this slogan was a shrewd way of shifting the aims and definition of health intervention in this area from the prevention of cancer to the prevention of death from cancer (the slogan was later changed to 'early detection is your best protection'). It should be noted, in any case, that mammograms – the best-established means of breast cancer screening – can be insensitive instruments of detection. This was indicated by an American study which estimated that half of all women will experience at least one false-positive result from a mammogram during ten years of routine screening, and that a fifth of women who do not have a breast cancer will have undergone an unnecessary biopsy after ten mammograms.[62]

Mammography – the radiation from which is thought to increase women's chance of developing cancer by one per cent per screening – is also poor at detecting lesions in the more dense breast tissue of younger women, which is where the cancers associated with BRCA1 and BRCA2 mutations tend to appear. It is also known to identify some slow-growing tumours and biologically benign cell changes which would not, if left alone, develop into health-threatening cancers. This may be why researchers at the respected Nordic Cochrane Centre in Copenhagen found that screening has increased the number of mastectomies by twenty per cent, but has at the same time failed to produce a detectable decrease in breast-cancer mortality.[63] It should be pointed out that this latter finding was subsequently challenged by Swedish researchers whose own study suggested that regular screening had reduced breast cancer mortality by a fifth.[64]

Given that ICI's spin-off company, Zeneca, remained, until its merger with Astra AB, a leading manufacturer of agrochemicals, and since AstraZeneca is itself the producer of tamoxifen – the world's best-selling anti-cancer drug, and one which, despite carrying further serious health risks,[65] is increasingly being prescribed as a preventative treatment for healthy but 'at-risk' women – it is small wonder that some health commentators continue to see Breast Cancer Awareness Month as an ideological attempt to mask the social and environmental origins of this disease, and to allow commercial gain to be made from the personal anxiety this campaign generates.[66]

Other Multifactorial Diseases

Cancer is not the only multifactorial disease that has been the subject of the genetic propaganda machine. Similar efforts have been made to identify genetic causes of chronic diseases like diabetes and hypertension, both of which have been shown by epidemiologists to be more strongly associated with intergenerational nutritional and socio-economic conditions than with genetic mutations.[67] Down's syndrome, to take another example, is a genetic condition characterised by clinically variable symptoms which are in most cases related to the presence of an extra copy of chromosome twenty-one. This is turn derives from an error in the process of cell division (meoisis) by which the founding gamete was formed, where two copies of the chromosome fail to part company and migrate to separate germ cells and are instead carried into a single egg or sperm. Since the risk of giving birth to a child with Down's syndrome increases with age, and since the rate of pathological copying and cell division errors rises as such mutations accumulate over the life-span, these genetic mistakes have been commonly regarded as an unavoidable malady of ageing cells. Yet this conventional view is accompanied by mounting evidence that the risk of having a baby with a chromosomal abnormality such as Down's syndrome increases – in the most recent study by *forty percent* – for women who live within two miles of a toxic landfill site.[68]

Probably few people were surprised to hear in the late 1990s that genetic mutations had also been identified as an inheritable cause of Parkinson's disease, a neurodegenerative condition caused by the loss of dopamine-secreting brain cells in a particular part of the brain. Damage to the alpha-synuclein gene on chromosome four was said to affect the expression of a protein thought to facilitate neuronal plasticity in healthy individuals,[69] while the mutation or deletion of a gene on chromosome six which normally codes for a cell-cleaning protein – triumphantly called 'parkin' – was later associated with both early- and late-onset Parkinson's.[70] Subsequent research found, however, that out of a hundred typical Parkinson's disease patients studied not one possessed the supposedly pathogenic mutation to the alpha-synuclein gene.[71]

Meanwhile public and charitable organisations like the US Parkinson's Disease Foundation and the Parkinson's Institute repeatedly stress that, except for in a small handful of families, the disease is not considered to be an inherited disorder.[72] A far more significant cause of the disease is likely to be exposure to environmental toxins, which is precisely what occurred in 1981 when severe Parkinson-like symptoms were observed in drug users in California who, believing they had been sold a new 'synthetic heroin', had injected themselves with the chemical compound MPTP.[73] Although MPTP is not found in the environment, its active metabolite is known to be structurally homologous to a number of agricultural chemicals, including the common herbicide paraquat. Though we should recognise the limitations of using animals as reliable models for human diseases, it is worth pointing out that when researchers exposed mice to low doses of paraquat combined with a common fungicide in an attempt to simulate long-term human exposure to the kind of agrochemical mixtures routinely applied to food crops, the mice exhibited behavioural and biochemical alterations analogous to Parkinson's disease.[74] Similar motor and postural impairments were observed in rats that were continually exposed to the common garden pesticide rotenone ('derris'), a natural compound widely believed to be a safe alternative to synthetic pesticides.[75]

Manufacturing Health in a Pathogenic World

We should now take stock of these observations and reflect on their implications for our understanding of the relationship between humans and their environment. First, it must be evident from the preceding discussion that the vast sums of money being poured into genetic science, and the huge public interest in and hope for the promises of the Human Genome Project, is out of all proportion both to the therapeutic interventions that new genetic knowledge is likely to yield, and to the contribution which conceivably correctable genetic mutations make to human suffering.

In the former case, we must see past the extravagant claim that extensive knowledge of the human genome will have obvious and direct clinical benefit. As Lewontin points out, the major advances in medical practice over the last century have been accomplished with little help from the Crick–Watson revolution in molecular biology. Instead they

> consist in greatly improved methods for examining the state of our insides, of remarkable advances in microplumbing, and of pragmatically determined ways of correcting chemical imbalances and of killing bacterial invaders. None of these depends on a deep knowledge of cellular processes or on any discoveries of molecular biology. Cancer is still treated by gross physical and chemical assaults on the offending tissue. Cardiovascular disease is treated by surgery whose anatomical bases go back to the nineteenth century, by diet, and by pragmatic drug treatment. Antibiotics were originally developed without the slightest notion of how they do their work. Diabetics continue to take insulin, as they have for sixty years, despite all the research on the cellular basis of pancreatic malfunction.[76]

As for the role played by genetic mutations in causing human suffering, we have already noted that monogenic diseases make up a fraction of the total human disease load. In the wealthy countries of the world, the greatest burden of disease (measured in terms of years of life lost to premature death and reduced quality of life due to disability), derives from cancer, cardiovascular disease, and neuropsychiatric disorders. In the world as a whole, on the other hand, the effect of these medical conditions is overshadowed by the consequences of communicable diseases, maternal and perinatal illnesses and nutritional deficiencies, which are together responsible for forty per cent of the global disease burden. Infant mortality exhibits a similar pattern, with more than two thirds of childhood deaths in the world caused by either pneumonia, diarrhoea, measles, malaria or malnutrition.[77] What is required by most of the world's suffering populations is not genetic intervention, but rather mosquito nets, clean water, nutritious food, basic perinatal medical services, and condoms.[78]

Medical genetics is also helping perpetuate the ideology of biological determinism, particularly when minor differences in individual susceptibility to social and environmental pathogens is presented as more significant than the pathogens themselves. New forms of genetic medicine may, in the future, 'correct' a genetic mutation predisposing a woman to breast cancer, but once delivered from the fate of this defect the woman will still be as susceptible as any other woman to the remaining ninety per cent or more of breast cancers that are not linked to heredity, as well as to the scores of other cancers whose cause is overwhelmingly environmental.[79] The fact that even people's susceptibility to infectious diseases is now thought to be subject to some degree of genetic variance, suggests that the current logic of medical genetics could eventually result in a complete abandonment of all concern for environmental sources of morbidity and a systematic attempt to make the least vulnerable genome the normative measure of human health.

We should in fact go further than this and reiterate Lewontin's observation that the ultimate cause of most ill-health and suffering in the world is neither internal genes nor external pathogens, but rather the structural inequalities and forms of powerlessness which allow those pathogens to inflict harm. Medical genetics, like the biomedical challenge to infectious disease that preceded it, essentially confuses the *agent* or *medium* of disease with its *cause*, and in doing so offers no credible means of resisting the replacement of old agents of disease with new ones. The genetic scientist thus 'isolates an alteration in a so-called cancer gene as *the* cause of cancer, whereas that alteration in the gene may in turn have been caused by ingesting a pollutant, which in turn was produced by an industrial process, which in turn was the inevitable consequence of investing money at 6 percent'.[80]

> Asbestos and cotton lint fibres are not the causes of cancer. They are the agents of social causes, of social formations that determine the nature of our productive and consumptive lives, and in the end it is only through changes in those social forces that we can get to the root of problems of health. The transfer of causal power from social relations into inanimate objects that seem to have a power and life of their own is one of the major mystifications of science and its ideologies.[81]

This point is also born out by Richard Wilkinson's extensive study of 'unhealthy societies'. Wilkinson shows that in affluent societies differences in health are indisputably correlated with social and economic positions, and that health inequalities not only increase in proportion to the degree of social inequality, competition and insecurity *as a whole*, but may also be greater amongst smaller, relatively privileged sectors of society where hierarchies constitute particularly visible and entrenched patterns of social reference and interaction. With death rates among junior civil servants in Whitehall three times higher than among senior administrators, for example, the evidence suggests that exposure to forms of subordination and stress may, once basic material needs have been met, be one of the major obstacles to human health and well-being.[82]

Finally, we need to relate this back to what we know about the capacity for adaptation and homeostatic regulation that characterises complex organisms like human beings. According to the theory of epigenesis, all complex organisms possess the ability to absorb environmental signals and produce adaptive ('homeostatic') behaviour which is not directly specified by inherited DNA – including behaviour which actively transforms the environment to better suit the needs of the organism. This variability of phenotype, however, is not infinite, and may be slow and limited in its accomplishments. From this perspective, ill-health and loss of biological function arise when environmental changes outstrip the adaptive and transformative capacities of individual human beings. Our genetic ability to store fat, for example, though indispensable to survival in harsh subsistence environments and hunter-gatherer societies, is incompatible with a hypercalorific environment where it instead confers increased 'susceptibility' to obesity and diabetes.

> When the world presents information for which the genome and its interactive epigenetic network have no adequate informational response – for example, an overly oxidative environment – the result is maladaptation, regressive-state change in cell behaviour, and finally end-state disease. Disease, then, is a result of the organism's frustrated attempts to adapt phenotypically to a hostile environment or set of elements for which there is no adequate response basis.[83]

In Strohman's account, the biomedical community should therefore 'focus not on genetic engineering designed to fit the organism to an increasingly hostile environment, but on environmental engineering designed to refit or return the world to a state consistent with evolved, conserved, stable genomic, and epigenetic capability'.[84] Yet changing society is, of course, a political project much harder to accomplish than changing ourselves, not least because collective solutions to shared problems are incompatible with the competitive logic of a capitalist society, its structural privileging of individual choice over common need, and its fierce refusal to allow any permanent sense of satisfaction, contentment or equilibrium amongst its members.

Consider the forced sedentarisation of existence in modern capitalist societies, the transformation of public space into corridors of cars, the stress of dangerous and competitive work environments, the chemical cocktails routinely consumed in the form of industrially farmed meat and vegetables. Consider the 'tyranny of choice' which faces the wealthy consumer in the developed world, the panoply of objects available to satisfy a single want or need, and the impossibility of sampling enough of those continually multiplying commodities to feel comfortable that one has made an informed choice. The overfeeding of consumers with refined, high-fat, sugar- and salt-enriched food is itself the product of a society which offers so much in the way of 'consumer choice' that, in order to keep up with the proliferating objects of wealth, we have rationalised consumption as if it were work. In the fast-food society of today, we seem to have forgotten Erich Fromm's reminder that 'consumption should be a concrete human act, in which our senses, bodily needs, our aesthetic taste – that is to say, in which *we* as concrete, sensing, feeling, judging human beings – are involved',[85] and instead strive in each act of consumption to free as much of ourselves (of our time, effort and care) for the consumption of something else, thus accumulating goods and experiences by dividing and diluting our taste for and interest in them.

In this society, health and fitness are rarely interpreted as a relationship of dynamic equilibrium between bodies and their environments, by which humans continually produce, destroy and refashion their worlds in accordance with their developing needs. Instead they become a package of commodities which, fragmented and specialised like the society that made them necessary, perpetuate the disintegration and alienation of people's bodily existence. It seems as if the commodification of well-being can only proceed by taking apart our cultural and biological lifeworlds, by excising health from life, sex from love (and reproduction from sex), by removing nutrients from food, meaning from work, exertion from mobility, and then selling these needs back to us in the form of nutritional supplements and dietary aids, as psychotropic medication and counselling, and of course as the ubiquitous fitness gyms where, arrayed on rows of mechanical treadmills like rehabilitated spinal injury patients, people exercise their biological capacity for locomotion by literally walking nowhere.

Genetic science is a conspirator in this process, persuading us that it is easier to adapt organisms to a hostile world than to make that world more habitable. Its project is persuasive because its goal is practically self-legitimising: to make 'human nature' a product whose authenticity or self-belonging can no longer be meaningfully ascertained, and for whom well-being is an unambiguous question of functional efficiency, strength, longevity and performance. Perhaps we have, in any case, already arrived at the borders of this post-human world, for the obsessive search for causal genes, the genetic gold rush, the hunt for the 'Holy Grail' of medicine or biology, is entirely consistent with a society which gives its members less and less opportunity to reflect, ponder, examine things from different perspectives, consider their feelings and experiences, and which instead turns scientists, and increasingly a spellbound public, into the fanatical accomplices of gene-detecting super-computers.

Future Trends

Despite the concerns expressed in this chapter, our appreciation of health and illness cannot do without an understanding of species-specific health, and we must recognise that the basic biological foundation of disease is an adverse departure from the range of variations typical of the normal functioning of the species. Endorsing this definition of disease, many will argue that a clear distinction can then be made between physiological pathologies on the one hand, and on the other non-pathological variations in species-typical functioning which, due to the prejudice and discrimination of the 'normal' majority, lead to mental suffering and, where environments reflect these prejudices, forms of material and physical disablement. From this perspective, it may be argued that forms of genetic intervention designed to adjust people to society *as a means of improving or restoring their health* – that is, enabling them to the escape the effects of prejudice and discrimination – are quite distinct from, and certainly less defensible than, genetic forms of treatment designed to improve people's health *in order to adjust them to society* – in the sense of enabling them to fulfil the rights and expectations of social membership.

In much of the literature on the new genetics, this distinction is expressed as the difference between, on the one hand, enhancing or improving uncommon or disadvantageous but non-pathological traits and features, and on the other, treating, correcting or curing pathological ones. I shall argue in the following chapters that this distinction is unlikely to survive the genetic revolution, and that acceptance of the latter will eventually lead to a positive eugenics aimed at improving the genetic resources of human beings in order to adapt them to social environments that increasingly exceed the sensory, cognitive, emotional and indeed moral resources of individuals.

At this point, two important factors may be considered as influential to this predicted trend. First, there is the fact that the market for genetic treatments for serious inherited diseases is likely to be small. The biggest need for relief from physical suffering lies amongst the poorest people in the world. The ideal solution to this suffering is cheap – basic resource redistribution – but offers little opportunity for the sale of profitable high-tech treatments. For this reason the future growth of the biotech health industry – if forecasts of growth are correct – will depend not on the treatment of ill-health, but on the strategic exploitation of wealthy people's fears, anxieties and prejudices, and of course the ever-expanding domain of 'multifactorial' disease.

Second, the most effective and efficient form of human genetic modification is likely to occur when the object of this intervention is an embryo or foetus. This will remove a vital ethical obstacle to the unilateral power of society to define and then procure the 'healthy' genetic constitution of the individual – the subject's right to refuse consent. The implications of this for the genetically modified person him or herself is a subject I shall return to in chapter 10.

7

From Pharmacogenetics to Gene Therapy

Pharmacogenetics

What, then, are the practical prospects for genetically-based forms of medical treatment, which are after all the main justification for the public financing and vaunting of the Human Genome Project? Medical supporters of the project believe it will deliver three main clinical benefits. The first of these, genetic testing, raises important ethical and political questions which will be considered later. From a strictly scientific point of view, the immediate benefits of genetic testing will be limited for the simple reason that most genetically-related conditions are incompletely penetrant or multifactorial, which means that the results of genetic tests have only modest predictive value. Many sceptics are concerned that the main consequence of the growth of genetic screening for 'susceptibility genes' will be the creation of a proto-morbid population for whom feeling well and being well have become two different things, and who see themselves either as the inevitable carriers of yet-to-be-discovered diseases (as 'healthy until further notice'), or as pre-diseased individuals living their lives in anticipation of predicted illness. 'In effect', Dorothy Nelkin points out, 'genetic risk for a disease has been reified as the disease itself, even in the absence of obvious manifestations of illness'.[1] These so-called 'pre-symptomatic ill' will of course be the prime target both of the insurance industry and of the marketing machines of the new life-science companies, which trade on people's anxieties and exploit their suspicion that health is a commodity that one either owns or lacks, and that possessing this commodity requires continual consumption of prophylactic medicines and lifestyle drugs, regular medical tests and examinations, and access to the constantly expanding range of 'healthy living' accessories.

If genetic testing expands at the rate that some predict, medical consultation will become less and less a matter of understanding the cause of the patient's presented symptoms, and more a question of predicting, minimising and preparing for the symptoms of an identified genetic flaw. For those unfortunate enough to test positive for a single-gene disease, in many if not most cases little in the way of improved treatment or prognosis can currently be delivered by virtue of this detection. Moreover, because 'presymptomatic screening' – a practice which of course is not restricted to and in fact historically predates genetic screening – fosters the belief that both ill-health and well-being are *medical diagnoses* rather than felt experiences, one notable danger is that over-reliance on formal certifications of health will lead people to ignore or misrecognise the physical signs of illness.

The second area of medical practice likely to be affected by the mapping of the human genome is pharmacogenetics, or what, in recognition of the fact that the whole human genome is now the object of study, is increasingly being referred to as pharmaco*genomics*. This is the customising of drugs and doses to suit the genetic profiles of different individuals, or more likely groups of individuals, thus exploiting the knowledge we already possess about wide variations in people's ability to metabolise important drugs. Millions of people, for example, possess mutated alleles of one or more genes necessary for the production of a family of enzymes, called cytochrome P450s, that are together responsible for metabolising most of the drugs used in clinical medicine. Six per cent of Caucasians and fifteen per cent of Nigerians, for instance, are unable to synthesise sufficient quantities of the liver enzyme debrisoquine hydroxylase, due to mutations in the gene CYP2D6. As a consequence, they cannot convert codeine into pain-relieving morphine, and are also at risk of adverse reactions from a range of pharmaceutical products – according to some accounts as much as a quarter of all prescription drugs[2] – used in the treatment of psychiatric, neurological and cardiovascular conditions. A fifth of Ethiopians and seven per cent of Spaniards also have extra copies of the CYP2D6 gene, as a consequence of which they metabolise these drugs so quickly that they require vastly increased doses to achieve any therapeutic effect. In many Scandinavian countries, where one in seventy-five people have such amplifications of the CYP2D6 gene, testing for mutations in this sequence in patients treated for psychiatric conditions is a widely accepted practice.[3] Today, gene chips designed to test for polymorphisms of the main cytochrome P450s are already commercially available.[4]

Nucleotide variations in the genes that produce aldehyde dehydrogenase and alcohol dehydrogenase, the enzymes responsible for metabolising (and thus detoxifying) alcohol, lead to a biochemical intolerance of alcohol that is particularly common in Oriental populations, affecting seventy per cent of Japanese people.[5] Another example is the 400 million people worldwide with an X-linked recessive condition called Glucose-6-Phosphate Dehydrogenase (G6PD) deficiency. Although the mutated form of the G6PD gene can once again confer resistance to malaria (in G6PD deficient red blood cells there are insufficient quantities of the metabolite required for the survival of the malaria-causing parasite), it also causes – when the affected individual consumes certain medicines (including aspirin and sulpha drugs) – potentially fatal hemolytic anemia (the premature death of red blood cells). Adverse reactions to drugs has in fact been ranked between the fourth and sixth largest cause of death in the US, with over two million hospitalised patients suffering serious adverse reactions from properly prescribed and administered drugs every year.[6]

Seen in this context, research in this domain certainly seems to be a worthy development, promising safer and more effective treatment of illnesses based on the well-marketed model of 'personalised medicine'. But we should not be under any illusions as to the motives of the pharmaceutical companies in driving the growth of genetic pharmacology. We are unlikely to see drugs designed for individual patients, but will enjoy at best a more successful fit between patients and the products available to treat their conditions. Drug research and development can be a costly business, with between eighty and ninety per cent of drugs entering the trial stage never reaching the market, often because a small number of people suffer adverse reactions to them.[7] Recovering this wasted investment, essentially by relaunching unsuccessful drugs as treatments 'customised' to those patients who are genetically susceptible to their ameliorative effects (and immune to their toxic ones), will be an attractive strategy for the industry. Yet it is one whose efficacy presupposes both that drug sensitivity is a single-gene phenotype, and that the results of pre-prescription genetic tests will always tell the truth. Since incorrect, misread or mislabelled test results are always possible, and since our knowledge of the complexity of genomic functioning suggests that it is highly unlikely that differing responsiveness to all types of drugs can be conclusively traced to and explained by single sequences of DNA, independent of the particular metabolic, digestive, and environmental circumstances, giving potentially dangerous drugs to supposedly protected patients could involve unacceptable levels of risk.

The development of pharmacogenetics may therefore signal a decisive shift in the balance of influence between patients and the medical industry, with pharmaceutical research ceasing to be a quest for remedies to human afflictions and becoming instead the search for the afflictions most suitable for the industry's products. Moreover, for such an enterprise to achieve even moderate success, extensive genetic screening and profiling of society's members may well be necessary, bringing with it the inevitable threat of genetic discrimination and loss of civil liberties. Critics may well wonder, in fact, whether the much-vaunted benefits of pharmacogenetics will serve as the Trojan horse which, by legitimising the identification and collection of genetic information, opens the door to a new eugenic structuring of society.

Gene Therapy

The third medical practice which genetic scientists believe will benefit from the mapping of the human genome and advances in molecular biology is 'gene therapy'. This involves the modification of gene expression by the addition of supplementary or replacement genes, or of DNA sequences designed to inhibit the expression of, or even to delete, defective genes (an intervention sometimes separately distinguished as 'gene surgery'). Gene therapy is, as Hubbard and Wald point out, already an ideologically-loaded term, since it confers therapeutic status on techniques which are still largely experimental.[8] These techniques are divided into two categories of clinical treatment. On the one hand, there is germ-line 'therapy', which modifies either the reproductive cells of individuals, or which targets the cells of embryos before the germ cells have formed and differentiated. Taking its lead from the genetic engineering of animals, this procedure is likely to involve the microinjection of DNA into the pronuclei of a single-celled fertilised egg,[9] the retroviral infection of early embryonic cells, or the transplantation of genetically modified spermatogonial stem cells – the parents of sperm cells – into the male's testes.[10]

Since this form of genetic manipulation is aimed at permanently altering the genome of individuals, and all the subsequent generations who inherit, via the germ cells, their DNA, genetic manipulation of the germ line is outlawed in most countries where it is seen as an unethical violation of the principle of consent, and an infringement of the belief that the human gene pool is the collective property of humanity. It also poses potentially enormous health risks which cannot be properly investigated except by allowing long-term trials (i.e. several generations of study) of genetically modified human beings.

The other form of 'gene therapy' developed for clinical practice is targeted at the somatic cells of individuals – specifically the cells of the organ or tissue in which the disease is located. This is seen as less controversial because the effects of this intervention should not be inherited by the patient's offspring. Although geneticists believe that ultimately all forms of disease, including non-genetic conditions such as infections, bone fractures and wounds, can in principle be treated by altering gene expression in the relevant cells, most research has to date been focused on serious medical conditions such as cancer, 'monogenic' disorders such as cystic fibrosis, haemoglobinopathies and immunodeficiency syndromes, and AIDS. One reason for the focus on serious, often life-threatening diseases, is that human gene transfer may carry considerable risks for the recipient, which many regard are entirely unacceptable in cases where the patient's life is not already at stake.

The first clinical trial of somatic gene transfer for a genetic disease involved a four-year-old American girl suffering from severe combined immunodeficiency disease (SCID). The girl was one of the thirty per cent of SCID sufferers whose condition is caused by an extremely rare autosomal recessive condition called ADA deficiency. This occurs when an individual inherits two mutated alleles of the ADA gene, resulting in loss of the enzyme adenosine deaminase, which normally plays a key role in the maturation of antibody-producing white blood cells. Because it prevents the mounting of an effective immune defence against infections, ADA deficiency is potentially fatal, and until fairly recently sufferers had to be kept in protective isolation unless or until they could receive a compatible bone marrow transplant. The prospects for sufferers in the US improved, however, when in March 1990, after four years of trials, the FDA approved the use of the bovine form of the ADA enzyme, wrapped in a polyethylene glycol (PEG) sheath to prolong its circulation in the bloodstream, as a drug replacement therapy for this condition.

With weekly or bi-weekly intramuscular injections of PEG-ADA costing around $100,000 per patient per year, the US National Institutes of Health

(NIH) was also keen to explore other possible approaches. Because four-year-old Ashanthi DeSilva's responsiveness to PEG-ADA had begun to decline, and because no lymphocyte antigen-matched sibling existed as a possible bone marrow donor, she became eligible for a gene therapy trial proposed by W. French Anderson and his colleagues at NIH and initiated in September 1990. Anderson had been an advocate of gene transfer since the late 1960s, and had founded the world's first gene transfer company, Genetic Therapy Incorporated (GTI), in 1986, his initial aim being to manufacture the vectors required for a full-scale clinical trial of the technique as a treatment for ADA deficiency. Failing to win regulatory approval for such a trial in the late 1980s, Anderson had astutely switched the focus of his research to the less controversial area of cancer treatment, successfully pioneering a protocol for the transfer of genetic markers to track cells in cancer patients. As a result, the ethical taboos and technical reservations surrounding gene transfer dissipated, and one year later Anderson's ADA-deficiency treatment proposal was given the green light.[11]

In the experiment, the researchers isolated and removed T-lymphocytes (a type of white blood cell) from the young girl, stimulated them to divide and proliferate, then incubated them with a murine (mouse) leukaemia retrovirus into which a 'healthy' copy of the human ADA gene had been spliced. The retrovirus, having been genetically altered to prevent it from replicating and causing disease, was the vector which, by inserting DNA copies of its own RNA into the genome of the T-cells, delivered the normal ADA gene into the defective blood. The 'gene-corrected' lymphocytes were then returned intravenously to the girl's blood stream, with the transfusion repeated a dozen or more times over the next two years.

Following the procedure, Ashanthi's ADA levels rose while her weekly dosage of PEG-ADA was reduced.[12] Anderson was also delighted to hear in 1995 that the patent he had filed with the NIH for the *ex vivo* gene therapy protocol had been approved by the US Patent Office. Licensed exclusively to Anderson's firm, it was one of the broadest ever patents awarded in the medical field, covering any human gene therapy that involves the removal, genetic alteration and reintroduction of a patient's cells. When GTI was subsequently sold to the Swiss pharmaceutical company Sandoz (now Novartis), the patent rights went with it.

Since the treatment of Ashanthi DeSilva, Anderson's technique has been used with other young ADA-deficiency patients, with seemingly comparable levels of success. In April 2000 French researchers announced that, ten months after delivering genetically-altered purified blood stem cells to two babies suffering from a different type of SCID, the results suggested 'full correction' – i.e. cure – of the disease.[13] In April 2002, the British press also reported the first case of gene therapy in the UK, in which Welsh toddler Rhys Evans, in the absence of a compatible bone marrow donor, was successfully treated for the same disease as the French babies. This form of SCID is caused by a mutation in the gamma c gene on the X chromosome, and it was treated by cultivating stem cells from the child's bone marrow with a retrovirus carrying a normal copy of the gene, then returning the modified cells to the boy's body.[14]

Despite the apparent success of these latter cases, gene transfer trials for SCID sufferers have not yielded conclusive proof of their safety and efficacy. In Anderson's work, for example, safety concerns obliged the researchers to keep the patients on PEG-ADA injections during the treatment, thus making it impossible to ascertain the cause of the patients' improved health (indeed, without the enzyme replacement therapy the researchers would probably have been unable to recover sufficient T-lymphocytes from the patient anyway). In the case of a second patient included in the Anderson trial, the nine-year-old girl's immune system response was highly variable before the onset of the gene therapy, ranging from zero to nearly normal. Even the scientists who conducted the research conceded that the contribution of their intervention to the children's improved health was therefore unclear.[15] In a subsequent study of three infants with ADA deficiency, who had been treated at birth to infusions of umbilical cord blood stem cells containing the inserted ADA gene, scientists tried to distinguish the effects of PEG-ADA from those of the exogenous gene by stopping PEG-ADA injections (which were also given from birth) for the healthiest child. The resulting deterioration in the boy's immune system convinced the researchers that 'the total amount of ADA gene-transduced cells must provide far less ADA activity than was being provided by PEG-ADA therapy'.[16]

Safety Concerns

Some of the questions inevitably raised over the true effectiveness of somatic gene transfer have in many cases been overshadowed by fears about its safety. These concerns centre on risks such as that the random insertion of exogenous genes may, by disrupting the patient's genome, lead to cancer-promoting mutagenesis, that the viral vectors, subject to horizontal gene transfer and recombination with other encountered viruses, may recover their replication competence (a grave concern where HIV-based vectors are used), and that the level of expression of the new gene may become uncontrollable.

There is also an appreciable risk that the vectors will migrate to non-targeted tissue – including, given what we know about the potential permeability of 'Weismann's barrier', the germ cells (from which the foreign genes can be passed to subsequent generations). In a trial by the leading gene therapy company Avigen, Inc. in California, which involved carrying the Factor IX gene into the liver of patients with haemophilia B, DNA traces of the adeno-associated viral vector were discovered in a patient's seminal fluid. Though the trial was suspended in October 2001, the FDA gave clearance for its resumption a month later, when subsequent testing showed the foreign DNA had cleared the patient's body and had not penetrated the sperm. According to the FDA official overseeing the trial, its continuation was necessary to allow an informed assessments of the risks and benefits. To minimise any hidden risk the patient will nonetheless 'be required to wear a condom'.[17]

Another pressing worry over the safety of gene transfer treatments concerns the risk that the recipient's immune system will be stimulated to fight off the viral invasion and destroy the infected cells. This fear proved well-founded when, in September 1999, the human gene transfer trials claimed their first victim. Jesse Gelsinger was an eighteen-year-old high-school graduate from Arizona who suffered from a rare X-linked condition called ornithine transcarbamylase (OTC) deficiency. OTC is a liver enzyme which plays a important role in clearing the blood of ammonia, a toxin produced when the body metabolises protein. Gelsinger's condition was at the mild end of the disease's spectrum, and was controlled by a special low-protein diet and drugs that sequester and excrete excess ammonia. In the trial that he participated in, the vector used to transfer the healthy OTC gene was a patented version of a common respiratory tract virus, called an adenovirus, which was again stripped of the genes necessary for its replication. An adenovirus is regarded as a far safer vector for human gene transfer, because it does not normally integrate itself into the host's genome, and hence avoids the risk of iatrogenic rearrangement of DNA. This vector was pioneered in the development of gene transfer protocols for cystic fibrosis sufferers, in which nasal sprays were used to shoot the adenovirus, combined with a working copy of the CFTR gene, into patients' lungs. In these experiments, take up of the vector by the cystic fibrosis patients proved to be extremely low, however, leading to few ameliorative effects, while higher doses were associated with inflammatory and immunologic responses which limited the duration of transgene expression as well as being hazardous to the patient.[18]

An adenovirus was chosen for the OTC-deficiency treatment trial because retroviruses can only infect cells that are actively dividing (such as blood cells, which can easily be extracted and cultured *in vitro*, then returned to the host), whereas in this case the non-dividing cells of the liver were the target. Gelsinger was the eighteenth patient in the trial by scientists at the University of Pennsylvania. Most of the patients before him experienced fever and other moderate symptoms in response to the treatment, while the tenth and twelfth patients exhibited signs of liver stress. Despite these responses, and having already witnessed the toxic effects of high doses of the transgenic virus on the livers of baboons, the researchers progressed through the volunteers. When the transgenic viral particles were infused by catheter into a vessel feeding Gelsinger's liver, his body launched a massive immune response which resulted in systemic inflammation, flooding of the lungs, acute respiratory failure and, by the fourth day, brain death.

A post-mortem found the viral vector dispersed throughout the body, with significant amounts in the spleen, lymph nodes and bone marrow. The recovered vector also contained DNA sequences not present in the original, implying that some form of genetic recombination had occurred.[19] When the US Food and Drug Administration (FDA) and the NIH launched their enquiries, it was found that of the 691 'adverse events' suffered by volunteers in experiments using the OTC adenovirus at the University of Pennsylvania over the previous seven years, only thirty-one had been properly reported to the NIH as required by law.[20]

Research on the more powerful and efficient adeno-associated virus (AAV), which is currently under trial as a gene therapy vector favoured because it carries a greatly reduced risk of immune system rejection, has also found that this vector can cause 'significant incidence' of tumours in mice that have undergone 'gene therapy'.[21] Unlike adenoviruses and retroviruses, adeno-associated viruses integrate themselves directly into the host's genome, and although the specific site of integration (on chromosome nineteen in humans) is itself relatively predictable, experiments on human cell lines have shown the insertion of the vector to be associated with random deletions and rearrangements of flanking DNA.[22]

From Cure to Enhancement

Given the unproven efficacy and well-documented risks of gene transfer experiments, it is not surprising that the revolution in health treatment predicted by the supporters of the Human Genome Project has not been universally applauded. Justifiable scepticism amongst many scientists and medical practitioners has also been magnified by an awareness that advances in medical genetics are unlikely to remain confined to the clinical domain, and that genetic modification will eventually be proposed as a means of enhancing or adjusting healthy human features, of providing an alternative to cosmetic surgery for breast enlargement, for example, or of improving memory and speed of thought,[23] or 'treating' socially stigmatised and often ethnically specific attributes such as excessive blushing, female facial hair, dark skin, non-Caucasian eye and nose shapes, or shortness of stature. A San Diego company specialising in anti-cancer treatments is already developing a gene therapy product for baldness, for example, and hopes that by first gaining approval to use the product on chemotherapy patients, the company will eventually be licensed to market it to ordinary consumers.[24] And since neuropharmacology has already made spectacular gains creating markets for behaviour-altering drugs like Ritalin and Prozac, and with companies already attempting to turn treatments for Alzheimer's into lucrative memory improvement drugs,[25] a progression from 'cosmetic pharmacology', as Peter Kramer calls it, to prosthetic gene therapy, looks like a fairly sound prediction.[26]

Another sphere of life which is an attractive target for the non-medical application of human genetic engineering is the highly competitive (and lucrative) world of sport, where, for example, somatic cell gene transfer could, by increasing the expression of growth-promoting proteins or the production of red blood cells, offer athletes potentially undetectable improvements in muscle strength and stamina. When the oxygen levels in a body's tissues drop below a certain threshold, the kidneys excrete a hormone, called erythropoietin or 'epo', which instructs the body to increase its production of oxygen-carrying red blood cells. Having isolated (and patented) the epo gene, scientists have successfully transferred it to the muscle cells of mice, monkeys and baboons, raising their red blood cell counts significantly (some monkeys had to be bled to keep their blood pressure within manageable limits). The medical goal of this research is ostensibly to develop epo gene transfer protocols as a means of treating severe anaemia in humans. But since Epogen – the injectable form of epo which is synthesised using recombinant bacteria – is already one of numerous performance-enhancing drugs which, although banned, are widely used in a variety of endurance sports, the market for epo gene-enhancement therapy may be large indeed.[27] Small wonder the World Anti-Doping Agency (WADA) – which is funded by the International Olympic Committee (IOC) to investigate and combat drug cheating in sport – brought thirty leading scientists together to discuss how to tackle 'Genetic Enhancement of Athletic Performance' at a conference in New York in March 2002.

Policing the boundary between medical and non-medical applications of gene transfer techniques will also be made difficult by the cultural construction of health and disease, and by the fact that, when medical treatments are expensive commodities, the economic pressure to produce new consumers often overrides social and ethical reservations. After the manufacturer of Epogen, the biotech company Amgen, sponsored medical literature research by the National Kidney Foundation in the US, the patient advocacy group recommended that optimum levels of red blood cells be raised above current guidelines, in response to which doctors began prescribing significantly higher doses of the drug. Epogen is today the most expensive drug in the US Government's Medicare programme, with sales of the product totalling $1.8 billion in 1999 in the US alone.[28]

The same growth trend has occurred with the sale of human growth hormone (hGH), but in this case it was the market rather than the dose that has expanded. Extracted from human cadavers, hGH has been used since the early 1960s. Following the discovery in 1985 that three young men treated with hGH had died of Creutzfeldt-Jakob Disease, distribution of the natural hormone was stopped by the US Public Health Service, and a market was immediately created for a synthetic version of the hormone, produced from genetically engineered bacteria. Once Genentech were able to manufacture large quantities of recombinant human growth hormone (rhGH), and Eli Lilly also entered the market, the profile of those deemed to require the treatment in the US began to change, and doctors started prescribing it to short people with fully functioning pituitary glands. This was despite the fact that scientific studies had suggested a possible link between long-term hGH use and leukemia, and despite the lack of evidence demonstrating its efficacy in children who already produced their own growth hormone (in the latter, the additional supply of the hormone is thought only to shorten the time taken to reach the same adult height).[29]

By the early 1990s the market for this drug – which, because of the rarity of the disease it was meant to treat, had been granted the tax exemptions and government subsidies permitted by the Orphan Drug Act – was worth nearly $500 million.[30] In 1994 senior executives of Genentech and the drug's distributor, Caremark, were brought before a federal grand jury in Minneapolis to answer charges that they paid $1.1 million in illegal kickbacks to a Minneapolis doctor who was the most frequent prescriber of hGH in the country (he had 350 patients). The companies were also known to be funding a charitable foundation which publicised the 'disease' of shortness and trained teachers to recognise it, and they had also hired a paediatric nurse to conduct a study identifying prospective patient-customers in schools.[31] Genentech scientists went on to suggest that anyone whose height falls within the lowest three per cent of the population should be considered suitable for treatment.[32] By pandering to widespread prejudice towards and discrimination against shorter-than-average people, the company's cure for height deficiency was simultaneously a means of making the condition a permanent (but lucrative) feature of American society.

By 1996 the US FDA had removed its ban on the use of recombinant human growth hormone (rhGH) in adults. The market for the drug was now estimated to be well over $800 million.[33] Meanwhile, evidence was mounting that, because the production of hGH steadily declines as people age, treatment with rhGH may deliver potentially remarkable improvements in health and longevity by increasing people's bone density, lean body mass, skin thickness, and size of organs like livers and spleens.[34] In this process of the research, development and marketing of an important therapy, what was originally produced as a treatment for an uncommon medical condition has become, through its aggressive marketing and success, the producer of new categories of illness: namely, height deviance and ageing. Given the marketing power of the biotech companies, and the extraordinary lengths to which some people are willing to go to correct their bodily imperfections – witness the decision of a sixteen-year-old British girl to have her femurs repeatedly broken and reset in order raise her above her existing 4 ft 9in of height[35] – there is every likelihood that genetic manipulation will eventually be included among the list of available correction/enhancement techniques.

Technicising Moral Judgement

The encroachment of dangerous and invasive genetic technologies from the clinical domain to the world of lifestyle and culture is not the only social risk associated with the development of human 'gene therapy'. What is also under threat is the integrity of the moral boundary separating genetic self-modification from the modification of gametes and embryos, and hence of other, non-consenting human beings. This boundary is partly endangered by the progressive technicising of moral judgement that accompanies the expansion of specialist technological practices – an attitude of pragmatic calculation and utilitarian reasoning which seems to exempt no values, rights or feelings from its grip, and in which all moral factors or claims can in principle be matched and disqualified by a manufactured equivalent.

There is no doubt that the predominance of a pragmatic, instrumental approach to evaluating the ethical issues raised by the prospects of human genetic modification has also been fostered by the growth of assisted reproduction technologies. As the application of these technologies has spread, it has become logical in many contexts to think of the creation of life in scientific and technical terms, and in a language compatible with the principles of efficiency, cost-effectiveness, and quantitative measures of success. On one level, this seems entirely sensible: assisted reproduction is a complex scientific procedure whose success depends on the technical skill of experts and the precision and reliability of their tools. Yet our understanding of, and contentment with, the successful technical manipulation and unification of gametes on behalf of the sub-fertile, may also have broader cultural and moral ramifications, altering the way we think about and articulate what has been portentously referred to as the 'traditional' method of conceiving and gestating human beings.

As one scientific commentator explains: 'Human embryonic development is remarkably inefficient. Most embryos die before a woman is even aware she is pregnant'.[36] What is insinuated here is that the human body is a sub-optimal organism to which functional improvements, corrections and 'upgrades' can and should eventually be made. Unfavourable comparison with productivity rates in IVF is an inevitable consequence: 'Some fertility clinics are achieving such high success rates for IVF that the women they treat now have a better chance of getting pregnant in one cycle than fertile couples relying on plain old-fashioned sex.'[37]

The same concern about the inefficiencies of the human organism may extend to the foetal production of eggs. By the time a baby girl is born, around three quarters of the seven million eggs produced in the ovaries during the first few months of foetal life have been lost, and by the time the female body is a proper candidate for ovulation and fertilisation less than five per cent of the young woman's original full complement of eggs remain. Of this five per cent, or 300,000 eggs, only one will normally be released for possible fertilisation every month, leaving 299,000 to steadily degenerate in the adult ovaries.[38]

Should efficiency-dedicated scientists be endeavouring to recover these hundreds of thousands – and in the case of foetuses, millions – of wasted ova in order to put them to good use? Scientists have already taken the immature eggs of new-born mice, cultured them to maturity, then fertilised them to produce a new generation of pups,[39] suggesting that the use of foetal eggs in IVF faces no unsurpassable technical obstacles. As indicated earlier, research into male infertility has also enabled scientists to turn mouse testis stem cells – spermatogonia – into mature sperm by incubating them in the sterile testes of other mice, and the same research laboratory has transplanted *rat* spermatogonial stem cells into the testes of mice – an animal separated from its rodent cousin by some fifteen million years of evolution – and successfully produced healthy rat sperm as a result.[40] Combine these two techniques and a steady supply of fertile gametes could be produced from foetal eggs and foetal testicular cells, enabling scientists to create embryos on demand for important medical research into infertility and disease. Were one such embryo brought to term by a surrogate mother, incidentally, a child would be created whose genetic parents *were never born*.

If cost-benefit analysis and considerations of efficiency are also applicable to the longer-scale processes of evolution, then human cloning may also be sanctioned as a more advanced and speedier form of genetic reproduction. If we adopt Dawkins' theory of the selfish gene, 'which treats an animal as a machine programmed to maximise the survival of copies of its genes', then obviously sexual reproduction is 'more "unnatural" than cloning', as Dawkins puts it, the latter being an evolutionary advance because it enables genes to reproduce themselves without competition from and contamination by the genome of a sexual mate.[41]

The cloning of human beings and the harvesting of gametes from unborn children are practical propositions the ethical implications of which scientists insist we can no longer address in emotional or sentimental ways. Our relationship to and understanding of our bodies, in other words, *should not be an embodied one*. This demand for dispassionate analysis conspires with the apparent successes of science in analysing and manipulating the behaviour of organisms and their components in controlled, calculable and impersonal ways. As a result, our tacit awareness that living bodies are something more than the mechanical operation of their parts, and that our relationship both with our own bodies as well as the bodies of others can never be reduced to a matter of function or technique, may be censored, with few moral convictions, as Leon Kass implies, surviving the scientific repudiation of affect: 'Shallow are the souls that have forgotten how to shudder'.[42]

The Medicalisation of Conception

The prevailing cultural and political view, sponsored by the medical and scientific community, that a week-old 'pre-embryo' is merely a ball of undifferentiated cells, is a clear example of the way the practical successes of new biotechnologies can change the perceptions and expectations of the public culture. Strictly speaking, an embryo is a mere ball of undifferentiated cells only for the scientist who has removed the early embryo from its original context in order to examine and manipulate it as a laboratory object. In its natural environment, this minuscule cluster of cells is something else: a hidden possibility, a hope, a fantasy, an embryonic fear or surprise, a moment of magic or ecstasy which will only reveal itself later as a memory, an imaginative story or an object of wonder.

Pre-embryonic life cannot justifiably be accorded the constitutional rights of inviolability associated with the dignity of the human person. But this fact alone does not make the early embryo an object undeserving of respect and reverence. Human being is not a manufactured product, but is recognised as an organic achievement occurring somewhere on a developmental continuum. It is because the precise point at which the life of a person truly begins is a mystery to us that the dignity attributed to that person, whose origins lie in the unknown and ineffable, seems proper and warranted. The scientific claim, on the other hand, to have identified a precursory stage in the development of human life, when the embryo's cells can be manipulated without moral censure, implies disposal over the beginnings of that life, and thus threatens to erode the foundations of our moral respect for it.[43]

Since the first 'test-tube baby', Louise Brown, was born in 1978, assisted conception through *in vitro* fertilisation (IVF) and intracytoplasmic sperm injection (ICSI) has grown so rapidly that nearly one child in eighty born in Britain is today conceived in a laboratory.[44] Part of the explanation for this is the way fertility treatment has consolidated the medicalisation of human reproduction, transforming childlessness into a diagnosable *illness* to be mastered by clinical science. This is despite the fact that in a large percentage – in some accounts a majority – of cases the inability of couples to conceive remains medically unexplained,[45] while fertility treatment 'cures' this apparent aberration in at best only one in five cycles. Yet by rendering visible and manipulable the creation of human life, one also has to ask to what extent reproductive technologies have fostered the suspicion that babies are *products*, not gifts or surprises, that they are conceived for a *purpose* (to meet the needs, plans, career patterns and lifestyle choices of adults), and that one should not have to suffer the inconvenience or frustration of an unpredictable body (nor perhaps, in the future, of an undesigned child).

We must also consider how the new reproductive technologies, by separating the earliest moments of human life from their embeddedness in and belonging to the female body, replace the strong and inalienable claim of women over the integrity and control of their own bodies with demeaning legal disputes over the ownership and rights of biological products, including embryos and babies. One thinks here of the case of an American couple undergoing fertility treatment in Australia who died in an accident in 1983 without leaving a will, leaving some to argue that the frozen embryos they had left at an IVF clinic could, if donated to another couple, in the future claim inheritance over their deceased biological parents' $8 million estate.[46] Or there was the legal wrangling over the fate of seven cryopreserved embryos in Tennessee between 1988 and 1992, in which a divorced couple fought for custody of the embryos frozen during their marital attempts to conceive, the biological father claiming that his right not to be a biological parent would be violated if his ex-wife chose to carry an embryo to term or donate them to another couple.[47]

When the mother of another set of frozen embryos died, the parents of the deceased hired a surrogate to gestate a child which they had no intention of rearing, it being a duty in their eyes to grow their daughter's last possession into a grandchild (the surrogate eventually miscarried). Jaycee Buzzanca, in another famous case, was conceived from the gametes of two anonymous and unrelated donors, which were bought by a couple who, before the child was born to a surrogate mother, initiated a divorce. With the genetic parents of the child remaining anonymous, the surrogate relinquished the child according to the terms of the original contract. Because the male divorcee had now renounced his claim to be the child's legal father, she was designated a legal orphan for three years, until the original female client won parental custody (and her ex-husband was ordered to pay child support).[48] There is also the case of a Detroit couple who, having purchased the eggs and rented the womb of a surrogate mother who subsequently became pregnant with twins, accepted the girl but left the boy (they already had three male children), their decision apparently simplified by the commercial nature of the transaction, and the technological separation of pregnancy and parenthood.[49] An American couple who withdrew from their deal with a British surrogate when they discovered that she was carrying twins (conceived from the man's sperm and a donor's egg), found it equally easy – when faced with the surrogate mother's claim for damages – to arrange for another couple to replace them.[50]

As for the technology of human cloning, might this become the most effective means of producing biological descendants for people who are severely infertile, for homosexual lovers, or people who want to be sole and single parent?[51] And what about the new possibilities and choices which are imposed by the overproduction of eggs and embryos, itself an unavoidable by-product of assisted reproduction? Because they are after all the products of a technical venture, conceived in a laboratory and nurtured by sterile scientific hands, these surpluses are inevitably discussed in terms of their 'usefulness' (for medical research, for example) or, when destroyed, regarded as a wasted resource.[52] And since embryologists logically select the healthiest embryos to be transplanted in the mother's womb, it certainly seems a comparatively small step down an ever-so-gentle gradient to favour those that are most 'useful' to the interested parties. This step was first taken to the great benefit of six-year-old Fanconi anaemia sufferer Molly Nash in the US, whose baby brother was selected from twelve embryos to provide, on his birth in August 2000, matching umbilical cord cells to treat her serious genetic disorder. It may also seem an equally small step to 'de-select' embryos with identifiable genetic imperfections, including those which would only be *carriers* of mutations that would not directly affect them.

This latter development has indeed taken place. In October 2000 the Universitat de Barcelona distributed a press release announcing that its researchers had successfully performed sex selection on IVF embryos in order to eliminate the risk of the father passing on his haemophilia to hypothetical future *grandchildren*.[53] Haemophilia is an X-linked recessive condition, which means that, while women who have inherited the mutated gene are usually protected from its effects by the presence of a normal allele of the gene on the other X chromosome, men – who inherit an X chromosome from their mother and a Y chromosome from their father – will not, if they acquire the mutation, have its effects masked in the same way. Ironically, however, male haemophiliacs cannot pass the disease directly onto their children: if the child is male (XY), his X chromosome must come from his mother; if the child is female (XX), she will inherit her father's genetic mutation, but the normal gene inherited with the maternal X chromosome will, by compensating for the mutation, make her a carrier of the defect but not a sufferer from the disease. In this widely reported case of assisted reproduction, female embryos were screened out not because they themselves would have been unhealthy, but because they would have been healthy *carriers* of a genetic defect with a fifty per cent chance, were they to mature and later have children of their own, of giving birth to (male) haemophiliacs. We can clearly see here how the technologising of conception, by expanding the reach of scientists' power beyond test-tube embryos to the future descendants of those embryos, makes the instrumental treatment of existing forms of life more likely, and more specifically how the moral–legal principle prohibiting sex selection on *non-medical* grounds (currently upheld in the UK), can be circumnavigated by an overriding moral concern for the medical well-being of future generations.

8

The Road to Human Cloning

Cooking Ourselves to Death

An analogy that is sometimes made to convey the fatal implications of the slow, incremental changes to our cultural expectations and practices induced by the technological advances discussed in the preceding chapters, is the cooking of a crab. The temperature of the water that eventually boils the crab to death rises so imperceptibly that the crab is never moved to try and escape. Likewise, the steady, piecemeal loosening of the ethical boundaries to the manipulation of human beings occurs in such small, isolated, uncoordinated and scientifically specialised ways, that the desire to call a stop to these developments seems like misanthropic paranoia, while consent to each new advance gains immediate legitimacy from the way the boundaries already seem in practice to have been redrawn, as scientific and economic interests create *faits accomplis* which can no longer be prevented by countervailing regulatory interventions.

The argument that fears of a slippery slope in human biotechnology is the irrational product of scaremongering colludes in this process. It appeals to a principle of ethical independence which implies that the possibility of moral conduct is rooted in human beings' capacity for abstract and logical reasoning. This is the dominant perspective amongst those who work in the bioethics discipline, many of whom seem to regard their task as one of *updating* moral principles to adapt them to – and thereby provide the ethical justification for – the pathbreaking implications and pragmatic demands of modern technology. This is also the opinion of the esteemed group of humanist laureates (including Isaiah Berlin, Bernard Crick, Richard Dawkins, Kurt Vonnegut and Edward O. Wilson) who signed the 'Declaration in Defence of Cloning and the Integrity of Scientific Research' in 1997. They disputed the view

> that future developments in cloning human tissues or even cloning human beings will create moral predicaments beyond the capacity of human reason to resolve. The moral issues raised by cloning are neither larger nor more profound than the questions human beings have already faced in regards to such technologies as nuclear energy, recombinant DNA, and computer encryption. They are simply new.[1]

But moral judgements are not the product of abstract, procedural reasoning. Moral regard for human life depends on sensitivity to the experience of life, to the meaning it has for people, and to the ambiguities and uncertainties that are definitive of it. It is in fact the false reduction of morality to absolute rules and distinctions which allows scientists to argue that, since these distinctions – such as the idea of a virginal 'nature' – do not correspond to ontological categories, then morality is transcended by the superior precision and objectivity of science. The zealous proponents of scientific advancement routinely point out that what ethicists may depict as absolute moral prohibitions – the instrumentalisation of life, the utilitarian calculation of pleasure and suffering, the artificial reconstruction of nature – have long since been broken. Since we already have widespread 'environmental or educational dragooning of children in attempts . . . to create cultural copies of their parents', Dawkins points out, why should cloning be prohibited 'simply because it involves *genetic* rather than environmental manipulation?'[2] And since the instrumentalisation of people is invariably present to some degree in many if not most decisions to have children – to produce an heir, to create a sibling-companion, to balance the family, to have somebody to care for, to remedy an empty marriage[3] – as well as being a formal feature of capitalist society at large,[4] then why should we be so concerned about the instrumental view of life propagated by the proponents of human cloning?

Though I shall suggest in chapter 10 that there may well be something distinctive separating genetic manipulation from the social instrumentalisation of persons, we can see here how the yearning for categorical moral distinctions can so easily become an alibi for laissez faire demoralisation – that is, for the replacement of difficult and uncertain moral judgements with the robust functional ideals of improving technological capacity, increasing the volume of commodities and expanding consumer choice. The utilitarian logic of cost-benefit analysis is not morally neutral, however. Nor is technological growth an inherent good which, as the biotech enthusiasts claim, can be legitimately opposed only if adverse consequences, risks or side-effects of an equivalent magnitude can be identified and proven. Rather, technological rationality is inherently degrading in the way it treats the ambiguous, uncertain and inefficient peculiarities of moral feeling as a problem to be solved, as an obstacle to its own ascendancy, as an impediment to the smooth and linear functioning of the well-oiled machine. Let me stress this point: if moral 'laws' existed, then *there would be no moral subject*, no sense of commitment, responsibility and freedom to animate and confirm our existence. It is the elimination of this subject, pursued in the name of scientific exactitude and methodical reasoning, which is the true aim of the technocratic imagination.

The Medicalisation of Care

What should concern us here is thus the direction in which genetic technology is pushing society, the degree of public involvement in this process, and, most importantly, how our capacity for moral judgement and feeling is likely to be affected by the future consequences of today's decisions. Will limited forms of human cloning or somatic cell gene transfer, by changing the points of reference, revising linguistic categories and lines of distinction, and generally altering people's perceptions and expectations of science, nature, and humanity, make the genetic engineering of humans a prospect which no longer appears to us to be on the wrong side of an ethical boundary? The fact that many of the proponents of cloning and elective eugenics, who argue that there is no clear moral divide separating cloning from other forms of artificial conception, were the same people who previously defended the nascent IVF industry on the grounds that the boundary between IVF and eugenics could be protected, suggests that this particular slope is slippery indeed.

An example of this subtle and progressive redrawing of boundaries occurs via the extensive *medicalisation of care* that is characteristic of modern societies. The medicalisation of care involves both the atrophy of compassion in situations of therapeutic importance, and the treatment and redefinition of 'ordinary suffering' as an illness to be remedied by physiological, biochemical, or surgical intervention. The alarming growth in the proportion of babies delivered by caesarean section is an illustration of this process. From under three per cent of UK births in the 1950s, caesareans are today performed on around one in five women approaching parturition, bringing the UK almost level with the rate of caesarean sections in the US.[5] 'Performed on' is the right phrase here, since the delivery of the child ceases to be an accomplishment of the mother, ceases to be the consummating act of a nine-month narrative of devotion, intimacy and self-conscious care. Instead the struggle by mother and baby to initiate a new chapter in their relationship becomes a malady to be remedied by surgical expertise, which in turn transforms the new-born child into a medical product.[6]

One notable feature of the broader medicalisation of suffering and care is the way new clinical technologies and specialist practices replace and devalue the nursing of the ill and impaired, and displace society's moral concern that people who suffer be *treated well*, by promoting instead the technical engineering of well-being and the manufacturing of cures. No doubt the dominant incentive driving this process is economic, for palliative care is time-consuming and difficult to commodify. As Anne Karpf points out, well-nourished children nursed through infectious illnesses would ordinarily develop strengthened immune systems and improved overall health, but vaccines are cheaper and more profitable that personal care.[7] Yet personal care is also a moral challenge in itself, suggesting that the most attractive aspect of the medicalisation of health and suffering is that it promises to relieve society of the burden of concern – to cure people of their sense of obligation and sympathy towards people who suffer.

This may be illustrated in the US by the activities of the Cystic Fibrosis Foundation, a charitable organisation which raises funds for medical research with aggressive advertising campaigns highlighting the role of gene transfer experiments in bringing forward the prospect of a cure for the disease, but which offers little in the way of patient education and other support services for patients and families.[8] The particularly aggressive search for genetic causes of those conditions – notably psychological disturbances and Alzheimer's disease – which are characterised by communication impediments and behaviours which challenge our ability to comprehend, also shows how complicit medical genetics may be in the disavowal of meaning in the experiences of those who are ill.

The Slippery Slope

What happens, moreover, when, having committed ourselves emotionally to the prospect of a successful genetic treatment for cystic fibrosis or Alzheimer's disease, the pioneering intervention reaches its technological limit and falls short of its original goal? Scientists know, for example, that when viral vectors are used in somatic cell gene transfer protocols, the patients build up immunity to the vectors so that they will, eventually, become completely ineffective. Won't the cumulative results of this clinical practice contribute directly to the argument in favour of germ-line therapy, which, in preference to a course of treatment that will become increasingly ineffective and preclude more lasting forms of genetic intervention, could deliver one final and permanent correction to a person's genetic defect?

The technological penetration of the mechanisms of inheritance and reproduction is, after all, heading in one simple direction: *backwards*, indeed, as far back into the ontogenetic past as scientists can possibly go, for they know that the more abstract the object, the more inchoate and non-human the biological form, the less ethical objections there will be to its technological modification. If genetic reductionism is anywhere legitimate then it has to be in the context of an embryonic ball of cells, where the human being is sterilised and distilled to its atomic essence, and where genetic oddities or imperfections reveal themselves as pure biological *errors*, programming mistakes that must be corrected or erased.

It goes without saying that the moral burden faced by women undergoing preimplantation genetic diagnosis (PGD) – the genetic analysis, selection and de-selection of early embryos prior to implantation into the uterus – is much lighter than that faced by the woman who must consider the abortion of a foetus she has carried in her womb for sixteen weeks, which has been diagnosed prenatally with a genetic disease.[9] Similarly, few parents having witnessed the first few months of their child's development would contemplate, were a condition like Down's syndrome belatedly detected, the humane killing of their disabled child, or the giving up of that child to better prepared carers. Yet society still finds such children – and such uncompromising parental attachments – sufficiently undesirable to offer pregnant women the opportunity to avoid their birth in the first place. By normalising the overproduction of embryos – as necessarily occurs in laboratory fertilisation techniques – reproductive science eventually succeeds in removing altogether the dilemma of choosing between life and non-life which otherwise faces the pregnant woman, replacing it instead with a *surplus* of possible lives and the *necessity* of distinguishing the more from the less valuable. Even the seasoned moral theorist Ronald Dworkin believes that the use of selection criteria in this context crosses no moral Rubicon:

> We accept in vitro fertilisation as a reproductive technique because we do not believe that it shows disrespect for the human life embodied in one zygote to allow it to perish when the process that both created and doomed it also produces a flourishing human life that would not otherwise have existed. When one zygote has already been implanted, a decision to terminate its life because it is female shows a disdain for its life, because the question is then whether a single, isolate human life will continue or cease. But before implantation some zygotes must inevitably perish, and using sex as a ground seems no more to disrespect life than using chance.[10]

If somatic cell gene treatments were to become common and routine, it is clear that they too would engender new risks, new expectations and new definitions of normality and danger which would alter people's attitude to germ-line engineering. For example, the danger of unintended germ cell conversion resulting from somatic cell gene transfer would, if realised, have intergenerational ramifications – exactly what changes will have been accidentally induced will be unknown until they are manifest in subsequent offspring – much more problematic than those associated with the deliberate manipulation of the germ-line, which may in contrast be seen as a more safe and reliable form of genetic intervention. Lappé poses the obvious question: 'If secondary germ line changes would be acceptable as a secondary effect of an otherwise acceptable protocol, what makes them less acceptable if they were the *primary* objective?'[11] 'Nothing', is Ray Moseley's answer to this question, provided that the unintended germ-line effects of somatic alterations are medically positive (removing the offending mutation from the germ line, for example). 'If it is morally acceptable that good (but unforeseen) germ-line consequences arise out of somatic manipulations then it would seem equally acceptable (and even more ethically praiseworthy) for good germ-line consequences to arise out of direct and intentional germ-line manipulations'.[12]

'Why subject each generation to having to undergo major, invasive intervention, when elimination of the culprit genes from the germ-line is possible?', writes the bioethicist and CEO of Spectrum Medical Sciences, Burke Zimmerman, pressing this point still further. 'It is far more efficient, in terms of suffering and discomfort, risk, and cost, and technically, to correct such disorders at the beginning of life'.[13] Indeed, once the language of economic investment, efficiency and cost-benefit analysis has been deployed, and resources have been expended reaching a point which has delivered little but which promises, on the other side of the moral threshold, a great deal more, what policy maker would resist the temptation, like the gambler whose stake is trivial in relation to his debt, to push back the threshold a little further? As Gregory Stock reminds the sceptics: 'We have spent billions to unravel our biology, not out of idle curiosity, but in the hope of bettering our lives. We are not about to turn away from this.'[14]

As for the fear that the iatrogenic consequences of germ-line modifications (such as those caused by insertional mutations) may not manifest themselves until several generations down the line, won't the limited improvements and successes of gene therapy be used to demonstrate that the natural progress of science will have caught up with these problems by the time they arise? 'The same genetic technologies which allow the modification of the human germ-line would allow it to be returned to its pre-modified state or to any genetically useful state', is Moseley's sanguine assessment of the risks.[15] An engineered human genome is, after all, already a product, an artefact to be continually worked on and improved, so why worry about 'future generations' when the future is increasingly a product of human ingenuity and expertise?

Zimmerman also argues, with candid foresight, that the need to manipulate the germ-line will increase as a direct result of the success of somatic cell treatments. This is because the latter, by increasing the survival rates for sufferers of monogenic diseases, will also increase their reproductive opportunities and thus, in the long run, multiply the number of people who are homozygous for autosomal recessive disorders. Since they in turn cannot help but pass those genetic defects onto their children, the moral case of germ-line intervention will in Zimmerman's view become unassailable.

Zimmerman's position is also shared by the molecular biologist Johnjoe McFadden. Echoing the thoughts of innumerable twentieth-century eugenicists, McFadden points out that modern healthcare and medicine 'has brought about the greatest shift in selective pressure on the human species since we came down from the trees'. Defective genes – including those associated with heart disease, diabetes or cancer – no longer mean death before reproductive fitness has been achieved, and are no longer weeded out by the forces of natural selection. Even the new reproductive technologies may be contributing to the survival of defective genes (if weak sperm is caused, as many scientists suspect, by faulty genes, the assisted injection of this sperm into eggs may result, when successful, in the reproduction of the genetic defect[16]). And when genetic cures inevitably become redefined as 'needs' which no liberal welfare state can legitimately ignore, the pressure to deliver permanent corrections to mutations in the human gene pool will be irresistible. As we increasingly become 'enfeebled parasites of our health system', McFadden predicts a necessary revolt against the economic burden this places on tax payers and the eventual embracing of the new genetic technology.[17]

It should also be pointed out, finally, that while some steps on the slippery slope lead us inexorably beyond them, other steps are made impossibly sticky by the prospect of those that follow. I am referring here to the way our ethical horizon is forcibly pushed forward to consider contentious future technologies in order to make dubious current practices seem more acceptable. A good example of this is the argument *against* germ-line genetic engineering which asserts that 'selection of healthy embryos will always be preferable to gene therapy of defective ones'. The norm here has now become *in vitro* fertilisation, pursued in order to produce a selection of embryos where 'there will always be some healthy embryos alongside the defective ones'.[18] Here we can see how proposals for more controversial practices invariably provoke the uncritical acceptance or consolidation of current ones.

Stem Cells

Certain precedents have, in any case, already been set which suggest that the genetic engineering of humans is closer than most people suppose. French Anderson, who advocates 'corrective germline therapy' for serious medical conditions, but 'excludes enhancement interventions in both somatic and germline contexts',[19] has been pressing the NIH to develop a satisfactory protocol for *in utero* gene therapy, since gene transfer is thought to be much more efficient when the recipient is a foetus rather than a mature organism.[20] Because there is an appreciable danger that such an intervention would introduce genetic changes in incompletely differentiated foetal germ cells, however, critics have voiced concern that Anderson may be trying to deliver the genetic engineering of the human germ-line 'unintentionally', by deliberate risk-taking aimed at dispelling the power of the moral taboo.[21]

In a startling announcement from the Institute for Reproductive Medicine and Genetics in Los Angeles, researchers claimed in October 2001 that by chemically stimulating unfertilised mouse egg cells they had miraculously persuaded them to divide like an embryo with a full double set of chromosomes. The purpose of the research was to investigate an alternative source of cells for therapeutic transplant: when the mouse 'parthenotes' were cultured *in vitro*, pluripotent stem cells were produced which in turn differentiated into medically valuable neurons (nerve cells). But the scientists also transferred sixty of the novel embryos into mouse surrogates, a fifth of which were alive and developing normally after thirteen days of gestation.[22] Small wonder that some commentators saw the research as a critical milestone on the path towards fatherless human conception.[23]

Current legislation on embryo manipulation in the US represents a curious and uneasy compromise between the powerful religious and scientific lobbies, with the federal government forbidden from funding any research which involves the destruction of embryos (on the grounds that this would legitimise and encourage the aborting and donating of unwanted embryos), but the research itself permitted so long as it can find private backers.[24] In the UK, cloning embryos by nuclear transfer for the purposes of fertility research was – subject to (highly unlikely) approval from the Human Fertilisation and Embryology Authority – never illegal, though no one is known to have applied for a licence. In December 2000, however, the House of Commons voted to endorse new regulations explicitly permitting the use of embryos, provided they are no more than fourteen days old, for research not just into infertility and contraceptive treatments, and into detecting and understanding genetic and congenital diseases and the causes of miscarriage (all of which has been legal since 1990) but now also into the treatment of serious diseases. The call for the relaxation of rules came from medical scientists interested in the potentially groundbreaking uses of embryonic stem cells.

Stem cells, as the name implies, are cells which are the origin of more specialised cells. Instead of dividing to form identical daughter cells, stem cells can divide unequally, forming a replica stem cell and a more specialised cell, this capacity for simultaneous self-renewal and differentiation being essential if the body is to replace ageing or injured cells. Though scientists do not fully understand the chemical cues which prompt and regulate this process of asymmetrical cell division – it cannot, of course, be simply genetic, since every cell in the body contains identical DNA – they know that without stem cells there would be no complex, multicellular organisms.

A single fertilised human egg and its immediate descendants are 'totipotent' cells, so-called because they are the parents of every specialised cell produced during the development of the human being. Any such totipotent cell can in principle be separated from its sister cell (as occurs naturally with identical twins) and then continue to develop into a fully formed individual, and this is what allows for the removal of cells for genetic and chromosomal inspection prior to the implantation of IVF embryos in the uterus. Strictly speaking, totipotent cells are not stem cells because they cannot reproduce themselves indefinitely.

When a fertilised egg has undergone five or more rounds of cell division (normally three to five days after conception), it forms a mass of cells called a blastocyst. The blastocyst has an outer wall of cells which will go on to implant in the uterus and form the extra-embryonic membranes of the placenta, the umbilical cord and the amnion. Inside this hollow sphere is a smaller group of cells (the 'inner cell mass') which will develop into the baby. This is made up of 'pluripotent' stem cells, which are the progenitors of all the 260 or more differentiated cell types present in an adult human. Until these cells have begun to differentiate and form body cell types – a process thought to begin with the emergence of the so-called 'primitive streak', which is the precursor of the spinal cord and backbone, after fourteen days – a human life, in the eyes of most scientists, has not begun. It is hence referred to at this stage as a 'pre-embryo' or 'pre-implantation embryo' (when a single primitive streak appears, the point at which twinning can occur has passed, and the embryo is committed to develop into a single human being, or none at all).

As the embryo implants and grows, a second more specialised class of stem cells is born. These cells are 'multipotent', so called because they are already committed to producing a small range of fully-differentiated cells. Examples of this type of stem cell (sometimes referred to as 'adult stem cells' because they survive in some adult tissues) include blood stem cells in bone marrow, which perform the critical role of replenishing the body's supply of red and white blood cells and platelets, skin stem cells that give rise to the various types of skin cells, and neural stem cells, which are the progenitors of the various cells that form the nervous system.

What makes stem cells so interesting to biologists, therefore, is first of all that they represent a developmental stage in the life of an organism from which point the future is, to varying degrees, not predetermined – and hence is potentially manipulable – and, secondly, because they have a capacity to produce derivative cells that is itself inherently renewable. Given the pre-differentiated nature of stem cells, it is unsurprising that in experiments with nuclear transfer technology a significantly higher proportion of embryos cloned from embryonic stem cell nuclei survive to adulthood than those cloned from somatic cell types.[25] Indeed, in the original *Nature* article that announced the arrival of Dolly the sheep, Wilmut and his colleagues conceded that, given the tendency for breast tissue to grow in preparation for lactation, '[w]e cannot exclude the possibility that there is a small proportion of relatively undifferentiated stem cells able to support regeneration of the mammary gland during pregnancy'.[26] In other words, the cell taken from the mammary tissue culture which was cloned to produce Dolly – so named, with typical schoolboy humour, after the American country and western singer memorable to Wilmut and fellow male colleagues not for her voice but rather the size of her breasts – may actually have been 'reprogrammable' precisely because it wasn't a fully differentiated somatic cell after all.

Regenerative Medicine

In 1998, funded by the ambitious biotech firm, Geron Corporation, a developmental biologist at the University of Wisconsin isolated stem cells from the inner cell mass of a human blastocyst, and propagated them in culture. As a result, the cells gave rise to a unique class of cells – called embryonic stem cells – which both maintained the developmental potential of pluripotent stem cells, and proliferated and replenished themselves indefinitely (most somatic cells, in contrast, have a finite lifespan, and even when grown in culture only survive for around fifty to eighty doublings).[27]

Having cultured seemingly immortal stem cells and prevented them from differentiating, biologists now believe they are close to discovering how to control and direct the behaviour of such cells so that they can be used to grow specialised tissue – skin, muscle, nerves, blood cells – for transplantation into diseased or injured patients, or as rejuvenated replacements for ageing organs. In August 2001 this conviction appeared justified, as Israeli researchers announced they had successfully grown cardiomyocytes – the precursors of heart cells – from human embryo stem cells, raising the possibility that injections of cultured cardiomyocytes could be used to produce mature heart muscle cells to replace those that are damaged in people with heart disease.[28] At the same time, another team of Israeli scientists published evidence that they had persuaded pluripotent stem cells from human embryos to differentiate into insulin-secreting ß-cells, which could in the future be injected into the pancreas as a treatment for type 1 diabetes.[29] And in a widely reported breakthrough, the original team involved in the isolation of human embryonic stem (ES) cells announced in September 2001 that, by co-culturing such cells with cell lines from mouse bone marrow and yolk sac tissue, they had persuaded the pluripotent cells to differentiate into hematopoietic cells, and subsequently into blood cells displaying all the characteristics of those produced from human adult bone marrow cells. The ultimate goal, say the scientists, will be to create biological blood factories: 'Because ES cells can be expanded without apparent limit, ES cell-derived blood products could be created in virtually unlimited amounts.'[30] The future, according to companies like Advanced Cell Technology, Geron, and Human Genome Sciences, is 'regenerative medicine'.

Another prime target for stem cell researchers is Parkinson's disease, sufferers from which it is hoped will be able to receive transplants of neural stem cells, which in turn should give rise to dopamine-producing neurons. The portents for this kind of treatment are not especially good, however. In one experiment, which involved the transplantation of foetal brain tissue (thought to be rich in neural stem cells) from aborted embryos, the patients enjoyed one year of modest improvement, but after this period fifteen per cent of them suffered devastating side effects, as the levels of dopamine appeared to increase uncontrollably.[31] In another experiment performed on rats induced by amphetamines to display Parkinson's-like symptoms of repetitive turning behaviour, transplants of undifferentiated mouse embryonic stem cells into the rats' brains differentiated into dopamine-producing neurons. Though most of the rats recovered normal motor control as a result, a quarter of those that accepted the graft developed fatal brain tumours.[32]

Other concerns have been raised by the discovery that mouse embryonic stem cells, when multiplied *in vitro*, display a high degree of epigenetic instability, with the genetically identical cells exhibiting radically different expression patterns which persist in adult mice cloned from the same cell line. This seemingly random variation in gene activation, imprinting and methylation is also thought to explain why the majority of cloned embryos do not develop to term, and why those that do survive frequently die of a range of post-natal abnormalities, such as misshapen or wrongly located organs, or enlarged blood vessels.[33] As Ian Wilmut concedes, commenting on the discovery that cloned mice, in the absence of extra food intake, become obese, but then fail to transmit this phenotype to their offspring (thus indicating the presence of unstable epigenetic changes induced by the procedure), 'cloning by the present methods is a lottery', and 'it is questionable whether there are any clones that are entirely normal'.[34]

Despite the cautionary discoveries of instability in stem cell lines, a British biotech firm, ReNeuron, has already announced plans to conduct stem cell transplant trials on stroke victims, implanting neural stem cells into patients' brains in order to repair the damage caused by blood clots and interrupted oxygen supply. By splicing into the stem cells a temperature-sensitive viral gene, which should prevent the cells from continuing to proliferate once they encounter the hotter environment of the brain, the researchers believe they will be able to avoid the risk of uncontrolled (cancerous) cell growth or excessive dopamine production.[35]

Therapeutic and Reproductive Cloning

Why does the decision of the UK parliament to update the Human Fertilisation and Embryology Act of 1990, so as to permit the use of embryos for stem cell research, make the cloning of human beings likely? One reason is that if stem cell therapy is to realise its promised benefits – treating, for example, the 120,000 Parkinson's sufferers in the UK, the annual toll of 40,000 surviving stroke victims, the millions of people with diabetes, and the thousands (in the US, 200,000) who are paralysed due to spinal cord trauma – it probably cannot rely, as recently approved by the Dutch, Australian, French, Canadian and Japanese governments, on 'leftover' embryos from fertility treatment as its main source of stem cells. The problem here is not just the volume of supply, but also that, as one UK stem cell researcher complained, these embryos are highly unreliable in developmental terms. 'We're starting from a population which has been selected to have fertility problems . . . we're getting the worst possible set'. In his view, legislation should be relaxed to allow aborted embryos to be donated for research. 'We are not getting access to good embryos that are just being thrown away'.[36]

Even were such a ruling to be passed, the volume of embryos 'donated' on this basis may also fail to meet projected needs. One alternative is for egg and sperm to be donated in order to allow scientists to create embryos from which stem cells could be extracted. This eventually took place in the Summer of 2001, when researchers at the Jones Institute for Reproductive Medicine in Virginia defied the recommended guidelines of a number of advisory bodies and, using private funding which enabled them to circumnavigate US law, harvested the first stem cells from embryos created specifically for research.[37]

There is, however, a further obstacle to the likely success of stem cell therapy, which is the problem of immune-system rejection by the receiving patient of foreign cells. It is widely believed that so-called 'therapeutic cloning' offers the best solution to this problem – a solution the main legal obstacles to which have been removed by the reform of the UK Human Fertilisation and Embryology Act. With therapeutic cloning, somatic cells will be taken from the body of the patient being treated and a nucleus extracted and fused with the enucleated egg of a donor. Using the same technique first used in the cloning of sheep, the egg is then persuaded to cleave and develop to the blastocyst stage of embryo development, at which point stem cells will be isolated, the embryo destroyed, and the cells then grown into transplant-ready tissue virtually identical to the recipient's DNA. (Pharmaceutical companies also believe the same procedure will enable them to make significant treatment advances by testing drugs directly on the stem cells of people with genetic disorders.)

Of course, the 'therapeutic' value of this procedure is as yet unproven – aside from the practical problems associated with stem cell transplantation, the alarming rates of largely inexplicable congenital abnormalities in cloned mammals suggests that the medical use of human stem cells derived from cloned embryos may equally be laden with hidden risks. Until the therapeutic function of 'therapeutic cloning' has been shown to be safe and efficacious, the use of this qualifier is thus more than a little misleading, since what makes it acceptable in the eyes of its supporters is no distinctive or inherently beneficial feature of the cloning technique itself, but rather the worthy medical goals that it is intended to serve, combined with strict limitations on the permissible life-span of the embryos that are produced from it.

Despite being made legal in the UK by the revision of the Human Fertilisation and Embryology Act, human cloning for the purpose of medical research is likely to be hindered by the scarcity of donated eggs – a problem already faced by IVF clinics[38] – and real advances in this experimental area may well demand the harvesting of eggs from aborted embryos, or from philanthropically motivated or economically deprived women. There is, nonetheless, one moral line which most scientists agree will not be crossed, which is the implanting of a cloned embryo into a woman's uterus. Scientists and politicians distinguish this practice from 'therapeutic' cloning by calling it, again rather disingenuously, 'human reproductive cloning'. It is probable that in the coming months the creation of cloned babies will be made illegal in many countries that possess the requisite technology, with the UN already pressing for an international convention to ban the practice. In the UK, the exposure of a technical loophole in the current law quickly brought the issue to Parliament, which voted in December 2001 to make the placing of a cloned embryo in a woman's womb a criminal offence. This legislation also appears to rule out, for the time being at least, the use of mature foetuses as 'spare parts factories' from which whole organs may be extracted for transplantation.[39]

Despite the existence of such a ban, there are solid grounds for believing that the moral and legal divide separating 'therapeutic' from 'reproductive' cloning will, however, be difficult to sustain in the long run, and that countries where cloning for medical research is permitted will inadvertently facilitate the eventual birth of a cloned human being. One reason for this is that making legal the creation, by novel technological means, of human life, but criminalising the preservation, by natural means, of that same life, will violate most people's intuitive moral convictions, and the bringing to birth of cloned embryos may eventually therefore be seen as an ethical responsibility greater than their humane destruction. One should not in any case doubt the commitment of those scientists involved in developing therapeutic cloning to the creation and maintenance of human life. In a testimony to a US subcommittee examining nuclear transfer technology in July 2001, for example, the President and CEO of the leading stem cell and cattle cloning company, Advanced Cell Technology, made it clear that his 'pro-life' attitude and youthful opposition to abortion clinics were entirely consistent with his passionate belief in the life-saving benefits of therapeutic cloning.[40]

Once cloned embryos for medical research can be efficiently produced and stockpiled, it also seems inevitable that a clone will be secretly or accidentally implanted into a woman's uterus. We may also encounter constitutionally persuasive legal arguments asserting the right of women whose egg and tissue cells have been transformed into living embryos to retrieve and freely dispose of their own lawful property – or at least to preserve that property from destruction by the state. A woman who carries a cloned foetus should also win legal protection against society's efforts to interfere with her body, and once a cloned child has been born attempts to prevent the creation of further clones will be interpreted as a form of discrimination greater than that suffered by people with disabilities who oppose the termination of foetuses with genetic imperfections. Banning cloning will thus result in what Laurence Tribe believes will be a monstrous caste system, 'one in which an entire category of persons, while perhaps not labelled untouchable, is marginalised as not fully human'.[41]

The palpable risk that therapeutic cloning will become reproductive cloning – which in any case is itself 'therapeutic', as fertility experts like

Severino Antinori insist, in its potential to cure infertility – is all the more unacceptable because scientists have already made significant strides in isolating multipotent stem cells from umbilical cord blood, adult bone marrow, fat, and most recently adult skin tissue and ordinary peripheral blood cells, indicating that replacement cells may eventually be produced without the need to create and then destroy cloned embryos.[42] Many critics suspect that the poor publicity given to this promising line of research reflects the way the medical potential of embryonic stem cells has been talked up by scientists eager to explore the technology of human cloning.

A case in point was the 'advanced online publication' of two articles by the science magazine *Nature* in March 2002, which purported to show that the new found plasticity of adult stem cells is probably a misinterpretation of experimental results, and that rather than differentiating into new cell types adult stem cells simply fuse and hybridise with the cells – in this case embryonic stem cells – with which they mixed.[43] The reports, though unavailable for several weeks to most ordinary subscribers to *Nature*, were widely publicised in the media, which was keen to contribute to imminent US Senate debates on bills proposing regulations for human cloning. What the media coverage did not say is that the researchers were experimenting with cells from mice, not humans, and that these cell experiments were performed *in vitro*, not by tracing the fate of donor stem cells in live recipients – where of course there is no risk of them encountering embryonic stem cells. It was also not widely known that the author of one of the articles, Austin Smith, is a keen UK lobbyist for 'therapeutic cloning' with financial and research interests in the leading Australian biotech company Stem Cell Sciences. Along with Peter Mountford, chief science officer at Stem Cell Sciences, Smith is named as inventor on a controversial European patent on the 'isolation, selection and propagation of animal transgenic stem cells' – a patent which the Munich-based patent office later acknowledged it was mistaken to award since it could cover the cloning of humans.[44]

The Lust of the Eyes

With current scientific ambitions and the startling pace of technical innovation in the reproductive field, humans may anyway be cloned and genetically engineered before legislation can effectively prohibit it. Muted ethical or legal protest, for example, greeted the US scientists who in March 2001 claimed credit for completing 'the first case of human germ-line genetic modification resulting in normal healthy children'. The researchers had injected human egg cytoplasm ('ooplasm') from a healthy egg into the eggs of an infertile woman. With the ooplasm came the thirty-seven genes of the mitochondria, and it was this 'normal' mitochondrial DNA which apparently corrected the woman's infertility, resulting in the birth of two children – each with three biological parents. Curiously enough, the reason for this experimental modification of the human germ-line was the researchers' unproven 'supposition' that the patient's fertility problem was 'related to hitherto unknown cytoplasmic pathology'.[45] Later investigation of the research programme found that fifteen babies were born using this technique but, alarmingly, two other foetuses were created which developed a rare chromosomal disorder called Turner's Syndrome – a rate that far exceeds normal statistical expectations.[46]

If the clinical justification for such a project seems rather tenuous, what about the creation of the world's first genetically engineered primate, announced in *Science* magazine in January 2001, by scientists at the Oregon Regional Primate Research Centre? In this celebrated experiment the eggs of rhesus monkeys were injected with a replication-defective retrovirus containing a gene, derived from the DNA of a jellyfish, which expresses a green fluorescent protein. The eggs were then fertilised, and of the three live monkeys born from twenty embryo transfers one contained the gene in all its cells (though predictably the gene was not expressed, and the animal failed to glow).[47]

The scientists who conducted this experiment claimed to be furthering research into the development of reliable animal 'models' of human diseases – a programme which I have already suggested is dubious in validity. The results of this and other experiments will certainly encourage the efforts of maverick scientists to carry out cloning, whether out of vanity (for which the wealthy Chicago physicist Richard Seed is the prime candidate), fanaticism (like the US cult, the Raelians, who claim to have assembled a qualified team committed to cloning a ten-month old girl who died due to a medical mistake), or with the medical dedication apparently displayed by the fertility experts Panos Zavos and Severino Antinori in their determination to use cloning as a 'solution' to male infertility.[48]

Last but by no means least in the catalogue of motives driving scientists' fixation with the idea of genetically engineering human beings, is the attitude of 'pure curiosity' expressed by, amongst others, Richard Dawkins, who facetiously confesses that he would 'love to be cloned'.[49] This may be, as he says, a 'self-indulgent fantasy', but it captures well the willingness of scientists to promote their research and experiments purely on the grounds that the accumulation of knowledge is a good thing, while 'the onus is on those who would ban it to spell out what harm it would do, and to whom'.[50] This is also akin to the seemingly insatiable desire to replicate in animals the clinically observed effects of pathogens and drugs on human beings – a practice which, legitimised in the name of safety and caution, is essentially a fanciful game of cleverness and invention played despite adequate but imperfect knowledge of the object pursued.[51]

This attitude is also a tacit illustration of that moral vice which, following St. Augustine of Hippo, John O'Neill refers to as the 'lust of the eyes'. Its offensiveness is well captured by George Bernard Shaw in his acerbic response to H.G. Wells's defence of vivisection – the unrestrained thirst for knowledge eventually leading to a desire 'to ascertain the gustatory and metabolic peculiarities of roast baby and fried baby, with, in each case, the exact age at which the baby should, to produce such and such results, be boiled, roast, fried, or fricassed'.[52] The value of greater knowledge, certainty and scientific exactitude becomes, in Wells's account, an alibi for unrestrained experimentation, as if a brutalised object could be repaired, and brutality redeemed, by the detail and accuracy of our understanding of it.

What kind of understanding of suffering is yielded by the creation of animals that are genetically programmed to suffer, and by the hardening of human sensibilities which this scientific exercise requires? The knowledge pursued by the indefatigable advocates of genetic engineering does not promise pleasure or well-being by revealing previously unappreciated qualities of a desirable object. Instead it is aimed at uncovering foreign objects – the clone, the redesigned baby, the legless mouse or monkey that glows in the dark – in order to temporarily satisfy the itch of curiosity. Legitimised on the grounds that expanding knowledge is the foundation of human control over their environment, that 'knowing more' leads to greater precision, safety and predictability in our dealings with the world, this attitude ignores the way science increasingly yields such knowledge only by inducing changes to that world which accumulate and outstrip its technical power, understanding and competence, as well as by concealing the subjective convictions which give the scientific project its meaning. In this way science ceases to be an instrument of meaningful human action and investigation, ceases to be a tool to be taken up or left to rest as occasions warrant, and instead becomes its own self-legitimising method, a self-sustaining logic, a means that cannot be refused.

9

Genetic Discrimination: Information and Power

Risk and Insurance

We saw earlier how testing patients for genetic mutations is expected to play an important role in the future of clinical diagnosis and treatment. As testing becomes a routine element of medical care, there will be a progressive accumulation of genetic information pertaining to the populations of the wealthy societies. With this accumulation of information comes the inevitable danger that exploitable or misleading knowledge about people's health risks, personality traits, environmental susceptibilities, and private behaviour and movements, will leak into the hands of agencies – including the police, insurance companies, employers, and pharmaceutical firms – whose primary instincts and interests are not always consistent with civil liberties.

A case in point here is the disclosure in September 2001 that Crispin Kirkman, the chief executive of the pharmaceutical lobby group the BioIndustry Association, had privately counselled the UK government that unless the industry was allowed to use genetic data collected by the National Health Service to develop new drug treatments, Britain would become 'a third world genetics country'. The model arrangement applauded by Kirkman is the sale to the biotech firm deCODE Genetics by the Icelandic government of exclusive access to – and the right to patent any resulting discoveries from – a database containing the genetic profile of virtually all of the population of Iceland (deCODE in turn sold access to much of the data it had collected to the Swiss pharmaceutical giant Hoffman-La Roche for $200 million). Kirkman also recommends the genetic screening of every new born baby in Britain, with a follow-up test at the age of eighteen.[1]

From a health perspective alone, it should be recalled that, since the majority of identified genetic diseases are incompletely penetrant or multifactorial, most of the information produced by genetic screening will have speculative rather than diagnostic value. The function of such tests will be to reveal predispositions, susceptibilities and risks, and to make the sensible management of such risks – even statistically small risks will bear the stamp of medical certification – an inescapable responsibility for the self-regulating individual.

Risk, moreover, is big business, and in modern capitalist economies huge profits are made through trading on risks and by selling people a more predictable future and a reduced burden of anxiety. All forms of commodification depend on relations of power and dispossession in order to flourish, and in this respect the commodification of risk is no different. The emergence of risk as a product which is itself bought and sold in pursuit of private profit is in fact founded on a quite unique asymmetry: namely, *inequality of information*. Businesses that make money by taking risks do so by calculating those risks more accurately than their customers and competitors. This is especially true of the insurance industry.

The domain of genetic health is unlikely to remain immune to the risk industry for very long. Up until now, ignorance of genetic differences has in many societies generated a sense of common fate and a willingness to pool resources in order to protect victims from misfortunes which could in principle affect anyone. If genetic screening becomes routine, and individuals gain possession of their own genetic profiles, health and life insurance companies may be faced with customers who know more about their individual health risks than do the companies they contract with. This could potentially bankrupt the private insurance industry, as high-risk applicants are able to exploit unsustainably low premiums by making large claims,[2] while low-risk applicants would see less need for insurance altogether. Premiums would have to be raised to prevent losses, which would in turn lead to further withdrawal of low-risk and low-income customers, and thus more pressure for a new round of premium increases (a phenomenon called 'adverse selection'). Many observers therefore predict that, for purely economic reasons, private health insurers will be granted access to customers' genetic profiles and allowed to introduce new categories of claimant, new exclusions and graded premiums in order to profit from the accurate management of genetic health risks.[3] Previously undisclosed genetic assets and liabilities will henceforth be exposed to the merciless forces of the market.

In the US, where applicants have already been denied medical or disability insurance for sickle-cell anaemia, arteriosclerosis, Huntington's disease, type-one diabetes, Down syndrome and AIDS, the lack of a mandatory national insurance programme combined with the expansion of genetic testing may well result in the eventual emergence of an uninsurable and medically abandoned 'genetic underclass'.[4] Becoming part of such an underclass is certainly a fear that is widespread amongst families with genetic disorders. According to a study reported in the journal *Science* in 1996, of 332 members of genetic support groups consulted, a quarter of respondents believed they or affected family members had been refused life insurance, while more than a fifth believed they or affected family members had been refused health insurance.[5] In another survey conducted in the UK, a third of people surveyed from support groups for families for people with genetic disorders reported that they had experienced problems when applying for life insurance.[6]

Much of the critical literature on the threat of genetic discrimination in the insurance industry focuses on the injustice of people being denied insurance, or quoted prohibitively high premiums, as a result of being diagnosed with a genetic predisposition to a disease whose symptoms they are currently free from, or which they are able to treat successfully (the so-called 'asymptomatic ill'). The argument normally made here is that insurers are acting *unreasonably*, that they are making judgements on the basis of prejudice and ignorance rather than an objective and informed understanding of the actual risks involved. As President Bush reassured the public in his weekly radio address in June 2001, genetic discrimination 'is unjustified . . . because it involves little more than medical speculation'.[7]

Nelkin and Lindee illustrate this point with a case uncovered in a study by Paul Billings, concerning a middle-aged American diagnosed with a genetic predisposition to the clinically variable neuromuscular condition Charcot-Marie-Tooth disease. Despite being certified fit and healthy and having driven without incident for twenty years, the man found his car insurance policy revoked when it came up for renewal.[8] Another person with the condition was refused life insurance, despite the fact that the disease does not affect life expectancy.[9] In another case reported by Billings, medical insurance was denied to an eight-year-old girl with PKU but whose dietary regime kept her in good health.[10]

Because they are concerned with the impact of reductive genetic essentialism, Nelkin and Lindee refer to cases like these in order to challenge the ill-informed treatment of genetic predispositions as an irresistible seal of fate. These and other examples – such as discrimination against people with sickle-cell trait, or cases where insurance has been denied to people with entirely manageable genetic diseases or who come from families with a history of a serious genetic disease but who have not been tested themselves – suggest that they have good reason to be worried about the ubiquity of such prejudice and its consequences. But we should not let this distract us from grasping the functional logic of the private insurance industry, which does not profit directly from the exclusion of high risk customers (this would eventually become an unworkable policy, since everyone carries potentially pathogenic mutations in their genes). Rather, the industry's success ultimately depends on *inequality in the perception and evaluation of risk*, and on persuading its policy holders to pay more in premiums than they are, on average, likely to claim. Though the profits of private insurers are of course amplified by the investment of members' premiums in the stock market, it is important to recognise that the for-profit insurance industry is essentially founded not on exclusion – though this has always been a policy integral to its functioning – but on *unequal exchange*.

Insurance companies do not, moreover, simply trade on the superior knowledge they possess of their customers' risks. They also exploit the anxiety of prospective clients, an anxiety which typically derives from their inevitable subjective entanglement in their own future. In this respect private health insurance providers have a vested interest in the proliferation of genetic testing which, as more and more genetic predispositions to disease are identified, will generate widespread uncertainty over people's own future. Instead of concentrating exclusively on cases where ill-informed insurers have misapprehended the true risks and discriminated against applicants as a consequence, we should perhaps be shifting our attention to the way genetic testing is likely to create new fears and anxieties among the general public. Magnified from the perspective of the person deemed at risk, these new anxieties will be readily seized upon by the insurance industry, which profits by exploiting the discrepancy between perceived personal risks and the more modest and reliable risk assessments obtained through statistical ('actuarial') calculation and sociological analysis of collective trends. As genetic testing grows, private insurance companies are therefore likely to move ever further from the goal of protecting the genuinely vulnerable, towards the far more lucrative business of underwriting the 'worried well'.[11]

Partly in response to concerns about genetic discrimination in the insurance industry, a regulatory body, the Genetics and Insurance Committee (GAIC), was set up by the UK Department of Health in 1999, charged with the task of evaluating specific genetic tests and judging whether they were sufficiently reliable and relevant to be used by insurers in assessing applications. The first major decision of the committee was to approve the use of two genetic tests for Huntington's disease by life insurance companies, ostensibly on the grounds that this would enable people from affected families whose own test results were normal to gain insurance at standard rates.

In May 2001, however, it came to light that the insurers Norwich Union had been demanding that insurance applicants disclose results of tests for the genetic mutations (BRCA1 and BRCA2) that carry a predisposition to breast cancer – despite the fact that these tests have not been approved by the GAIC. Indeed, the Association of British Insurers (ABI) had earlier said that it would not wait for government approval but would continue to require disclosure of test results for seven genetic conditions – including myotonic dystrophy and familial early-onset Alzheimer's – unless they were legally prohibited from doing so.[12] It was also found that a member of the GAIC was simultaneously employed as genetics advisor to the ABI, helping the industry to draw up applications for the use of tests whose validity he was also responsible for judging.[13] Though the ABI subsequently announced a voluntary five year moratorium on the use of genetic test results by insurers, the growth in genetic medicine and screening – which critics believe will rapidly expand as fears over the misuse of genetic information recede – will certainly strengthen the insurance industry's case in the long run.

Unhealthy Work

A second area where the expansion of genetic information is likely to result in new forms of discrimination is in the labour market. Employers are normally free to look for the most suitable candidates for a given job, and certain physical and mental aptitudes may be legitimate requirements for the performance of occupational tasks. By the mid-1980s, however, around fifty genetic conditions had been identified as having the potential to increase an individual's susceptibility to the toxic effects of environmental agents,[14] raising the possibility that job applicants would soon be screened for genetic as well as physical and cognitive suitability for work.

According to an American Management Association study, this is already happening, with thirty per cent of large and medium-size businesses obtaining genetic information about their employees, and seven per cent admitting to using that information to inform hiring and promotion decisions.[15] In the UK, genetic testing for employment is restricted by the terms of the 1995 Disability Discrimination Act, but this legislation offers no protection to people with genetic mutations but no symptoms. This contrasts with the situation in France and Norway, where genetic testing for employment is illegal.

In 1992, the US Department of Defence launched an ambitious programme to collect and preserve several million DNA samples from its military personnel, including reservists and civilian employees, purportedly to enable the accurate identification of men and women lost in combat. In the legal controversy that followed, as two marines defended their refusal to give blood under the Fourth Amendment right to privacy, widespread fears were expressed that the genetic samples would be used for biomedical research to identify the best 'military genes', to weed out soldiers most susceptible to particular 'risks' (including homosexuality), to facilitate research into biological weapons, or to withdraw the right to claim discharge benefits from individuals with a genetic predisposition to disease.[16]

Often the critical issue raised by these debates is expressed as a warning that genetically vulnerable individuals will suffer discriminatory exclusion from occupations in which they should be entitled to earn a living. This is certainly likely, and not just because ailing and absent workers are a threat to productivity. In countries like the US, a sizeable proportion of the workforce receive medical insurance through their employment contracts, with employers either acting as direct insurers or else buying cover from commercial carriers at rates which rise when their employees fall ill and make claims. In these circumstances genetically susceptible workers may be perceived as too risky an investment by the employer. This issue is particularly serious because it may be those who are most vulnerable to illness who are being excluded not just from work, but also from this common source of medical cover.

Beneath the terms of this debate lies a more important danger, however. This is that employers recruiting workers for hazardous occupational environments will be able to minimise public criticism of their businesses by employing people who are least susceptible to their pathogenic effects. This was evidently the logic behind the publication in 1990 of a US Congress report listing twenty-seven 'high-risk groups' of people described as being 'hyper-susceptible' to environmental contaminants.[17] The same questionable logic is apparent in the consumer domain, where people are being encouraged to buy tests that may indicate an increased sensitivity to unhealthy diets, thus giving them extra reason to do what is widely recommended as beneficial to all.[18] Genetic tests may also allow employers in dangerous or stressful industries to screen for and exclude applicants with putative susceptibilities to alcoholism, depression, or behavioural disorders, thus silencing demands for more humane working conditions by making 'sensitivity' – to stress, brutality, injustice, tedium – an individual pathology.[19]

Alpha-1-antitrypsin (AAT) deficiency, for example, is a genetic condition which, like the more obvious case of cancer, has an important environmental dimension. AAT is a protein, produced by the liver and carried in human blood serum, which is responsible for controlling the protein-splitting activity of an enzyme released by a particular class of white blood cells to fight infection. Mutations to both copies of the AAT gene – the disease is autosomal recessive – normally leads to insufficient levels of the protein being synthesised, with the result that the enzyme, called elastase, begins to digest healthy cells, typically those of the lung walls.

Around three-quarters of all AAT deficiency sufferers, who together number up to 100,000 people in the US, develop the chronic lung disease emphysema. Classified as 'genetic emphysema' or 'AAT deficiency-related emphysema', they make up between one and three per cent of all emphysema cases. But emphysema is otherwise a disease with classic environmental aetiology: it is caused by chronic inflammation of the bronchial tubes either by infection or by a wide spectrum of possible airborne pollutants, ranging from tobacco smoke to coal dust and petro-chemical emissions. People thought to be at risk of emphysema include metal grinders, sand blasters, and workers who are regularly exposed to silica, asbestos, fibreglass, pesticides, iron filings, and dust from wood, cotton, talc, coal, coffee, cereal grains and drugs.

The fact that AAT deficiency does not always lead to emphysema suggests that the condition essentially renders people hypersensitive to the effects of air pollution. Instead of regarding the plight of the growing numbers of emphysema sufferers – rates of chronic bronchitis and/or emphysema in the US have nearly doubled since 1970 – as a warning of society's deteriorating working and living conditions,[20] employers and other parties interested in the proliferation of genetic information may use the occupational exclusion of people with AAT mutations as a way of deflecting attention from universally harmful working conditions, and of making it even more difficult to isolate single agents of environmentally-induced disease.

DuPont is one major company which regularly tests its employees for AAT deficiency, though relocation rather than redundancy is the most likely consequence of a positive test.[21] The company also tests for glucose-6-phosphate dehydrogenase (G6PD) deficiency. This is the most common human enzyme deficiency in the world, affecting around 400 million men worldwide. The relevant gene is X-linked, and critical mutations to it lead to defects in the actions of the oxidation-catalysing enzyme G6PD. One consequence of this is that the erythrocytes (red blood cells) of G6PD deficient individuals are sensitive to oxidative stresses, which may eventually cause severe anaemia. It is thought that such stresses may include prolonged exposure to oxygen-depleting industrial chemicals and a range of environmental and occupational pollutants, including ozone, nitrogen dioxide, chlorite and trinitrotoluene (TNT).[22]

As with AAT-related emphysema, the occupational exclusion of people with this genetic disorder is likely to be favoured as a means of minimising public criticism of toxic work environments. But there is an added dimension to this problem, which is that, as in many genetic conditions, it is concentrated among population groups which in the western world are socially disadvantaged minorities. Especially common in Africa and south east Asia, it also affects around sixteen per cent of African-American men.[23] The danger, therefore, is that the refusal of employers to make working conditions safe for G6PD deficient individuals will disproportionately penalise groups who are already disadvantaged by the effects of racial discrimination – and indeed that this very policy may be justified on racist grounds, and carried through due to the economic and political weakness of the victims.

Currents of racism certainly appear to have influenced the way people with sickle-cell trait – the unaffected *carriers* of one allele of the sickle-cell mutation – were treated in the US in the 1970s. In a person with two copies of the sickle-cell allele, the structural abnormality of haemoglobin molecules seriously depletes their oxygen-carrying capacity by causing them to cluster together in such a way that the erythrocytes are bent into sickle shapes. With around one in ten African Americans carrying the sickle-cell trait, and one in five hundred having both copies of the mutation, many liberal reformers and health and welfare workers in the late 1960s saw sickle-cell anaemia as a politically neglected 'racial disease' which exacerbated the disadvantages already faced by the black community in America. Despite the fact that white people from Greece, Sicily and Turkey also suffer from the disease, and some of the highest incidences of the sickle-cell mutation are found in India and Saudi Arabia, not Africa, the disorder gained cultural and political currency amongst black civil rights activists and Black Panther members as a symbol of African-American experience and suffering.[24]

After Republican president Richard Nixon had signed into law the Democratic Congress's National Sickle-Cell Anaemia Control Act in May 1972, however, what first appeared as a reformist political concession to a disadvantaged social group swiftly led to an entrenchment of racial discrimination, as the publicity surrounding the disease combined with technical limitations and medical ignorance to nourish popular fears and prejudices. The common test for sickle-cell anaemia at this time, and which was widely used until the early 1980s, was the 'sickledex' assay, in which a drop of blood is deoxygenated by a reducing agent which then induces sickling in susceptible cells. Critically, however, the sickledex test does not discriminate between those with sickle-cell *trait* and those who are homozygous for the mutation – the blood of both turns cloudy under these conditions. In an atmosphere of widespread ignorance about the disease – including amongst members of the medical community – the clinical insensitivity of the sickledex assay exaggerated the number of afflicted individuals, which in turn fed both latent white racism and the political paranoia of the black community. (Even the preamble to the 1972 Act reproduced the conflation of the trait with the disease, erroneously stating that two million Americans suffered from the 'disease'.[25])

Despite the fact that most cases of sickle-cell disease were at that time already identified in children and young adults through the clinical diagnosis of their symptoms, and in the absence of any definitive treatment for the disease or immediate reproductive relevance for affected young children, some fourteen US states proceeded to implement screening programmes for black Americans, making the test a mandatory requirement for school enrolment, a marriage licence, and in some cases for certain jobs. In 1972 DuPont began a decade-long programme of screening all black workers and job applicants for sickle-cell disease and sickle-cell trait. Although this was apparently a response to requests from a group of black employees, the tests were initially performed without the knowledge or consent of the individuals.[26] Many African Americans who were identified in their medical records as being sickle-cell carriers were charged higher rates for health and life insurance, while the US Airforce and some commercial airlines either grounded or fired personnel with sickle-cell trait, claiming they were medically vulnerable to the effects of high altitude. One military trainee who was forced to resign from the Air Force Academy when he tested positive for sickle-cell trait was Stephen Pullens. Pullens was a mountain climber, a four-sport athlete, and a former state champion high hurdler who had excelled in races at high altitude.[27]

Even today, some employers still screen black job applicants or employees – and not necessarily with their knowledge or consent – for sickle-cell trait, believing that it may make them more vulnerable to oxidising industrial chemicals, heavy physical exertion, or stress on the circulatory system, or simply in order to avoid the risk of having to pay the health care costs of a sickle-cell child born to an employee with the trait. Why workers at the Lawrence Berkeley Laboratory at the University of California were surreptitiously tested for sickle-cell disease and syphilis during the 1980s and early 1990s is unclear, but the fact that black workers were tested for syphilis more frequently than their white colleagues (the only white worker repeatedly tested was married to an African-American woman), suggests that this policy also had a sinister basis.[28] The ease with which the geneticisation of illness and identity can lead to racist discourse and practices should serve as a warning to those who believe the principles of genetic health and hygiene can be surgically disentangled from racial reductionism and prejudice.

Medicalising Childhood

A third arena of genetic discrimination is likely to be childhood, and especially the social domain of the school. In some states in the US, programmes have been introduced to test children at school for genetic diseases. One such disease is fragile-X syndrome. This is a form of mental retardation linked to a break or fragile site on the X chromosome, which is in turn the result of extensive mutations to the gene associated with the protein FMR1. Not only is the genetic anomaly incompletely penetrant – a fifth of males and more than two-thirds of females with the mutation show no symptoms – but it also displays a varying phenotype, ranging from mild intellectual and behaviour problems to severe mental disability. Despite the fact that the treatment for fragile-X is no different to that recommended for children with learning difficulties, testing programmes were launched in elementary schools in Colorado and Georgia in the early 1990s. One predictable result was that children diagnosed with the genetic mutation, but lacking any clear symptoms, were thought by their parents to have been given reduced opportunities and encouragement at their school.[29]

The hunt for genetic causes of cognitive and behavioural impairment should of course be considered in its broader context, for there has been a concerted effort over the last twenty years to biologise the difficulties faced by children in environments which are not congenial to their flourishing, and to extend the logic of social normalisation and control into infancy and schooling. The search for genetic explanations for unusual child behaviour is partly an attempt to legitimise the medicalisation of childhood that is otherwise already well advanced, essentially by uncovering irrefutable evidence that the problems of troublesome children are biological in nature.

Genetics is likely to accelerate this trend towards the medical categorisation of children with learning difficulties and challenging behaviour. In doing so it threatens to reduce the responsibility of schools, parents and other relevant parties in facilitating the educational and personal development of such children, transferring this responsibility to the eager hands of the pharmaceutical industry. Because genetic reductionism is the apotheosis of biological determinism, condemning people to a destiny that is 'written in their genes', any future testing of children for genetically-linked learning disabilities and personality disorders is likely to raise new psychological and interpersonal hurdles to the healthy development of many children, hardening the perceptions and narrowing the expectations of their carers.

Yet once again the issue here is more than one of discrimination – it touches on the relationship of children to a hostile world, and society's desire to adapt children to that world, however high the cost. It would be an omission to ignore in this respect the issues raised by so-called Attention Deficit Hyperactivity Disorder (ADHD), a putative psychiatric condition first listed in the *Diagnostic and Statistical Manual of Mental Disorders* in 1980, and which has apparently reached epidemic proportions among children in the US, and increasingly in the UK. The biochemical treatment of ADHD – which has been a goldmine for Novartis, the manufacturer of Ritalin – illustrates how biological essentialism, though superficially exposing the vagaries of an immutable 'human nature', so easily serves the interests of power and domination, reducing the biological to a form which can be manipulated in the name of health and normality.

ADHD is defined by a cluster of symptoms which are said to indicate inattentiveness, impulsiveness and hyperactivity. Many psychologists believe it is a 'state deficit' disorder, where people – mostly children – actively seek to arouse themselves in compensation for what they experience as inadequate sensory stimulation. It has generally risen to prominence in the context of the classroom, where restlessness and loss of concentration can lead to disruptive behaviour and conduct which challenges the authority and patience of overstretched teachers.

Despite decades of research, and one famously flawed claim to have traced the origins of ADHD to lowered cerebral glucose metabolism,[30] no reliable evidence of neurological dysfunction has been found to indicate an organic explanation for this disorder. Nonetheless, drugs are the most popular form of treatment for ADHD, the market leader – used in up to ninety per cent of diagnosed cases – being methylphenidate. Paradoxically, methylphenidate is a stimulant with biochemical effects very similar to those of amphetamines and cocaine. It works by inhibiting the normal recycling and removal of extracellular neurotransmitters, such as dopamine and norepinephrine, thereby increasing their concentration in the brain. Research shows that while a normal dose of cocaine blocks the actions of around half of the brain's dopamine-cleaning transporters, a typical children's dose of methylphenidate blocks seventy per cent of them.[31] By providing a chemical alternative to the stimulation derived from novelty-seeking and risk-taking behaviour, methylphenidate can lead to a reduction in hyperactive conduct, greater docility and obedience, improved concentration, and better performance of simple or repetitive tasks.

Widely available as an illegal recreational drug (designated Class A in the UK, and Schedule II in the US), the illicit use of methylphenidate by high school seniors in the US (to improve performance as well as to heighten pleasure) increased from three to sixteen per cent from 1992 to 1995.[32] When crushed and snorted or injected for an immediate high, the drug becomes addictive. Its effectiveness is also known to decline with extended use, creating the need for higher doses over the long term. Like all toxic drugs, it may also be accompanied by a range of unpleasant side effects, including insomnia, anorexia, abdominal pains and dizziness.

Nearly three per cent of Americans between the ages of five and fifteen – almost three million children – are on methylphenidate, which until the patent expired in 1996 was only available under the brand name Ritalin. The 1990s saw a huge increase in the number of Ritalin prescriptions issued in the US, with domestic production of the drug (the US consumes ninety per cent of the world's supply) rising eight-fold between 1990 and 1998.[33] This trend was also accompanied by a startling growth in the consumption of psychotropic medication by children who are not old enough to attend school. Researchers found a three-fold increase in methylphenidate prescriptions in Canada, and a ten-fold increase in the prescription of Prozac-family anti-depressants in the US, for children of five years and younger between 1993 and 1997. French statistics similarly indicated that twelve per cent of children starting school were receiving psychotropic medication.[34] In one survey of two Virginia school districts, it was found that twenty per cent of fifth-grade boys were being administered stimulant drugs during school hours.[35] A study of young children in two Medicaid programmes and a managed care organisation in the US also found that the proportion of children between the ages of two and four using Ritalin had increased three-fold between 1991 and 1995, raising concerns that the drug was being used to induce social compliance amongst the socially disadvantaged.[36] In Britain, an estimated 21,000 children are taking methylphenidate, with Ritalin prescriptions exploding from 3,500 in 1993 to 157,000 in 1998. It is now estimated that nearly 200,000 UK children are currently prescribed psychotropic drugs of one kind or another.[37]

Mobilisation and Inertia

How should we understand this phenomenal growth in the psychopharmacological treatment of children? Francis Fukuyama believes that one important factor is the increasing desire of ordinary people to medicalise their own and their dependants' behaviour so as to reduce their responsibility for their actions. This desire is willingly met by pharmaceutical companies keen to promote their products. In 2000, separate class action lawsuits were filed in three US states by leading tobacco and asbestos litigation attorneys. Inspired by Peter Breggin's book, *Talking Back to Ritalin*, they charged that Novartis, together with the American Psychiatric Association and a huge national parents' organisation partially funded by Novartis, Children and Adults with Attention Deficit/Hyperactivity Disorder (CHADD), committed fraud in conspiring to over-promote the diagnosis of ADHD and its treatment with Ritalin, repeatedly violating Article Ten of the United Nations Convention on Psychotropic Substances.[38]

From Breggin's perspective ADHD is a diagnosis manufactured in order to medicalise the disruptive behaviour of children frustrated by over-regulated and under-stimulating environments. The proliferation of Ritalin thus meets the dual interests of social order in the under-resourced classroom, and increased profits in the pharmaceutical industry. In *Ritalin Nation*, Richard DeGrandpre takes a somewhat different approach. He agrees that there is insufficient evidence to suggest that ADHD is an organic medical disorder, and an equal lack of evidence proving the efficacy of Ritalin as a long-term treatment for the behavioural traits it describes. In DeGrandpre's view, children diagnosed with ADHD exhibit the most extreme manifestations of a phenomenon of 'sensory addiction' which he believes is becoming increasingly prevalent in the developed world. Sensory addiction has its roots in the normal neurological adaptation of human beings to steady levels of sensory stimulation, this process of adaptation creating a continually renewed desire for stimuli which have a greater impact on the senses. According to DeGrandpre, this natural tendency is being transformed into a pathological addiction because of the desensitising conditions of modern 'rapid-fire' culture, conditions which reward impulsiveness and a state of hurried consciousness by satiating people's sensory appetite without the need for effort, patience, or improvisation. This culture is characterised by an unprecedented intensity of visual, auditory, kinaesthetic and other perceptual signals, and by the increasingly aggressive bombardment of individuals with new information, sounds, images, music, and even fragrances. Both these sensory stimuli and their pharmacological equivalents are increasingly targeted by adults at children as a substitute for more intensive and time-consuming forms of care, thus enabling the young to feel contentment and self-esteem without participating in formative interpersonal struggles for recognition.[39]

Our culture is, in addition, suffering an extraordinary addiction to speed, the consequences of which have been astutely described by the French social thinker, Paul Virilio. In his account, the constant speeding-up of life, the dramatic compression of space and duration by new communication technologies, leads to the paradoxical situation in which the growing pursuit of novelty and stimulation results, by virtue of its very success, in increasing passivity and inertia. New sources of sensory stimulation are so ubiquitous that they no longer have to be sought out, but rather solicit our attention before we have acquainted ourselves with restlessness and boredom. When new experiences are available instantaneously – and, as is the intended function of communications technologies like mobile phones, often *simultaneously* – the individual is no longer required to journey in search of an object, and must no longer suffer the anguish of deciding which object to pursue (and which to renounce). Instead the hyper-stimulated subject suffers a 'negative behavioural involution leading towards a pathological fixedness',[40] as the close proximity of a multitude of possibilities makes self-mobilising curiosity and perseverance unnecessary.

For his part, DeGrandpre stresses that, though sensory addiction may justifiably be treated as a developmental problem, it is context specific. This is why behavioural differences between ADHD children and children regarded as normal usually disappear when the level of sensory stimulation is raised to meet the heightened needs of the sensory addicted child, and why hyperactive children who encounter for the first time the novel environment of a doctor's consultation room often exhibit acceptable behaviour that confounds the expectations of parents and clinicians.

Yet DeGrandpre does not argue that active and more dynamic learning environments are the best treatment for the hyperactive child. In his view, both this solution and the Ritalin solution only deepen children's dependence on sensory stimulation, whether environmental or biochemical. In doing so they inhibit the individual's developmental capacity to adapt to and enjoy conditions of low-intensity stimuli and slowness – conditions which even today's harried civilisation cannot fully eliminate, and which in any case provide the opportunity for the expression of values and qualities (contemplation, perseverance, loyalty, patience, sympathy, imagination, tolerance, restfulness and relaxation) which are vital to the formation of human relationships and to the health and rounded flourishing of individuals.

Notwithstanding DeGrandpre's call to readapt the child to conditions of slowness, it also seems possible that the hyperactivity of the ADHD child is a form of rebellion or protest against a social and technological environment which, like the dispensing of Ritalin, exacts a kind of Foucauldian imprisonment by hyper-stimulating the individual into a condition of inertia. Of course, the hyperactive child is most demonstrative in conditions of slowness, and most sedated by conditions of passive stimulation (such as TV watching). But it may be an error to mistake the trigger of this rebellion for the object it opposes, which may not be slowness as such, but rather the more seductive and irresistible condition of agitated immobility which unhurried environments may recall. As people's lifestyles in the affluent societies become increasingly sedentary, we should probably take heed of the disruptive protests of the hyperactive child, and consider whether they represent a defence, however misguided, of a right to vitality and self-animation which is increasingly under threat.

Unsurprisingly, geneticists and molecular psychiatrists now claim that there is a 'substantial genetic component' to ADHD, suggesting that inheritable mutations to the D4 dopamine receptor gene, the dopamine transporter gene, and the D2 dopamine receptor gene, may all predispose individuals to the disorder.[41] These claims converge with research into the genetic bases of 'novelty-seeking', which has led some scientists to conclude that ten per cent of the population's variation in novelty-seeking – as measured by personality tests – is the result of mutations to a gene which governs the ability of cell receptors to bind dopamine (an impaired ability means the release of dopamine due to sensory stimulation yields limited pleasure, and thus the desire for new sources of satisfaction).[42] In one popular account, it has even been suggested that the high rates of ADHD identified in the US may be the result of a hyperactivity gene that was positively selected for by the challenging conditions sought out and encountered by the first European settlers (the exploits of Benjamin Franklin, these authors suggest, 'gives us ample ground to speculate that he may have had ADD and was the happier for it').[43]

What may have initially appeared as an unnecessary digression on a pseudo-psychiatric disorder, should now indicate to us what is at stake in the medicalisation – and with it, the geneticisation – of childhood. With cosmetic pharmacology already so prevalent among infants and teenagers, the use of genetic screening to deselect faulty embryos or to earmark hyperactive children for gene therapy may not be far behind. Once again the biological reductionism that is encapsulated by the genetic paradigm promises, by reconciling stimulation with inertia, to adapt people to a society which is maladapted to themselves.

Further evidence that this is indeed an attractive and plausible programme for the biomedical research community can be gleaned from examining the studies on (and the predictable patenting of) the mammalian genes which are thought to govern the molecular mechanisms of the circadian clock system. This refers to the twenty-four-hour cycles of biological activity and rest which characterise the physiological functioning of most living organisms. By inducing mutations in the relevant genes in rodents, scientists have been able to create mice with altered biological clocks which are no longer synchronised with or responsive to natural environmental cues.[44] The identification of structural polymorphisms in the human *period3* gene in people suffering from sleep disorders has convinced many scientists in the field that there is a genetic explanation for common forms of sleep disturbance such as insomnia, restless legs syndrome, and feelings of sleep deprivation. More significantly, research in this area is being promoted with the promise that it will yield therapeutic products – ranging from the biochemical to, presumably, the genetic – which can adapt the human circadian clock to societal demands and rhythms which are no longer synchronised with it. Chief among such demands are those of night shifts – the increasing prevalence of which is characteristic of the non-stop, 'twenty-four-hour society' – and the crossing of time zones, which is equally central to a society of accelerated mobility and restless communication. The pharmacological or genetic resetting of people's biological clocks will thus enable them to be adjusted to conditions which few organisms are naturally equipped to bear.[45]

Commodification of the Body

The development of medical genetics has also generated a fourth category of political concerns, which revolve around the way the genetic paradigm is opening new pathways through which the logic of property relations and commercial exchange can penetrate deeper into the domains of personal well-being and selfhood. The commodification of the body is of course already a well-established trend in the modern world, and its essential vices – notably the degradation of the moral status of persons, and the introduction of power and inequality into people's relations to their own and others' bodies – have been well-documented by Andrew Kimbrell[46] and Andrews and Nelkin.[47]

The commodification of genetic information and derivative medical practices has, as we have already seen, been made possible by the loosening of intellectual property law so as to permit the patenting of DNA. Much of the outrage that has greeted this trend has been fuelled by the way profitable patents are being granted to individuals, institutions and companies which have benefited directly from publicly-funded research. As with the development of the US military-industrial complex, the scientific and intellectual groundwork for corporate enterprise in the biotech industry is being subsidised by tax-payers, with nearly two-thirds of all gene patents awarded in the US based on research funded with federal money.[48]

The expanding portfolio of patents on cancer susceptibility genes and related antibodies and proteins held by Myriad Genetics, for instance, would never have developed without years of international collaboration between research groups and co-operative members of the public around the world. Much of the work on the BRCA2 gene was performed at the Sanger Centre in Cambridge (financed by the Wellcome Trust and the publicly-funded Medical Research Council) and by the Institute of Cancer Research (ICR). The ICR still claims, in fact, that it was the first to discover the gene (Myriad filed its patent application hours before the publication of ICR's findings in *Nature*). The US patent won by French Anderson and his colleagues for *ex vivo* gene therapy was also the product of nearly three decades of research for the NIH. Like many biotech patents, it is, as noted earlier, scandalously broad, covering *any* future technique for genetically altering human cells *ex vivo* and then returning them to the body. Now held under licence by Novartis, the patent will be a fabulous money-spinner for the company, with the market for all gene therapy products expected by some analysts to exceed $45 billion by 2010.[49]

The patenting of genetic information with apparent medical value also seems unjust because it implies the right to a commercial monopoly over potentially life-saving clinical knowledge, and because the basis of that knowledge is normally an already existing organism – not an invention or 'new composition of matter' (as the US patent statute requires). Where specific disease-linked alleles have been patented – this includes the mutations associated with Huntington's disease, cystic fibrosis and breast cancer – private companies are also able to charge hospitals and health providers for a licence to test for those mutations. In one study of genetic testing for the iron-overload disease haemochromatosis, for example, it was found that thirty per cent of US laboratories had discontinued the performance or development of such tests in the wake of the awarding in 1998 of patents for the alleles most commonly associated with the condition.[50] Such patents may also require researchers in the field to buy permission to study the particular sequence of the human genome which the company claims to have isolated, copied, and recreated in a 'purified' form which does not exist in nature. In other cases researchers may have to sign a material transfer agreement (MTA), which commits them to surrender all property rights covering any discoveries they may make using the company's proprietary material.

The increasing commodification of the human genome is, however, only the tip of the commercial biotech iceberg. In the US and elsewhere, money is made from markets in blood, serum, organs, cell lines, eggs, sperm, embryos and wombs, and in virtually every case these economic exchanges presuppose and perpetuate deep social inequalities. Although today less than one per cent of the US blood supply comes from paid sources, for example, attempts to prevent the commercial exchange of human blood have often faced powerful legal and political opposition, such as the court ruling in 1966 which charged a non-profit community blood bank in Texas with conspiracy to restrain trade in blood, and the licensing by the US Food and Drug Administration (FDA) of a plasma centre run by the Nicaraguan dictator Anastasio Samoza in the 1970s, which collected blood from poor Nicaraguans and political prisoners and exported the purified plasma to the US and Europe.[51]

Though the sale of humans organs is prohibited in most developed countries, the commercial exchange of body parts still occurs in poor regions of Asia, Africa, Latin America and Eastern Europe. In Third World countries the sale of a kidney or cornea can fetch the donor more than a lifetime's wages in a single transaction, thus allowing the impoverished to trade less essential parts of their bodies so as to guarantee the basic nutritional needs of themselves and their families. The pressure to find new organ donors in the wealthy world has also led to medical calls to widen the definition of 'brain death', so as to enable the harvesting of body parts from comatose patients who have lost their higher brain functions (so-called neocortical death) but retained sufficient brainstem activity to continue breathing unassisted.[52]

With many advances in biotechnology centred on the field of biological reproduction, it is hardly surprising that one of the most prominent and controversial trends in the commodification of the body is the exchange of reproductive cells and services, and even, as the technology of genetic modification and cloning advances, the commodification of human life itself. I shall consider the political and ethical debates surrounding this trend in the next chapter.

10

Making Babies: The Appropriation of Life

The Reprotech Revolution: Surrogacy in Britain and the US

Despite having the highest per capita living standard in the world, the US has an infant mortality rate double that of the country with the lowest rate (Singapore), and stands in twenty-sixth place in a table of international comparisons assembled by the World Health Organisation.[1] With infant health apparently so low on the US public agenda, it may seem strange that baby making in America is today a multi-billion dollar business. The gamete industry has been especially quick to capitalise on the seductions of genetic reductionism, as models, scientists, athletes and successful business people have begun advertising their supposedly superior genetic inheritance to prospective parents, using private clinics such as the Fertility Research Foundation in Manhattan, Cryobank in Boston and Palo Alto, and the Repository for Germinal Choice in California, as brokers.[2] Though successful men who donate their sperm may not be drawn to such clinics for financial reasons, the situation is different for women. With payments for the invasive, painful and dangerous extraction of eggs averaging around $3–5,000 – but rising as high as $80,000 where premium maternal traits and achievements are up for sale – the economic incentives for women who consent to this procedure are thought to be particularly strong.

It is logical to assume that the unpleasant nature of this procedure – including days of hormone injections to stimulate super-ovulation, and the surgical removal of eggs – combined with the limited number of children (compared with male donors) that an egg donor is likely to produce, will increase the likelihood that women who contribute eggs to infertile couples will want to trace their genetic offspring at a later date. An even greater sense of maternal attachment and loss – stronger, even, than that experienced by mothers who gave up their children for adoption – has been observed amongst women who are hired to gestate children for paying clients. This is because, unlike the decision to have one's baby adopted, the choice to carry a baby for somebody else is made before the woman has experienced the long nine months of being pregnant with that child.

For a country which has one of the most liberal regulatory frameworks for reproductive technology in the European Union, restrictions on surrogacy arrangements in the UK, established under the recommendations of the 1984 Warnock Report, and formalised in the Surrogacy Arrangements Act of 1985 and the Human Fertilisation and Embryology Act of 1990, certainly appear sensible. It is estimated that at least one baby a week is born to a surrogate mother in the UK, all of whom must be registered at birth as the child of the birth mother and her partner or husband. The transfer of legal parentage to the commissioning couple cannot occur any earlier than six weeks after the birth of the child, at which time the consent of the 'surrogate' parents is still required – thus guaranteeing the right of the birth mother to change her mind after the birth. The commissioning couple must also be married to each other, and the child must be genetically related to at least one of them. The child cannot be exchanged for money – which would be a violation of the 1976 Adoption Act – but the surrogate mother is entitled to compensation for the inconvenience and expense of pregnancy. In agency-arranged agreements this seems to involve payments of around £10–15,000 – in some cases well beyond the common interpretation of what the 1990 Act refers to as 'expenses reasonably incurred', though as yet not high enough to have provoked any legal challenges.

The agencies themselves are prohibited from charging for their services, though commissioning couples usually have to pay to be members of the agency-run organisation. Advertising to recruit surrogates is also illegal, and any agreement between the parties involved cannot constitute a legally enforceable contract. It was the general wishes of the majority in the Warnock Report that surrogacy, while not made a criminal offence, should be strongly discouraged, and that excessive regulation of the surrogacy industry should therefore be avoided for risk of giving the state's moral blessing to legal versions of the practice.[3]

In the US, on the other hand, where thousands of babies have been born to surrogate mothers and millions of dollars have been made by surrogacy brokers, surrogacy laws are patchy and inconsistent. In most, but not all, states, surrogacy contracts are not recognised in law and therefore cannot be enforced, but in the absence of explicit regulations some disturbing judicial precedents have been set which will certainly influence future legislation and the resolution of surrogacy disputes. The most important of these was an Orange County Superior Court judgement in 1990, subsequently upheld by the California Court of Appeal, and later by the California Supreme Court, which stripped surrogate mother Anna Johnson of parental rights over the child she had carried.

Maternity for Sale

By 1986, reproductive technology in the wealthy world had advanced to the point where surrogate mothers no longer had to have their eggs fertilised by a stranger's sperm, but could have implanted in their wombs embryos with no genetic relation to themselves (such 'host' or 'gestational' surrogacies make up around forty per cent of surrogacies in the UK). Anna Johnson in the US was one such non-genetic surrogate, and it was on these grounds that the trial judge, following the breakdown of the agreement between the surrogate and the commissioning couple, ruled that the age-old definition of motherhood had been superseded by the effects of the new technologies, and that the birth mother, being a 'genetic hereditary stranger' to the baby she had borne, was merely 'the gestational carrier of the child, a host'. In conformity with the dominant DNA-centric paradigm, the long, unique and indispensable physiological contribution of the birth mother to the creation of a viable human being was deemed inferior to the genetic contribution of the couple whose gametes were combined in a laboratory petri dish. 'Anna's relationship to the child', the judge continued, 'is analogous to that of a foster parent providing care, protection and nurture during the period of time that the natural mother, Crispina Calvert, was unable to care for the child'. Judge Parslow also referred to Johnson as the baby's 'wet nurse'.[4]

Johnson's claim for custody was also weakened by testimony in court that she was a single mother with few financial resources and who had difficulty holding a job.[5] She was also black, and the child she gave birth to was white (the genetic mother, Crispina Calvert, was in fact a Filipina, but in the binary terms of racial discourse this effectively made her white). In the subsequent hearing in the appellate court, the trial court's decision was upheld by a verdict which reaffirmed the primary and undisputed power of the gene. In its judgement:

> As evidence at trial showed, the whole process of human development is 'set in motion by the genes'. There is not a single organic system of the human body not influenced by an individual's underlying genetic makeup. Genes determine the way physiological components of the human body, such as the heart, liver or blood vessels operate. Also, according to the expert testimony received at trial, it is now thought that genes influence tastes, preferences, personality styles, manners of speech and mannerisms.[6]

Though the California Supreme Court later also upheld the original decision, it did so on rather different grounds. Drawing on principles derived from contract law and intellectual property rights (rather than family law, which privileges the interests of the child), the majority of the Supreme Court judges agreed that it was the commissioning mother who had 'intended to bring about the birth of a child that she intended to raise as her own', and this intentional act of authorship made her the 'natural mother'.[7]

This principle of intentionality seemed to address the contradiction created by the earlier genetic definition of parenthood, since the latter implied that anonymous gamete donors with no original intention to be parents would have the right to make subsequent claims on their genetic offspring. As bioethicist Joseph Fletcher put it, urging ethicists to keep up with the effects of the new reproductive technologies, we have to 'embrace a moral rather than physical definition of parenthood'.[8] This idea that intention should be considered the primary determinant of parenthood had in fact already been promoted by a number of legal commentators, including some who believe that, by de-biologising the definition of parenthood, it prevents the claims of many fathers to be nurturing parents from being culturally relegated, often with legal sanction, by the assumed strength of the 'gendered' gestational bond.

As Marjorie Maguire Schultz writes: 'Determination of parental status on the basis of expressed and bargained-for intentions would create greater gender equity in access to child-nurturing roles than does the current system.

Unlike biologically-based variables, the capacity to form and express intentions is gender neutral.' Developing this logic further, Schultz argues that attempts to outlaw the commodification of surrogacy as a contractual exchange perpetuate the sexist convention that 'women, being women, should do their woman-things out of purity of heart and sentiment. Women are too delicate, too pure, to be tainted by filthy lucre.' A woman's role as biological mother is consequently 'reified' through its separation from the marketplace and, by exalting 'a woman's experience of pregnancy and childbirth over [the] formation of emotional, intellectual and interpersonal decisions and expectations', policies are defended which suppress gender-neutral definitions of parenthood and 'unnecessarily and unwisely entrench rather than mitigate the sex-based biological dependency of men on women'.[9]

The difference between the reasoning employed in the original court decision, and the later discourse of contractual 'intentionality' employed by the Supreme Court and Schultz, is of course not as stark as it may seem. Both geno-centricity and the idea of intellectual authorship take as the primary and indisputable progenitor of life what is in effect an ideological abstraction, whether this appears in the guise of the 'master molecule' that is DNA or the sexless, disembodied agent that is intellectual intention. Indeed, the repudiation of the generative powers of nature and the denial of the sexual specificity of the female body seems perfectly congruent with the patrilineal history of dualistic Western cultural thought, in which mental 'conception' is elevated to a founding role in order to legitimise the treatment of women's bodies as passive material vehicles for the realisation of male designs and incubators of men's seeds.

One may also note that this logic mirrors the legal protection of the 'separation of conception and execution' that characterises capitalist employment relations, where workers are equally denied ownership of the products they create because their original intention – an income – is external and indifferent to the commodities they produce. It also fits perfectly with the dominant legal paradigm of intellectual property law, which grants monopolistic powers of ownership to those who claim authorship of productive ideas. And just as the earliest forms of capitalist industry used technology and the division of labour to sever the attachment between craftworkers' intentions and their creations,[10] so reproductive technology, by taking the idea of breeding factories to its logical conclusion, may eventually produce using a programmed machine what surrogate mothers will only deliver with economic and legal persuasion: babies as pure and alienable commodities.

What reproductive technology makes possible, of course, the market economy consolidates and legalises. Schultz's argument that 'the capacity to form and express intentions is gender [or class] neutral' is ideological precisely because it conjures away the material and legal circumstances under which people formulate their intentions. After all, Mark and Crispina Calvert had the intention to use a surrogate mother to produce a child for them to parent because they had the money to rent, along with the requisite technology and expertise, the womb of a woman like Anna Johnson, and because Johnson was sufficiently lacking in resources that this seemed like a worthwhile exchange.[11] Because sensible people do not, ordinarily, have intentions which they lack the power to realise, the neo-liberal argument that intentions constitute legal grounds for ownership simply functions to preserve the existing distribution of resources and opportunities.

The Domination and Exploitation of Surrogates

By separating the genetic and gestational aspects of motherhood, the reproductive technologies associated with surrogacy now also enable people who approach the baby brokers – and who, in the US, need not be infertile, but may simply have the money and desire to avoid the inconvenience of pregnancy – to ignore the genetic profile of prospective surrogates. More accurately, it encourages the normally wealthy clients to suspend any prejudices, preconceptions or expectations which would otherwise place expensive restrictions (racial type, social class, appearance) on their definition of an appropriate surrogate, and to elect the cheapest carrier for their baby.[12] Once the biological features of the female host are no longer important, and the surrogate has become an exchangeable environment for the gestation of a foreign product, then market forces will ensure that it is the poorest, most desperate women who trade themselves into surrogacy. The greatest demand for this service may in future come from high-earning women who refuse to choose between career and motherhood, and who know, given the abundant supply of young wombs for hire, that they can pay to be a parent at any stage of their lives.

John Stehura, president of a surrogacy brokerage called the Bionetics Foundation, indicated the ambitions of his industry in an interview with Gena Corea in 1983. There he complained that the current prices for surrogate mothers were too high in the US, and that prospects for the future expansion of the industry lay in the recruitment of women from developing countries.

He also outlined his plans to establish a surrogacy clinic in Mexico to produce babies for US clients.[13] What he did not foresee was the break-up of the Soviet Union and the subsequent economic collapse of the Eastern European economies, which created a whole new black market for the baby brokers. There is now a growing trade in the smuggling of young women from Eastern Europe and the Balkans into countries such as Italy and Germany, where they are paid to gestate and give birth to the babies of respectable couples. In one such case a Moldovan nurse, earning less than $220 a year, was smuggled into Italy, impregnated by a stranger, cared for by the man and his infertile wife until she gave birth to a boy, then sent home on a plane with $6,000 in payment.[14]

Poor women are also often selected as surrogate mothers in the US because it is assumed that their need for money will override any inconvenient feelings of attachment they may feel towards the baby, and because they are less likely to have the resources to fight expensive legal battles if they change their mind. According to a psychologist working for a Philadelphia surrogacy firm, 'candidates with an element of financial need are the safest. If a woman is on unemployment and has children to care for, she is not likely to change her mind and want to keep the baby she is being paid to have for somebody else'.[15] Indeed, surrogacy agencies normally always perform extensive psychological screening of surrogates in order to select women most capable of surrendering the child they intend to bear. Not only is there a growing body of literature produced by the surrogacy industry providing instructions to surrogates on how to dissociate from a foetus during the course of pregancy, but some agencies even stipulate in the contract that the surrogate mother should not attempt to form a bond with the baby she carries.[16] And while the surrogate mother is expected to suppress her affections for her baby, she is at the same time likely to be encouraged to act in a spirit of generosity and moral commitment towards the needs of the contracting couple – more so, in fact, when the legal rights of the commissioning couple are weak. As Elizabeth Anderson points out, the moral pressure exerted on the surrogate to feel a personal dedication to remedying the misfortunes of the infertile couple is fraudulent because such feelings are exploited and unreciprocated: once the terms of the contract have been fulfilled, the birth mother's continuing interest in the couple and the child will not be welcomed by the new family.[17]

The idea that some pregnant women should not feel a physical and emotional attachment to the baby they are carrying and nourishing is not only problematic from the point of view of *in utero* child development – it is now well-known that the gestational setting, attitude, behaviour and emotions of pregnant women can have a substantial impact on the well-being, temperament and personality of the developing foetus – but it also sanctions a form of violence inflicted on the surrogate mother by deliberately censoring her affective and relational sensibilities.[18] Granting the surrogate mother the option to retain parental rights after the birth of the child – which is effectively the way the law works in the UK – does not necessarily address this problem, and may actually increase the pressure on surrogacy brokers to select vulnerable women and to use psychological and legal strategies to undermine the pregnant woman's sense of maternal attachment.[19]

Such contractual strategies may include the use of specific conditions in the surrogacy agreement which regulate the woman's day-to-day relationship to her body and foetus. By providing instructions on lifestyle, diet, exercise and rest, these terms of agreement are designed to make the experience of pregnancy as akin to formal employment as is possible, and to maximise the submissiveness and dependency of the child bearer. In the contract signed by the surrogate mother Mary Beth Whitehead, as revealed in the famous *Baby M* trial in New Jersey that began in 1987, it was stated that she would abstain from alcohol and cigarettes, that she would undergo amniocentesis, that she would have to terminate the pregnancy if a genetic defect were discovered and the genetic father demanded it, and that she could only choose an abortion if her life was endangered by continuing to carry the child.[20]

Disputed Parentage

The problems which commercial surrogacy may pose for the healthy development of children conceived in this way should also merit our attention. Perhaps the most striking feature of the stories of scandal and legal wrangling that inevitably surround the baby brokering business is the way the technological separation of genetic and gestational motherhood, by allowing multiple claims to a child's parentage, divorces the child from pre-contractual bonds of endearment and devotion and transforms him or her into disputed property. And with so many *partial* stakes involved in the creation of a child – who may, after all, have two genetic parents, two commissioning parents, and one gestational mother – there is plenty of latitude for a change of heart or withdrawal of commitment, for the maternal surrogate to choose new clients (as UK surrogate Karen Roche attempted to do in 1997, having decided that her Dutch clients were unsuitable), or for the commissioning parents to withdraw interest in a child because it is the wrong sex, colour, or suffers health problems.

This was precisely what happened in the Stiver-Malahoff case in 1983, when a Michigan surrogate, Judy Stiver, gave birth to a child with congenital microcephaly. Microcephaly is a generic term used to describe a variety of medical conditions associated with an unusually small head. This normally indicates an underlying reduction in the size of the brain, and leads to symptoms varying from minor intellectual impairment to cerebral palsy and serious developmental delay. Delivered what he perceived to be a defective product, Alexander Malahoff declined to accept the child, and asked the hospital to take no steps to treat the baby for a potentially lethal streptococcus infection. The hospital went to court and won permission to treat the child and to allow Michigan Department of Social Services to arrange adoption care.

In a further degrading development, Mr Stiver and Mr Malahoff began a dispute over who the real genetic father was. The results of a paternity blood test were then revealed to millions of viewers live on the Phil Donahue talk show, where the two men had agreed to appear. They showed that the husband of the surrogate was in fact the real biological father (Judy Stiver later complained that she had not been instructed to abstain from sexual intercourse before artificial insemination took place, only afterwards). In the litigation farce that ensued, Malahoff sued Stiver for not producing the child he had contracted for, while the Stivers sued Malahoff for invading their privacy by making the matter public, as well as their doctor, lawyer and psychiatrist for not advising them properly about pre-insemination sex. They also alleged, in a final *coup de grâce*, that the child's illness was caused by a virus transmitted in Malahoff's sperm.[21]

If cloning became an accepted fertility treatment, this displacement and complication of parenthood would of course be even greater. Imagine two heterosexual lovers, neither of whom has viable gametes, who want a child genetically related to one of them.[22] Imagine also that the woman is incapable of bearing a child – she may, for example, have had a hysterectomy to prevent the spread of a tumour. A cell from the man is cloned with the help of an enucleated egg from an anonymous donor, and the resulting embryo is implanted in the uterus of a surrogate mother. The boy would grow up with a birth mother who is not his genetic parent, a father who is his genetic twin brother, a mother who is his genetic sister-in-law, grandparents who are his genetic parents, and possibly siblings who are, genetically speaking, his children.

One may agree with Leon Kass that what is endangered here is the vital sense of identity young people search for and ideally gain through clarity about their origins.[23] Yet genetic essentialism is also, curiously, on the minds of the parents. Being as committed as they are to conserving and reproducing a specific genome, they may be equally intolerant of the novel confusion of genetic ancestry which is the result of human cloning. Will it really be so easy for the adults involved in this possible scenario to ignore the complicated implications of the same genetic mindset – the enchantment with a linear genetic 'bloodline' – that animates them? Can they want so dearly to reproduce the purity of a specific human genome whilst remaining indifferent to the jumbled mixtures of genetic kinship and heredity that are the effects of the new reproductive technologies?

Can the fifty-one-year-old American woman who gave birth to a baby created from a donor egg fertilised, in fulfilment of her desire to continue the family ancestry, with her brother's sperm,[24] dismiss as insignificant the genetic fusion of uncle and father that it implies? Or there is the case of the sixty-two-year-old French woman whose brother masqueraded as her husband at a Californian fertility clinic. The French siblings received one child who was successfully born to a surrogate mother whose egg had been fertilised by the man's sperm. But the French woman also enjoyed a successful pregnancy herself, giving birth to a boy whose genetic mother is the birth mother of his sister, but whose own birth mother was the surrogate wife and genetic sister of his father.[25]

In reviewing these cases I am reminded both of the booming cosmetic surgery industry, which preys upon and consolidates people's uncritical acceptance of current cultural definitions of beauty and desirability, and of the attitude implicit among those men (and some women) who believe that the surgical transformation of their anatomy can deliver them from the trials and alienations of their gender. Instead of rebelling against the cultural conventions attributed to their sex, such men seek refuge in the imaginary essentialism of a physically reconstructed femininity. Instead of changing the world, they change themselves, desiring a purity of sexual identity – to *be* a woman – that even naturally born women cannot authentically claim to feel. The revolutionaries of the new reproductive technologies seem equally conformist, unable to challenge the cultural denigration of childless women, or to question the way the gene has become the supreme mark of intergenerational continuity and attachment.

Where legal statements, contracts and signatures have replaced sexual procreation and childbirth as the principle determinants of parenthood, inevitable administrative mistakes or misunderstandings can also lead to disputes over parentage. In one unique and complicated case in the UK, a thirty-eight-year-old man was confirmed as the legal father of a child born to his estranged girlfriend, but created using the sperm of an anonymous donor. The couple had separated after several unsuccessful attempts at IVF, but on establishing a relationship with a new partner the woman returned to have one of the frozen embryos implanted. Because she did not inform the hospital that her partner had changed, hoping to keep the matter secret from her ex-boyfriend, the latter's signature remained on the relevant consent forms. Though he has been denied direct contact with the child until the infant is three years old, the 'intentional father' has been advised that he will get full parental responsibility for the child so long as his current commitment – apparently measured by doting gifts and cards – is maintained. His strong conviction that the boy is truly his own is strangely reinforced by the fact that the sperm donor was selected for maximum physical resemblance to the prospective father.[26]

The Poverty of Property

Assisted conception has also led to disputes over the ownership and control of reproductive material between would-be mothers expressing legal claim to their biological property, and other interested parties such as estranged husbands who do not want to be forced into biological fatherhood, and the medical scientists whose expertise has produced that particular object in its current form. A frozen embryo or egg is not a natural product of human procreation or biological functioning, and until recently it was illegal in the UK to use frozen eggs in fertility treatment, with fertility experts concerned about the low chance of successful live births from defrosted eggs (approximately one per cent, according to the HFEA), and about the risk of producing babies with chromosomal defects. Dozens of women in the UK have exercised their right to freeze their eggs nonetheless, many of them doing so before enduring infertility-inducing cancer treatment or surgery, while others have simply paid to postpone motherhood to a later period of their life when their chances of conceiving naturally will have declined. The demand that women be entitled to reclaim their own biological property subsequently became a feminist argument that resonated with the interests of fertility clinics. With improved freezing and defrosting techniques on the horizon, the ban was lifted in January 2000.[27]

Challenging the liberal feminist argument, advanced in particular by Lori Andrews, that the right to rent out, buy and sell one's body and its component parts is a crucial ingredient of women's reproductive autonomy, Maria Mies suggests that the likely result of the liberalisation of reproductive alternatives is in fact the antithesis of freedom: the invasion of personal life, relationships and bodies by the juridical arm of the state.[28] When a woman claims that her body, egg or foetus is her *property*, she is simultaneously alienating part of herself and allowing the rights of ownership, the computation of value and the definition of use (and misuse) to be regulated and brokered by external parties. Even when the exchange and remuneration of bodies and body parts is given over to the forces of the market, the state is inevitably required to insinuate itself between contracting parties, to supervise and enforce commercial agreements, to monitor product quality, to help stabilise the forces of supply and demand, and to police the boundaries – however generously they have been redrawn – between the permissible and the profane.

As Mies points out, there is in fact a startling similarity between the liberal defence of reproductive commercial freedoms and the assertion of the rights of the unborn child by the Right-to-Life movement. In both cases a symbolic antagonism is constructed in the woman's body between herself and her embryo or foetus – an antagonism reinforced by ultrasound images of the foetus as a fetishised, miniature astronaut, floating free in space.[29] And in both cases the state has to be called in, either to protect the mother from competing claims to her property, or to protect the foetus, via a new technological panoptics of the womb, from the competing interests of the mother. 'It is this new type of economic and scientific cannibalism, based on the bourgeois property concept and the "progress" of reproductive technology to which both positions, the liberal one and the conservative one, converge'.[30]

All this is to say nothing about what it means for a child to grow up knowing that his or her current parentage was the result of a (theoretically reversible) commercial transaction, nor what it means for society and culture in general to permit the selling of children and the trading of parental rights. Proponents of commercial surrogacy typically argue that, because payment is made for services rendered rather than for the sale of the child, surrogacy does not involve the commodification of children.[31] This was of course the opinion of Judge Sorkow in the *Baby M* case who ruled that, since William Stern had paid Mary Beth Whitehead 'for her willingness to be impregnated and carry his child to term', then the baby could not be regarded as another person's property that was exchanged: 'He cannot purchase what is already his', was the judge's insightful conclusion.[32] This reasoning is also shared by many liberal feminists, some of whom see surrogacy as no different from post-natal forms of child care, and who believe pregnancy is nine months of labour that should be properly remunerated rather than surrounded in romantic mystique.

Feminist legal scholar Lori Andrews, for example, maintains that banning surrogacy 'interferes with the couple's constitutional right to privacy to conceive a child who is biologically related', and that, since 'it is permissible to pay all sorts of surrogate childrearers, e.g., baby sitters, nannies, day care centres', then 'payment to women who serve as surrogate mothers should similarly be allowed'.[33] Christine Sistare similarly argues that women's right to engage in commercial surrogacy is a feminist issue. It is her belief that curtailing this right serves the interests of men, whose real concern is not that children will be commodified but that men will have to pay for what they otherwise tend to get through coercion:

> What is truly feared, here, is not that babies will become commodities but that women's reproductive services will no longer be cheap or available on demand. The most dreaded process of commodification is the one through which women's reproductive role will become a commodity valued in the best capitalist style: one for which the buyer must go asking and pay well.[34]

The belief that the full commodification of women's reproductive function is an advance over their patriarchal enslavement is also shared by the Israeli feminist lawyer Carmel Shalev. For Shalev, the marketisation of social relationships which tore apart the tradition-bound hierarchies of the feudal order initiated an emancipatory process that will only be completed when motherhood itself has become a commercial transaction. 'A movement from status to contract in defining parent–child relations would be a progressive step towards freeing ourselves from the yoke of traditional gender relations', she argues. By transcending the unique and incommensurable features of sexual biology, and replacing them with 'an androgynous mode of human being' expressed through the universal medium of exchange-value, wages for reproduction will enable women 'to reclaim the procreative power that has been subsumed under patriarchy as a mark of their inferiority'.

> The idea of wages for reproduction poses an essential challenge to the public–private/market–family division that is the patriarchal foundation to our

> postindustrial economy. Woman as conscious, moral, social, and political being is also woman as economic being . . . [T]he failure to acknowledge the economic value of female reproductive labour is blind folly for those who wish for equity in women's social situation.[35]

This argument that capitalist market economies are inherently and beneficially corrosive of patriarchal relations ignores two important facts. The first is that women's unique reproductive capacity has always functioned as an obstacle to men's control and definition of productive resources (unlike maternity, the determination of biological fatherhood – at least until 1978 – has always been a matter of convention, trust or conjecture). It has also limited the extent to which women's reproductive faculties can in fact be *treated* as a productive resource, rather than an intrinsic and inalienable component of selfhood. Destroying this obstacle by making reproduction a form of paid labour is, as Pateman points out, fully consistent with patriarchal interests, since it allows men to claim private ownership and control over women's bodies.

> The logic of contract as exhibited in 'surrogate' motherhood shows very starkly how extension of the standing of 'individual' to women can reinforce and transform patriarchy as well as challenge patriarchal institutions. To extend to women the masculine conception of the individual as owner, and the conception of freedom as the capacity to do what you will with your own, is to sweep away any intrinsic relation between the female owner, her body and reproductive capacities. She stands to her property in exactly the same external relation as the male owner stands to his labour power or sperm; there is nothing distinctive about womanhood.[36]

The second fact that is ignored by the liberal defence of surrogacy contracts is the way the commodification of women, while conceived as a logical solution to their exclusion from and disadvantage in the capitalist market place, also changes people's cultural perceptions and expectations of 'non-commodified' women, and of those dimensions of women's existence which many believe should not be treated as alienable commodities. Margaret Radin calls this the 'domino effect'. 'To commodify some things is simply to preclude their noncommodified analogues from existing'.[37] An understanding of this process was apparent in Richard Titmuss's argument against permitting a market in human blood: if blood can be exchanged for payment, donating blood becomes a gift of money forgone rather than a contribution to the health or life of a stranger.[38]

This is why the critique of surrogacy made by feminists like Barbara Katz Rothman and Sara Ann Ketchum stresses precisely the way the technological and economic colonisation of the biological and interpersonal processes of pregnancy and childbirth results, by virtue of its apparent technical success, in a corollary transformation of human sentiments into a form which is inherently compatible with it.

> We are encouraging the development of 'production standards' in pregnancy – standards that will begin with the hired pregnancy, but grow to include all pregnancies. This is the inevitable result of thinking of pregnancy not as a relationship between a woman and her foetus, but as a service she provides for others, and of thinking of the woman herself not as a person, but as a container for another, often more valued, person.[39]

Or as Ketchum puts it: 'once there is a market for women's bodies, all women's bodies will have a price, and the woman who does not sell her body becomes a hoarder of something that is useful to other people and is financially valuable. The market is a hegemonic institution; it determines the meanings of actions of people who choose not to participate as well as those who choose to participate'.[40]

The argument that commercial surrogacy is not baby selling is in any case undermined by the fact that, in most surrogacy arrangements in the US, at least half – and in some cases all – of the fee is withheld until custody of the child has been turned over to the contracting couple. Under the conditions of an ordinary employment contract, an under-performing worker who produces insufficient or defective products may lose his or her job, but is still entitled to payment for the labour time exchanged during the life of the contract. A surrogate mother who refuses to hand over her child, on the other hand, or who miscarries late in pregnancy, or whose baby is stillborn, is not likely to receive her fee, despite the fact that she has performed her 'prenatal baby-sitting', as one legal advocate of surrogacy has called it. This also explains why couples sue for specific performance of the contract when the surrogate mother decides to keep the child.[41]

Nor does the empirical evidence that children born through surrogacy arrangements are normally reared in loving and stable families make the selling of children morally acceptable. The same defence could of course be made of black market adoptions or slavery. During the Civil War era in America there were, as Scott Rae point out,

> undoubtedly instances in which slaves were treated humanely, without having to suffer the indignities that were forced upon many other slaves. But that did not change the fact that they were still slaves. They were still objects of barter that could be bought and sold, even if they were bought and sold into environments in which they were not treated as slaves . . . Child selling, like slavery, is inherently objectionable, whether or not it is accompanied by objectionable circumstances.[42]

Child-selling and womb-renting is inherently objectionable – that is, unacceptable from a 'deontological' ethical stance – because it contravenes the Kantian principle that human beings are ends in themselves, and thus devalues and degrades our appreciation of personhood. As Ketchum rightly argues, if bodies are to be treated as alienable forms of property, ownership and control over which can be contractually exchanged, the modern ethical tenet that an assault on the body is an assault on the person is weakened, and respect for the physical integrity of individuals loses its moral foundation.[43]

Patriarchal Reprotech: the Artificial Womb

The development of reproductive biotechnology thus appears inseparable from the commodification of bodies and selves, and this is both a violation of the modern cultural principles protecting the dignity and integrity of human beings, and a process which magnifies and exploits existing market-generated inequalities in wealth and power. And it is women who, on the whole, are the principal losers in this exchange. Even the well-established technologies of IVF bear the mark of male domination: despite the fact that a significant proportion of fertility problems among couples are attributed to low sperm counts, abnormal sperm morphology, or poor sperm motility – all of which make the male a prime candidate for injection-assisted conception – the laboratory unification of gametes presupposes an extraction process which is as long, painful, invasive and dangerous for women as it is as brief and inconsequential for men.

Probably the most startling future development likely to come from advances in reproductive, obstetric and prenatal medicine and technology, is, as I intimated earlier, the invention of fully mechanised *in vitro* gestation, or 'ectogenesis'. Ectogenesis may well be favoured as a solution to some of the problems raised by current reproductive technologies and practices, including the refusal by some surrogate mothers to hand over their children to the contracting clients. It may also offer a means of reconciling fiercely divided opinion in the abortion debate, enabling unwanted foetuses to be transferred to artificial wombs, gestated to maturity and then adopted by welcoming adults.

The ability to gestate embryonic human beings outside of the maternal womb is in fact a widely predicted, though rarely expressly intended, outcome of current advances in the medical care of premature babies. Premature babies can now be incubated and sustained from the moment their lungs are functional – which may be as early as twenty-one weeks – and pregnancies have been sustained in, and babies delivered from, the wombs of mechanically respirated women who have suffered brainstem death, including one case of 'post-mortem pregnancy' that was prolonged for sixty-three days.[44] While scientists at Cornell University's Weill Medical Centre have made significant progress in their attempts to build a replacement womb lining from regenerating cells grown on biodegradable scaffolds,[45] research by a Japanese gynaecologist, Yoshinori Kuwabara, has gone a step further, enabling goat foetuses to be successfully incubated in tanks filled with laboratory-made amniotic fluid. Fed with oxygenated and nutrient-enriched foetal blood, the animals can survive in this artificial womb without lung respiration for three weeks.[46] At the other end of the gestatory life-span, embryos are routinely cultured for several days in IVF clinics before implantation, and all of the major steps involved in IVF have been automated by biomedical engineers in the US, who have built a computer-driven prototype of a female reproductive tract.[47] Though experiments aimed at prolonging the life of human embryos *in vitro* are in most countries illegal, by the early 1980s scientists had already succeeded in culturing mouse and rat embryos to a level of post-blastocyst differentiation equivalent to about four weeks of human embryonic life.[48]

There are also influential medical arguments in favour of ectogenesis. Harvard professor and expert on biomedical ethics, the late Joseph Fletcher, expressed sentiments he believed were shared by 'most of those in responsible roles – embryologists, placentologists, foetologists', when he welcomed the future opportunity 'to monitor foetal life in the light, out of the darkness and obscurity of the womb.' 'We realise', he continues, 'that the womb is a dark and dangerous place, a hazardous environment. We should want our potential children to be where they can be watched and protected as much as possible'.[49]

Fletcher's vision of a clinically faultless replacement for the dangerous uterine environment is a valid indicator of where today's hunger for 'perfect babies' is likely to take us. This depiction of the female interior as a dark and primitive continent that must be illuminated and colonised by the rationalising efforts of a benevolent male science, lies at the heart of the patriarchal–technocratic project to dominate women's bodies and selves. Under the pretext of improving health and biological functioning – most human embryos, recall, do not survive beyond the first few weeks of pregnancy – it provides ideological legitimation for dispossessing women of control over their bodies, and for making those bodies into an object of medical science and contract law. An artificial womb would, of course, not only free the foetus from the deleterious effects of women's imperfect biology, but would also protect that foetus both from the hazardous consequences of irresponsible maternal behaviour, and from increasingly toxic working and living environments. 'Fifty or 100 years from now,' reproductive physiologist George Seidel predicts, 'our in vitro procedures for parts or even all of pregnancy may end up being safer than dealing with the various things that occur in the body – in terms of viruses that the mother comes across, toxins, and so on.'[50]

Fletcher, who also foresees the eventual transplantation of women's reproductive organs into men, as well as the deliberate creation of animal–human hybrids to perform dangerous or demeaning work, writes confidently of the prospects for ectogenesis.

> The glass womb is after all nothing more than an extension of the 'extracorporeal membrane oxygenator', the incubator which already feeds 'preemies' and babies with hyaline membrane disease. An artificial placenta, like a heart–lung machine, is a substitute for a natural function; it provides amniotic fluid chemically. Glass wombs are a radical version of early Caesarean sections. With artificial placentas we could save a foetus which might otherwise have to be lost in a medical abortion; both the patient *and the baby* could be saved. If we can save people with kidney failure by putting them on machines (artificial kidneys), and people with heart disease by putting machines in them (artificial hearts), why not do the same for *potential* people. When cloning becomes fully operational for humans, ectogenesis would in some situations eliminate the reimplantation stage, to advantage.[51]

Of course, it is not women's bodies pure and simple which are the target here, but the personal relationship between mother and child which is fashioned during nine months of continual and intimate physical connection. This experience is ordinarily the primordial basis for an affective and emotional bond between mother and child which refracts society's attempts to bind individuals, by the processes of socialisation, to the dominant norms, roles and expectations of that society, including norms of domination and inequity. By virtue of this maternal bond, which all totalitarian regimes have sought to sever by collectivising parenting or turning offspring against their parents, even the socially useless or unlovely child has a strong chance of being loved, and of subsequently articulating its personal opposition to the society that has no use for it.

Paradoxical as it may sound, there can in fact be no firmer basis for resistance against the economic commodification and social control of human beings than the maternal possessiveness born of pregnancy and childbirth. The argument made by some feminists – and personified in Shulamith Firestone's classic and prescient demand that women use ectogenesis to liberate themselves from the 'tyranny of reproduction'[52] – that pregnancy and maternal care is an oppressive function performed for male society which should be properly remunerated or mechanised, in practice only conspires with the technocratic–authoritarian instincts of modern society. Such reasoning amounts, as André Gorz put it, 'to *de-feminising* the biologically, corporeally and affectively specific dimensions of motherhood'.[53] By reducing pregnancy and childbirth to abstract labour, and babies to interchangeable products, it repudiates women's perception of and attachment to their children as *uniquely their own*, and in doing so weakens the dynamics of love and rebellion through which the social integration of the child is otherwise mediated, personalised, and thereby rendered incomplete.[54]

Even at its current early stage of development, the phenomenon of ectogenesis also serves as a reminder of the way our mechanistic perception of living things invariably leads to the construction of surrogate machines, and how the successful functioning of these machines deepens the mechanisation of thought and perception in turn. This preference for the superior functionality of the intelligent machine has its roots, as Edmund Husserl intimated in his critique of the Galilean mathematisation of nature, in the increasing 'superficialisation of meaning' brought about by the transformation of science into technique. By formalising thinking to enable its detection of the pure and objective boundaries, shapes, magnitudes, properties and causal relations of a geometrically idealised world, the natural-scientific method conceals both the fluid and indeterminate qualities of the 'world of sensibility' in which we permanently dwell and consequently the *act of idealising* that world which gives the scientific project its only meaning. 'Are science and its method not like a machine, reliable in accomplishing obviously very useful things, a machine everyone can learn to operate correctly without in the least understanding the inner possibility and necessity of this sort of accomplishment?'[55] It is this mechanisation of thought and feeling which ultimately makes possible our misrecognition of machines as surrogates, substitutes, or competitors for ourselves.

> The capacity to conceive these machines finally comes to conceive of itself as a machine; the mind which has become capable of functioning as a machine recognises itself in the machine which is capable of functioning like itself – without realising that in reality the machine does not function like the mind but only like mind when it has learned to function as a machine.[56]

Negative Eugenics: Aborting Concern

I want to conclude this chapter by returning at length to one issue that has exercised the imagination and drawn the ire of numerous commentators on the genetic revolution: the threat of eugenics. It has to be said that the eugenic practice of preventing or discouraging people with supposed physical, intellectual and psychiatric disabilities from reproducing has a long and ugly history of support amongst many members of the liberal intelligentsia in twentieth-century Europe and America, including members of the British Fabian movement such as Sidney and Beatrice Webb, George Bernard Shaw, and H.G. Wells, supporters of the Progressives in the US, and prominent Social Democrats in Denmark and Norway.[57] Though a private member's bill on eugenics, introduced to the British parliament by a Labour MP, was blocked in 1934, policies of compulsory sterilisation are thought to have claimed several hundred thousand victims in the Western capitalist democracies as a whole. Such policies, which were pursued on grounds as diverse as class snobbery, racial prejudice, Marxist utopianism, and a social democratic concern to reduce demands on the new universal welfare state, were abandoned in most US States in the late 1930s, but in some European countries (including Austria, Denmark, Finland, Italy, Norway, Sweden and Switzerland) sterilisation laws remained on the statute books until the 1970s.[58]

Despite Charles Murray's provocative claim that, with the distinctive philosophical commitment of socialists to the interventionist role of the state, 'eugenics will become a cause for the left',[59] it is in fact the 'laissez faire' form of eugenics – what has been called a 'home-made' or 'consumer' eugenics – which most critics believe will be the greatest threat posed by the expansion of genetic technology in the coming decades. Instead of the state exposing itself as the coercive agent of genetic cleansing, the societal goals of reducing the burden of health care and improving the performance capacity and functional uniformity of the population may well be realised through the exercise of 'consumer choice' in an atmosphere of diffuse prejudice and discrimination.

We should recall the peculiarity of a genetic research paradigm which has become locked into a search for genetic aberrations which have little practical relevance to modes of treatment and the healing arts. The significance of the findings of most genetic research – findings which are increasingly represented as insights into 'behavioural diseases' as much as those of pure biology – is their contribution not to the care and treatment of the ill, but to a programme of 'prevention'. Unless and until *in utero* gene therapy becomes an acceptable practice, this of course means abortion. As the critique of medical paternalism and demands for freedom of information gain ground, women may in the future have the constitutionally protected right to deselect embryos or foetuses with genetic features indicating, however tenuous the evidence, a wide range of socially maligned characteristics.

As noted earlier, what is most distinctive about advances in prenatal genetic testing is the inexorable trend towards ever-earlier stages of investigation. These investigations are aimed at identifying disorders which cannot yet be treated *in utero*, the age of onset and severity of which often cannot be determined, and which are therefore pursued with the sole purpose of providing the pregnant woman with the option to terminate. In some cases – and this is likely to be the trend in the future – foetuses have been aborted because genetic or chromosomal irregularities have been detected whose phenotypic effects are disputed or unknown.

Amniocentesis, which is the extraction, by syringe through the pregnant woman's abdominal wall, of foetal cells that have been shed into the amniotic fluid, was initially only possible at around eighteen weeks of pregnancy, but can now be done at fourteen weeks. A new and safer test for Down syndrome, which involves removing cells from the placenta, has now been piloted which can be conducted at eleven weeks. Chorionic villus sampling, in which cells are removed (either transabdominally or transcervically) from the hair-like membranes enveloping the early embryo, can be performed even earlier. And with new 3D-ultrasound technology, minor or cosmetic congenital dysmorphologies such as cleft-lip, malformed ears, polydactyly (extra toes or fingers), club-foot and claw-hand, can now be detected in the first trimester of pregnancy, thus allowing for the termination of such pregnancies before a strong foeto-maternal bond has developed.[60]

Pre-implantation genetic diagnosis of course takes this process back even earlier. Although it is routinely used on embryos that have been formed *in vitro*, the embryos of fertile couples deemed at risk of having babies with genetic diseases can also receive pre-implantation genetic diagnosis, and not necessarily via IVF. In a relatively new process called 'uterine lavage', naturally formed embryos are flushed from the uterus and tested *in vitro*. Because the viability of the embryo deteriorates the longer it is left out of the body, but the more mature the embryo the more likely it is to survive the removal of a cell for testing, uterine lavage allows embryos to be tested that have the robustness of age but have not been weakened by excessive laboratory culturing.

Though prenatal testing is normally depicted as a means of improving prenatal care and increasing parental choice, it is also expanding as a result of defensive legal measures designed to protect physicians from later charges of negligence.[61] That the vast majority of foetuses diagnosed with genetic diseases are aborted suggests, in any case, that the options made available to pregnant women in these circumstances are not perceived by the latter as having equal (or neutral) value. Indeed, society communicates to the woman *its* preference for termination by offering the tests in the first place, and justifying them – in economic and legislative debates – on the grounds of cost-effectiveness (prenatal screening and abortions are far less costly than social support and medical care for people with disabilities).[62] Add to this the social stigma which the pregnant woman knows a disabled child will face, the inadequate welfare support she is likely to receive, and the likelihood that she herself will have to shoulder the lion's share of domestic care for the child, and the choice to terminate is barely a choice at all.[63]

Though moral objections to the trend towards ever-earlier testing of foetuses and embryos are often wrapped in the rhetoric of 'foetal rights', what rightly concerns the more perceptive commentators on this issue is how the growing phenomena of prenatal testing represents 'disability' to the rest of society – how it essentialises and medicalises impairment, exonerates society from disabling the modestly competent and able, and relieves the rest of us of the burden of concern.[64] What is at issue here is the prejudicial treatment not just of those people with genetic anomalies that may slip through the net of prenatal screening, but also of the vast majority of disabled people whose impairments are acquired – through accident, illness or inescapable physical deterioration – rather than inherited. Will a society which is successful in its attempt to eliminate the possibility of childhood disability continue to care about, and even remain tolerant of, the progressive deterioration of physical and mental faculties in the elderly? And what about those fatally disabled babies, a small number of whom will always be born, no matter how sophisticated and extensive the genetic testing? Will their fate be a medical question of 'post-natal abortion', of the destruction of 'defective organisms' with no meaningful future or value, or will society have the moral courage to face their suffering in a spirit of mercy and grace?

These questions recall Hans Jonas's observations on the prevailing medico-scientific view that a person without brainstem activity is 'already dead', the function of this viewpoint being to 'let no ancient fear and trembling interfere with the relentless expanding of the realm of sheer thinghood and unrestricted utility'.

> The responsibility of a value-laden decision is replaced by the mechanics of a value-laden routine. Insofar as the redefiners of death – by saying "he is already dead" – seek to allay the scruples about turning the respirator off, they cater to this modern cowardice which has forgotten that death has its own fitness and dignity, and that a man has a right to be let die.[65]

Jonas's point is that the existential disturbances of death, disability and suffering can only be technologically eradicated by eliminating our capacity to care. While disability activists and their critics discuss the rights of people with disabilities, his analysis prompts us to ask whether the right of 'normal' children to be loved, cherished and respected – for their idiosyncrasies, their deviations, their failures, misfortunes and rebellions – will continue to be met in a world where health and normality is perceived as a rationally planned and pre-selected asset, a genetic programme chosen and paid for by demanding consumers.

Positive Eugenics: The Perfect Baby

Eugenic pressures in modern societies have not only become softer and indirect, but are also moving towards a *positive* or 'melioristic' framework in which gametes or embryos will be selected for desirable traits, and even modified and improved by genetic engineering. Cloning is perhaps the most controversial manifestation of this trend, since it offers a means of circumventing the ordinary lottery of genetic recombination and ensuring instead the exact replication of the superior genomes of successful athletes, musicians, intellectuals, and so on, as well as the resurrection of dead family members or friends, or the recreation by adults of themselves as children.

Perhaps the most compelling philosophical argument against human cloning, the positive selection of embryos, and indeed the genetic engineering of embryos for the purposes of 'enhancement', is that it removes a primordial barrier to the social determination – or 'predetermination' – of human beings. This barrier is the *contingency* of human life – what Hilary Putnam refers to as 'the "right" of each newborn to be a complete surprise to its parents',[66] and what in legal debates contesting the absolute freedom of parents to shape the preferences, values and talents of their offspring, has been called 'the child's right to an open future'.[67] As the French molecular biologist Axel Kahn puts it, 'part of the individuality and dignity of a person probably lies in the uniqueness and unpredictability of his or her development. As a result, the uncertainty of the great lottery of heredity constitutes the principal protection against biological predetermination imposed by third parties, including parents'.[68]

Kahn's argument that 'One of the components of human dignity is undoubtedly autonomy, the indeterminability of the individual with respect to external human will',[69] is not without ambiguity, however. Does he mean that the genetic uniqueness and unpredictability of children created by natural fertilisation is the reason why such children *should* be treated with respect and dignity, or the reason why they *are* treated so? Or to put this another way, is the cloned child, reduced to an artefact of human design, *undeserving* of love and respect, an inappropriate object of dignity and moral consideration in Kahn's eyes, or rather is such a child *unlikely to receive* this consideration in a society which has permitted his or her creation?

It is the first interpretation of Kahn's argument which appears to have gripped most people who have reflected on this issue, and most of whom have subsequently dissented. The point is, of course, that only someone who subscribed to a form of complete genetic reductionism would see cloning as a complete form of predetermination. As critics of Kahn's position – and both advocates and opponents of cloning – have pointed out, different uterine environments, random cellular processes ('developmental noise'), interaction with diverse post-natal personal and social environments, and of course the spontaneous choices and projects of cloned individuals themselves, mean that the resemblance between the latter and those with whom they share a genetic identity is not likely to be overwhelming.

> To produce another Wolfgang Amadeus Mozart, we would need not only Wolfgang's genome but his mother's uterus, his father's music lessons, his parents' friends and his own, the state of music in 18th-century Austria, Haydn's patronage, and on and on, in ever-widening circles . . . If a particular strain of wheat yields different harvests under different conditions of climate, soil, and cultivation, how can we assume that so much more complex a genome as that of a human being would

yield its desired crop of operas, symphonies, and chamber music under different circumstances of nurture.[70]

It is the second interpretation of Kahn' argument, therefore, which makes most sense. The effects of eugenic technologies like human cloning are mediated by a cultural attachment to genetic determinism which both underpins and is consolidated by those same technologies. The problem is not that the autonomy and uniqueness of individuals will be lost, and hence that they will be undeserving of the respect and dignity normally accorded to human beings. It is, rather, that the respect, love and recognition ideally expressed by adults towards the child will be subverted by their *expectation* that they have ordered a predetermined product, and this expectation will in turn lead to the misrecognition or repression of the child's attempts to assert its autonomy and uniqueness. The issue, as Lewontin puts it, 'is not whether genetic identity per se destroys individuality, but whether the erroneous state of public understanding of biology will undermine an individual's own sense of uniqueness and autonomy'.[71]

Rothman tells a salutary anecdote about a couple of intellectuals who reconciled themselves to the birth of their disabled son. Diagnosed with a neural tube defect, his wheelchair-dependency would, they reasoned, pose much less of a problem to their cerebral values and lifestyle than if they were keen athletes. Fortunately, they were willing to revise their expectations, as the young man became a national wheelchair-basketball champion and they adapted their lives to his game schedules.[72]

Rothman's argument is not just that biology is not destiny, but also that the assumption that genes are determinants of identity is deeply entrenched at all levels of society, and it provides the cultural matrix in which eugenic technologies and practices can flourish. She also discusses the case of sex selection, a phenomenon that can no longer be legitimately treated as an exclusively 'Third World problem'.[73] In a 1998 study of nearly 3,000 geneticists and genetic counsellors, a third of those practising in the US reported they would perform prenatal sex identification if requested by a hypothetical couple with four girls who want a son and would abort a female foetus. This percentage was exceeded only by Israel (68 per cent), Cuba (62 per cent), Peru (39 per cent) and Mexico (38 per cent). Comparing this data with the results of a 1985 study by the same team, the two surveys suggest 'a clear trend in most of the world towards greater willingness to perform or refer for sex selection.'[74]

Sex selection for non-medical reasons is legal in the US and Canada (though in the latter there is a badly-observed moratorium on the practice), and prohibited in France and the UK. There is no reliable record of the number of pregnant women who have used foetal testing for the purpose of sex selection, nor of those who have terminated their pregnancies as a result (though in the study mentioned above, nearly half of all respondents reported that they had received outright requests for prenatal diagnosis solely to select the parents' preferred sex). What is certain, however, is that both sexual inequality (which diminishes the cultural value and material life-chances of women) and the stark differentiation of gender roles (which leads to the fetishisation of the 'other' sex and makes families feel 'unbalanced' without their fair share of both sexes), together create significant incentives for the selective abortion of foetuses because they are the less favoured sex. These incentives are, moreover, strengthened by society's rightful endeavour to restrain scientific paternalism and defend the reproductive rights of women,[75] as well as by the development of new, safer and less invasive selection methods, such as the pre-fertilisation sorting of sperm into separate X and Y chromosome-containing gametes.[76]

Because the medical risks involved in sex selection are declining, and because the dominant reason for seeking sex selection in the wealthy world – the desire for a 'balanced' family – is not seen as a threat to the population's sex ratio, David McCarthy has argued that the option to choose the sex of one's child should be legally available in a liberal democracy.[77] Yet the trouble with sex selection, as Rothman points out, is not that it involves the deliberate creation of an irresistible biological identity – an outcome which is in any case of no concern to McCarthy so long as there is an even spread of boys and girls, with both sexes valued equally. The problem, rather, is that sex selection nourishes adults' belief that the physiological and anatomical characteristics of a particular sex determine the social behaviour, temperament, values and beliefs of *gender*. Parents who choose the sex of their child are unlikely to tolerate the 'unbalancing' of their family by a boy who wants to learn ballet, or a girl who plays football, eschews the conventional trappings of femininity, and curses like her brothers.

What is true of sex selection is no less applicable to genetic enhancement or cloning. A child designed for maximum physical attractiveness is likely to be a source of considerable parental displeasure if it finds intimacy difficult, is acutely shy or shows no interest in sexual relationships. A child selected or manipulated to grow to a generous height may be deeply flawed in the eyes of its parents if it fails to corroborate the well-known link between tallness and economic success, preferring to work with the homeless or hang around on the dole.

The point is that the desire for designer babies is not, in fact, a desire simply to bestow one's children with advantageous biological attributes, but rather to guarantee that those children follow a socially esteemed life-plan – that they exploit those apparent biological advantages in an acceptable and conventional manner, and live out the future that is expected of them. Despite the claims of some sociobiologists, there is no possibility of delivering this degree of social control and predictability by simply altering the human genome (and even if there was, incidentally, the competitive advantage resulting from this control would be devalued in the long run by the employment of the same technology by others). One may therefore reply that the experience of parents whose genetically selected or engineered children fail to perform as anticipated will be a sobering and worthwhile education, and that the ideology of genetic reductionism will consequently not survive the empirical test of time. But we should not blind ourselves to the fact that such experiences are shaped by the intentions that beget them. Many parents of identical twins encounter the spirited efforts of their children to assert their individual uniqueness, but nonetheless still corral, prime and groom those children to behave like carbon copies of each other. The truth is that gratifying the desire to predetermine the genetic inheritance of one's children can only consolidate the same prejudices of biological determinism, and will inevitably result – even, and perhaps especially, where the product falls short of expectations – in attempts *to programme and control those variables which genetics leaves untouched*.[78] How indeed can a procedure such as the cloning of human beings justify the extraordinary risks, uncertainties and investments involved if not by making the complete abolition of risk and certainty – the transformation of the human being into a programmed product – its ultimate and redeeming goal?

As Jonas writes of cloning:

> Note that it does not matter one jot whether the genotype is really, by its own force, a person's fate: it is *made* his fate by the very assumptions in cloning him, which by their imposition on all concerned become a force themselves. It does not matter whether replication of genotype really entails repetition of life performance: the donor has been chosen with some such idea, and that idea is tyrannical in effect. It does not matter what the real relation of 'nature and nurture', of genetic premise and contingent environment is in forming a person and his possibilities: their interplay has been falsified by both the subject and the environment having been 'primed' . . . [E]xistentially significant is what the cloned individual *thinks* – is compelled to think – of himself, not what he 'is' in the substance-sense of being. In brief, he is antecedently robbed of the *freedom* which only under the protection of ignorance can thrive; and to rob a human-to-be of that freedom deliberately is an inexpiable crime that must not be committed even once.[79]

The Self-Understanding of the Species

Jonas rightly enjoins us here to consider the effects of genetic engineering on the child itself. Liberal theorists and commentators who believe we should defend adults' right to exercise freedom of choice over their children's genetic inheritance, invariably refer their opponents to the freedom which capitalist democracies grant to all parents to mould the character and enhance the aptitudes of their offspring by selecting the particular domestic, educational, social and dietary conditions under which they are raised. The unease this 'liberal eugenic' position generates amongst its critics reflects, as I have indicated, the concern that the technology of human genetic engineering adds a kind of scientific legitimacy and psychological encouragement to existing forms of parental despotism which, because they cannot easily be legislated against, must be restrained by the norms and expectations of the public culture. But the genetic programming of unborn lives represents not only the deepening of a capacity for parental domination which many would prefer to see curtailed. It is also a power of an entirely new order, which brings with it unique ethical problems.

The socialising efforts and ambitions of parents, though exercised on a dependent and unequal subject, are normally always in principle *contestable*. By critically re-evaluating the interpersonal and communicative processes which brought about the adaptation and consent of the child to its parents' demands, the reflective adolescent can reappraise and break free from the restrictive expectations of those parents, and from a socialisation process which necessarily *comes to an end*. This end is marked by the accession of the youth to the world of autonomous and equal adults. With the cloned or genetically enhanced child, however, the parents' intentions are fixed in a past which predates the offspring's experience and memory, as well as projected into a future from which the genetically engineered adult will never be free (the modified genetic programme, in the eyes of the zealous programmers, does not fall silent when the child reaches physical maturity, and the child's adult life will be as much the intended object of the parents' instrumental [and experimental] intervention and curiosity as his or her childhood). As Habermas explains:

> The programme designer carries out a one-sided act for which there can be no well-founded assumption of consent, disposing of the genetic factors of another in the paternalistic intention of setting the course, in relevant respects, of the life

history of the dependent person. The latter may interpret, but not revise or undo this intention. The consequences are irreversible because the paternalistic intention is laid down in a disarming genetic programme instead of being mediated by a socialising practice which can be subjected to reappraisal by the person 'raised'.[80]

The problem is not only that the intentions of the programming parents are fixed in a mute and unanswerable form that cannot be addressed and challenged, but also that the authors of these intentions do not appear as opponents inhabiting a shared world. The struggle to be and be recognised as a subject is originally a struggle to transcend the dependent state of childhood, and it is the mutual recognition by parent and offspring that the latter's dependency has come to an end which gives substance to the moral–political belief in the fundamental rights of all persons. With genetic programming, however, an asymmetry prevails which precludes the genetically designed being from exchanging roles with its designer, from inhabiting a symmetrical relationship of reciprocity and mutual recognition.

The development of identity and autonomy – of a sense of being a coherent and stable centre of initiative and originator of meaning – may require more than participation in a world of reciprocity and generalised norms. It also presupposes a continuity of self, a sense of being to which one can 'be true', which is jeopardised by genetic intervention, the logic of which leads imperiously to the dissolution of the boundary between the natural and the made. Following Hannah Arendt, Habermas suggests that the capacity to see oneself as the irreducible origin of one's own actions and judgements requires the opportunity to locate this origin in a beginning which eludes human disposal, which reaches back beyond the vagaries and contingencies of culture and socialisation:

It is only by referring to this difference between nature and culture, between beginnings which we cannot dispose of and the plasticity of historical practices, that the acting subject may proceed to the self-ascriptions without which he could not perceive himself as the initiator of his actions and aspirations. For a person to be himself, a point of reference is required which goes back beyond the lines of tradition and the contexts of interaction which constitute the process of formation through which personal identity is moulded in the course of a life history . . . We can achieve continuity in the vicissitudes of a life history only because we may refer, for establishing the difference between what we are and what happens to us, to a bodily existence which is itself the continuation of a natural fate going back beyond the socialisation process. The fact that this natural fate, this past before our past, so to speak, eludes human disposal seems to be essential for our awareness of freedom.[81]

By dissolving the boundary between nature and nurture, and by preventing the genetically engineered person from facing his or her progenitors as an equal, the selection and enhancement of human genomes is likely to obstruct the self-understanding of persons as autonomous and responsible selves. More than this, human genetic engineering threatens to destroy the anthropological self-understanding of the species which, elaborated in our world religions and in our metaphysical and humanistic traditions, portrays all humans, whatever their cultural situation, to be heirs to the biological inheritance of 'man', and thus born equal in dignity and freedom. In Lee Silver's dystopian vision, human genetic modification will eventually result in the biological division of human beings into two non-interbreeding populations.[82] But reflection on the cultural consequences of a modern liberal eugenics suggests that it is the normative foundation of society – the collective understanding of human beings as members of a shared community of equals – which is truly and more immediately at stake, and that a society in which human life is at the disposal of people's contingent and perishable preferences and tastes will not survive as a moral community for very long.

One reason for this, as Fukuyama points out,[83] is that by replacing the lottery of natural heredity with deliberately enhanced genetic endowments, the parents of genetically designed children will deprive their offspring of the sense of good fortune which normally enables those dealt a strong hand to feel sympathy for the less fortunate. Instead of attributing their strong start in life to the favourable accidents of birth and upbringing, they may identify their talents and opportunities as benefits merited by the superior choices and planning of their parents. And the aristocratic mentality nurtured in such people may be all the more imperious if their sense of superiority can be justified by reference to their superior genes.

In the form of 'genetic communitarianism' considered by Allen Buchanan and his colleagues, on the other hand, parents would be free to transmit to their children by means of genetic engineering the definitions of human virtue and the good life shared by the social and moral communities to which those adults belong. Were such traits to become amenable to genetic manipulation, the members of one benevolently minded community may choose to enhance their offspring's capacity for self-sacrifice, empathy and generosity, while another community may pursue the perceived virtues of physical strength, competitiveness and individualism. Where the transmission of a particular way of life is the overriding goal, adults may even seek to *limit* their offspring's capacities – as at least one deaf couple has already done in the

US – in order to create the biological conditions which best prime the child to internalise the parents' culture.

One foreseeable result of this scenario, as Buchanan and his colleagues point out, would be a declining commitment to the liberal principle of mutual tolerance and respect, based as it is on the conviction 'that we are all "reasonable people" who have come, for complex reasons having to do with the limits of human judgement and facts of history, to believe different things'. As Buchanan *et al* summarise the dangers of this liberal-communitarian eugenics:

> Genetic communitarianism might result in different communities coming to view their differences as no longer the result of commitment and persuasion, but of their different 'natures', with the result that these differences come to be regarded as irreconcilable. Under these conditions any suggestion of compromise with the values of another community might be regarded as literally a threat to the identity and hence the survival of one's group . . .
>
> The threat of locking-in is thus not just a threat to the individual and his or her rights to an open future. It is a threat to the basis for political cooperation in a liberal society that involves a respect for individual liberties and toleration for those who are different. The threat is that people will come to think of themselves as different in ways even more fundamental than they do today.[84]

Of course, a social consensus over what constitutes 'improvement' to the human condition may well be possible, particularly if the protagonists of human engineering appeal directly to the objective and measurable standards of functional performance, speed of thought, physical strength, biological longevity, powers of communication, metabolic and immunological efficiency, and so on. These performance criteria figure prominently in the increasingly popular discourse of the cyborg, and they gain currency the more the complexity and inhospitableness of society disables healthy human beings from playing a full and fulfilling role in the production and reproduction of the social world. By turning to the concept of the cyborg, I hope to expose even deeper problems associated with the prospect of what Edward O. Wilson calls the 'volitional evolution' of humankind.[85]

11

The Cyborg Solution

The End of Metaphysics?

'There is a strong and, it seems, almost irresistible tendency in the human mind to interpret human functions in terms of the artefacts that take their place, and artefacts in terms of the replaced human functions.'[1] We have seen this tendency at work in reproductive biotechnology and the genetic reductionism that accompanies it, and it is central to the increasingly popular view that people, being nothing more than sophisticated biochemical-informational machines, will in the future be able to re-engineer their minds and bodies, and thus make human nature itself the deliberate product of a transcendent or supremely intelligent will. This 'power to change the nature of humankind', as Lee Silver describes it, will involve transferring adaptive genes from other species to ourselves, enhancing our sense of smell, sight and hearing, giving us the ability to see in the dark, to generate organic electricity or light, to produce and sense high frequency sound waves, and even to send and receive information as radio waves (so-called 'radiotelepathy').[2]

This enthusiasm for the self-enhancing power of human beings is shared by the contemporary German philosopher, Peter Sloterdijk. For Sloterdijk, the flourishing of genetic science, in which the building-blocks of life are understood and manipulated as the reprogrammable elements of complex information systems, marks the long-anticipated decomposition of the false metaphysical distinction between culture and nature, subject and object, spirit and matter. In the information age, both sides in this opposition – on which rests the whole of Western humanism and the forms of domination that are, for Sloterdijk, its inevitable nemesis – can no longer sustain the characteristics that previously divided them. In particular, formerly soulless matter is now understood to possess a 'parasubjective' intelligence: 'in the fundamental material structure of life, as represented by the genes, nothing "material" in the sense of the old culture / nature divide is to be found any more. Rather one finds the purest form of information, for genes are nothing but "commands" for the synthesis of protein molecules'.[3]

As electronic and biological information systems exert their supremacy in advanced societies, the fetishised self in conflict with an unyielding nature is a picture that Sloterdijk believes must be abandoned in recognition of the fusion of the subject with the pliable materials and technologies of a post-metaphysical world. The domination of nature, its treatment as raw material or 'standing reserve' in Heidegger's parlance, 'can only be found where raw subjects – call them humanists or other egoists – apply raw technologies to them'. But purposeful action in today's world is, according to Sloterdijk, ceasing to employ the old 'allotechnologies' – technologies which executed a violent and counter-natural restructuring of matter in order to meet ends that were indifferent or alien to it. Instead such actions are increasingly accomplished through a 'non-dominant form of operativity' which he calls 'homeo-' or 'anthropo-technology'. This technology, which is exemplified by genetic engineering, is a product of unprecedented insight into the functioning of nature, and is finely tuned to the language of matter itself. Indeed, since genetics *is* a linguistic code of sorts, it can only be spoken to in its own native language. 'By its very nature, [anthropo-technology] cannot desire anything different from what the "things themselves" are or can become of their own accord. The "materials" are now conceived in accordance with their own stubbornness [and are integrated] into operations with respect to their maximum aptitude.'[4] The result is 'co-operation rather than domination', as the struggle to master the Other gives way to the self-manipulation of the human essence, to the triumph, as Hegel might have seen it, of Absolute Mind, of knowledge materialised in nature aware of itself.

Because humanity is, in Sloterdijk's account, defined by the technologies it constructs and uses, because the human is *made* rather than given, then with the maturation of that technology and the deepening understanding of our biological and evolutionary condition, the turning of that technology on ourselves is entirely consistent with what we are.

> If there is man, then that is because a technology has made him evolve out of the prehuman. It is that which authentically brings about humans. Therefore humans encounter nothing strange when they expose themselves to further creation and manipulation, and they do nothing perverse when they change themselves autotechnologically, given that such interventions and assistance happen on such a high level of insight into the biological and social nature of man that they become effective as authentic, intelligent and successful coproductions with evolutionary potential.[5]

Sloterdijk's vision is echoed in the more populist account of Gregory Stock, who also believes that the technological manipulation of humanity is ceasing to be a form of violence 'because biology itself is becoming more and more like our technology'.[6] The primary tool of what Stock calls 'germinal choice technology' will be the artificial chromosome, a synthetic molecule of chromatin added to the nucleus of the cell which, though manufactured 'blank', will provide a multitude of vacant sites into which can be inserted 'genetic modules' manipulated and selected for the traits they express. Pioneered in 1997, the artificial chromosome has been successfully replicated in dividing cells, and has been transmitted through the germ-line of successive generations of rats. By persuading the cells to synthesise proteins from these auxiliary chromosomes, scientists will be able to treat the human genome like electronic software that may be continually 'patched', updated and revised, whilst at the same time minimising the risk of disturbing endogenous DNA sequences on the natural chromosomes.

The problem of transferring these genetic changes to non-consenting future generations will also be avoided by a technique similar to that used in the 'terminator' technology discussed in chapter 4. Here the central locus of the artificial chromosome which is necessary for its successful replication – the centromere – would be sandwiched between sequences identified as sites to be cut and recombined by the enzyme produced by the Cre gene. The latter would be driven by a promoter which only functions in the germ cells, and which is activated by the presence of a drug. Before trying to conceive, the genetically engineering individual would consume the drug, thus precipitating a molecular chain of events which ends with the deletion of the artificial chromosome from the person's egg or sperm cells. This technique for the spontaneous removal of DNA sequences from the germ-line has already been accomplished in mice.[7] It could also be used to control the activation of the genes on the auxiliary chromosome, thus allowing the effects of the chosen genetic modules to be delayed until the child is old enough to give consent.

Cyberbeings

Whether scientists' level of insight into the genetic constitution of organisms is as deep and penetrating as Sloterdijk and Stock claim, is a question which, in the light of the preceding discussions in this book, only a genetic reductionist could answer in the affirmative. The idea that human beings can become knowing authors of their own evolution, that humanity itself can be the direct target of our instrumental interventions in the world, is, however, not restricted to the enthusiasts of genetic science. Since genetic engineering seems particularly limited as a means of adult *self*-transformation, and since the genetic manipulation of embryos carries disturbing social and political ramifications, the possibility of changing and enhancing our own human nature has also begun to attract the interests of philosophers and scientists working in the fields of robotics and artificial intelligence.

Confidence in people's ability and right to transform themselves into humanly constructed and improved artefacts became politically and academically fashionable following the publication in 1985 of Donna Haraway's abstruse and delirious text, 'A Cyborg Manifesto'. In its more lucid moments, this essay appears to celebrate, not unlike Sloterdijk's valediction to metaphysical dualisms, the deconstruction of the divisions between the natural and the artificial, organism and machine, reproduction and replication, and to propose as a feminist strategy the replacement of the Western ideal of the integrity and sincerity of the self with a 'postmodern' world of information flows, networks and interfaces. 'Intense pleasure in skill, machine skill, ceases to be a sin, but an aspect of embodiment. The machine is not an *it* to be animated, worshipped and dominated. The machine is us, our processes, an aspect of our embodiment.'[8]

According to the pioneering robotics researcher Hans Moravec, there will be a natural progression from molecular biology to computer technology because genetic engineering offers limited opportunities for humans to expand their powers and capacities, with DNA-synthesised proteins providing fragile and inflexible homes for upgraded forms of consciousness. Protein, Moravec writes, is an imperfect material:

> It is stable only in a narrow temperature and pressure range, is very sensitive to radiation, and rules out many construction techniques and components. And it is unlikely that neurons, which can now switch less than a thousand times per second, will ever be boosted by the billions-per-second speed of even today's computer components. Before long, conventional technologies, miniaturised down to the atomic scale, and biotechnology, its molecular interactions understood in detailed mechanical terms, will have merged into a seamless array of techniques encompassing all materials, sizes, and complexities . . . At that time a genetically engineered superhuman would be just a second-rate kind of robot, designed under the handicap that its construction can only be by DNA-guided protein synthesis. Only in the eyes of human chauvinists would it have an advantage.[9]

Genetic programmes must, in Moravec's view, be replaced by their silicon equivalents, which will eventually be maintained by self-replicating nanobots built at the atomic level and incorporating microscopic motors derived, for example, from rotational enzyme reactions. The limitations of genetic biology will thus be transcended in an evolutionary shift which Moravec believes is analogous to that which, according to one theory, marked the beginning of life on earth. In this account, certain microscopic species of clay crystals gained a critical adaptive advantage by supplementing the chemical machinery of crystal growth and replication with the synthesis of long carbon molecules of DNA, a polymer of sufficient stability and complexity to make the original crystal scaffolding eventually redundant. Now the evolutionary goal in Moravec's view is to complete the emancipation of culture from biology that has been augured by ten thousand years of human development – to liberate the human mind from its finite material infrastructure by transmitting its contents to artificial neurological networks.

Ray Kurzweil, a leading authority on artificial intelligence, also thinks that the 'plodding type of circuitry' characteristic of genetically regulated organisms will have to be replaced by 'a computational technology a millions times faster than carbon-based neurons'. Like Moravec, Kurzweil is confident that the exponential growth in computer power – the 'law of accelerating returns' – will deliver computers with the memory capacity and processing speed of the human brain by around 2020. Once the advantages of artificial intelligence – in terms of speed, storage capacity and accuracy of computation – become clear, humans will begin replacing their ageing neural circuits with silicon implants, enhancing their brains with electronic prostheses until thought itself can migrate to 'far more capable and reliable new machinery'. Micro-electronics, not genetics, will pave the way to immortality.

> Up until now, our mortality was tied to the longevity of our *hardware*. When the hardware crashed, that was it . . . As we cross the divide to instantiate ourselves into our computational technology, our identity will be based on our evolving mind file. *We will be software, not hardware* . . . Just as, today, we don't throw our files away when we change personal computers – we transfer them, at least the ones we want to keep. So, too, we won't throw our mind file away when we periodically port ourselves to the latest, ever more capable, 'personal' computer . . . Our identity and survival will ultimately become independent of the hardware and its survival.[10]

Because the human brain follows the laws of physics, it must, in Kurzweil's eyes, 'be a machine, albeit a very complex one'. The two essential functions of this complex machine are the processing of information (computation) and its storage (memory). With a hundred billion neurons, and an estimated one thousand connections between neighbouring neurons, the human brain is capable of making around one hundred trillion simultaneous calculations – far in excess of today's silicon circuits. The speed of these calculations, however, is dramatically outshone by microprocessors, and the steady growth in computers' processing speed will soon make them, even in the absence of parallel processing, fiercely competitive with the human brain. The million billion 'bits' that is thought to represent the average human memory will also be matched by electronic memory, the capacity of which has been doubling every eighteen months, and which will equal its human counterpart in two decades.[11]

The information-model of organic functioning thus makes plausible the replacement of the genetic and neurological networks of the human organism with more sophisticated and powerful electronic systems. According to Kurzweil, the resulting cyborg will even be capable of enhanced emotional or spiritual experiences:

> When we can determine the neurological correlates of the variety of spiritual experiences that our species is capable of, we are likely to be able to enhance these experiences in the same way that we will enhance other human experiences. With the next stage of evolution creating a new generation of humans that will be trillions of times more capable and complex than humans today, our ability for spiritual experience and insight is also likely to gain in power and depth.[12]

Lest these ideas appear to belong entirely in the realm of fantasy, we should note that experiments in this direction are already taking place. In August 1998, UK robotics expert Kevin Warwick had a silicon chip implanted in his forearm, enabling a computer to monitor his movements through his university, opening doors and turning on lights, computers and heaters when he approached. In a subsequent experiment which began in March 2002, a new chip was implanted and connected to the nerve fibres in Warwick's arm in order to send signals back and forth between his nervous system and a computer. Having recorded – and translated into digital format – the signals transmitted by the microchip during specific movements, sensations, emotions and moods, Warwick and his collaborators will attempt to induce or modify the original motions and feelings by sending signals back from the computer to the implant. If this is a success, and ethical obstacles are overcome, Warwick's wife will also undergo the operation, and they will attempt to link up their nervous systems through the medium of the computer.[13]

The Upright Posture

I shall consider in a moment where this infatuation with the cyborg is taking us. First I want to expose a central defect in the cyborg manifesto: namely, the disembodiment of the human will which the task of refashioning human nature both presupposes and intends.

In his justly celebrated essay, 'The Upright Posture', the phenomenological psychologist Erwin Straus suggests that the physical state of uprightness as maintained by human beings is more than merely an allegory for the moral connotations of the term (to be honest and just, dignified and noble, accountable for one's actions and convictions). Though non-human primates have the capacity to stand upright, Straus observes, the upright *posture* – in which the skull rests centrally on the vertebral column, and the hips are thrust forward to permit upright locomotion – occurs in no species other than our own.

While the shape and functioning of the human body is made by and for the upright posture, Straus points out that this posture is, nonetheless, both a developmental and existential accomplishment. For the child, being upstanding is a physiological achievement crucial to the development of its autonomy and independence, and in fact there are recorded cases of severely bow-legged infants who eventually brought about the self-straightening of their flared leg bones by spontaneously persevering with an upright posture.[14] But equally for the adult, who is dogged by the need for rest and sleep, and burdened by the force of gravity, the unique capacity for the upright posture – our human nature – must be constantly renewed by pitting nature against itself:

> Upright posture characterises the human species. Nevertheless, each individual has to struggle in order to make it really his own. Man has to become what he is . . . In getting up, in reaching the upright posture, man must oppose the forces of gravity. It seems to be his nature to oppose nature in its impersonal, fundamental aspects with natural means. However, gravity is never fully overcome; upright posture always maintains its character of counteraction. It calls for our activity and attention.[15]

Standing upright requires wakefulness, and wakefulness in turn presupposes sleep. Sleep gives the first moments of awaking their ponderosity and heaviness against which, arising upright and mobile, we excite our appetite for levity and freedom of action. Yet in rising and counteracting gravity we remain determined by it – indeed, it requires of us constancy and perseverance in our conduct: 'Held back by gravity to a precise point, we can overcome distance only in an orderly sequence. . . . Gravity, which holds us in line, imposes on waking experience a methodical proceeding.'[16]

Our upward struggle against gravity is symbolised in the positive meanings associated with elevation and high-standing, and we find dignity in this posture because we know it carries a heavy burden. Because the upright posture requires constantly renewed resistance to the danger of falling – an occurrence which is always tragic or comical in effect – it also expresses strength of character and firm convictions. 'Our task is not finished with getting up and standing. We have to "withstand". He who is able to accomplish this is called constant, stable.'[17] Hence also the existence of transitive and intransitive meanings to the verb 'to stand': to get up, but also to endure.

Yet the upright posture is also, Straus points out, a source of profound ambivalence. It removes us from the ground, and makes us long for the permanent stability and comfort associated with yielding to gravity and to the voluptuousness of rest. By extending the horizon of sight, it also distances us from things, separates seeing from our other senses and, by loosening our touching and grasping link to the world, allows the lengthening of this distance through the mediation of tools. The upright posture also distances us from each other. Its verticality signifies formality, aloofness, austerity, and mercilessness. Only by deviating from the vertical – by inclining towards one another, by bending, bowing, kneeling, nodding, stooping and so on – do we overcome our separation and express our social relatedness.

On a pure physiological level, the evolutionary capacity for upright posture enables the frontal extremities to be freed from the function of supporting the body. The hands and arms can now create through their horizontal extension an 'action space', and a territorial and possessive domain that may be empty or occupied. This space functions both to remove oneself from the world and to link one to it through throwing or embracing. In pointing, similarly, 'man's reach exceeds his grasp'. In referring to objects that escape our individual possession, pointing is a social gesture, an act of communication and inclusion as well as withdrawal. Hence the upright posture not only distances us from objects, and from fellow men and women, but it also provides the expressive tools to communicate the interior sense of this breach. The culturally universal gesture of shrugging the shoulders and baring empty hands, for example, expresses the felt inability to advise, help or be touched by the plight of another.

By removing the head from its proximity to the ground, Straus continues, the upright posture dethrones smell as the primordial sense, while seeing and hearing – the pre-eminent senses of distance – assume dominion. In non-human animals, which always move in the direction of their digestive axis, the jaws lie directly in the 'visor line' of the eyes, the latter dutifully leading the former to the animal's prey. With upright posture and the liberation of the forelimbs, the mouth is no longer needed for fighting, catching and carrying. Instead it drops below the field of view, allowing the eyes to direct the movement of the person, to rest on distant objects, and to marvel at their figurative appearance rather than their utility – to apprehend them with imagination and wonder rather than hunger. 'Bite has become subordinate to sight', Straus observes. 'Eyes that lead jaws and fangs to the prey are always charmed and spellbound by nearness. To eyes looking straight forward – to the gaze of upright posture – things reveal themselves in their own nature. Sight penetrates depth; sight becomes insight.'[18] The mouth also becomes a flexible instrument of language, the use of which again introduces a reflective distance between the speaker, speech, the recipient and the world to which language refers.

A Rational Nature

What is the significance of Straus's phenomenological account of the upright posture? First of all, Straus shows that our uniquely human way of being-in-the-world – our rationality and reflectiveness, our humility and sociality, our appreciation of form and of creature comforts, our sense of moral rectitude – cannot be explained as the expression of some large and sophisticated neurological apparatus (whose cognitive powers may be upgraded or replaced), nor of a disembodied spirit or consciousness in fundamental opposition to its animal flesh. We are, rather, *rational thinking bodies* – our humanity is always present in and conveyed by the totality of our physical and animal being:

> The traditional definition of man as a rational animal has frequently been interpreted to mean that one has to conquer the animal in order to be rational – that rationality has to be severed from animal existence. Considering awakeness, one may find that rationality originates in and issues from the animal nature, i.e., from corporeality and motility. While man transcends the boundaries of his here and now, he remains bound to the original situation.[19]

'The "rational" is', as Straus puts it, 'as genuine a part of human nature as the "animal".'[20] We acknowledge this fact, Leon Kass suggests, when we honour – as do all cultures – the lifeless bodies of the dead.[21]

Yet there is another side to our human nature, for our embodiment also makes us vulnerable and suffering creatures. Our rational humanity is a natural possibility but a precarious accomplishment, for it is continually menaced and interrupted by our perishable and needy bodies, by our finitude and mortality, by our material exigencies, appetites and inconveniences. To be a natural body is to be an object of nature. And here indeed lies the ambiguity of the human condition: to exercise our rational nature, we must defy nature in a natural way. For just as we owe to gravity the dignity of our upright posture, so we gather against the opposition of nature a ballast of interiority, a refusal of the object – and of being an object in its thrall – which enables us to reapprehend the world in its aesthetic and moral dimensions, and to perpetuate and embolden this awareness through cultural representations. Hence 'man's natural opposition to nature enables him to produce society, history, and conventions'.[22]

The adornment of the body is one such convention. The fig leaf, as Kant rethinks the biblical creation myth, is the means by which the transient natural impulses of sexual attraction 'can be prolonged and even increased by means of the imagination – a power which carries on its business, to be sure, the more moderately, but at once also the more constantly and uniformly, the more its object is removed from the senses'.[23] The concealment of our nakedness is not merely symbolic of our refusal of nature, for it also moderates, rationalises and prolongs the attraction of that nature, removing from sight the object of desire, but simultaneously stimulating the erotic imagination, the arts of love and romance, the thought and anticipation of sexual fulfilment. By clothing our bodies we not only hide our vulnerability, our blemishes and imperfections, our imperious physiological needs and desires. We also cultivate a subjective inwardness – an *awareness* of our physicality, which is often all the more enhanced by garments that must be *mastered* by their wearer, and whose mastery expresses, as itself a magnetic object of attraction, the human qualities of elegance, composure, balance and refinement.

Contrast this, then, with the casual brutality of the cosmetic surgeon, who is paid not to conceal the body's imperfections or to encourage the client's expressive use of his or her body, but rather to make that body into the shameless object of a disembodied desire. The surgically 'enhanced' body loses its affinity with selfhood, ceases to be a medium for the subtle arts of eroticism and intimacy, and instead becomes a form of manipulable and hard-earned property, an object *designed for display*. In the case of the now-growing fashion in the US for female cosmetic genital surgery ('labiaplasty'), we have an extraordinary example of the repressive 'deployment of sexuality' described by Foucault in *The History of Sexuality*.[24] Here the most intimate and unexamined of physical idiosyncrasies, the smallest and most concealed anatomical imperfections, no longer seem worthy of an assertion of privacy, an erotics of modesty, or a defence of undisclosed interiority, but are instead willingly exposed, under the pretext of sexual liberation, to the normalising discipline of the social and technocratic gaze.

Today this social immobilisation of the body, its disconnection from the expressive spirit of the person, is equally apparent in the prolific use of the botulinum toxin to lessen, by paralysing non-essential facial muscles, the decipherable wrinkles of ageing skin, as well as in new keyhole surgery aimed at preventing excessive blushing by severing the relevant nerve chains.[25] The ultimate function of such procedures, by erasing all signs of personal history and character, and by denying the natural form of the body any organic relationship to personhood, seems to be to remove all meaningful criteria for *being oneself*. The result is that satisfaction with one's commodity-enhanced body becomes an ever more elusive possibility, while the body itself ceases to express who one is, but instead conveys how much of who one isn't one can afford.

The transformation of our bodies into inert and imperishable objects of human design is, of course, the mutual concern both of cyborg theorists and of biotech companies like Geron Corporation and Human Genome Sciences – firms that are eagerly searching for genetic means to extend the human life-span.[26] If these efforts are successful, one likely political consequence will be chronic generational conflict, as older people refuse to make way for new social entrants by moving *down* the hierarchy they so painstakingly ascended.[27] Conquering death also has dramatic existential ramifications, for it is the very perishability of human life, and the futility of devoting ourselves exclusively to our own survival, which moves us to human acts of creation and moral judgement. Without the certainty of death, we would have no reason to live life passionately and to the full, for everything would have the status of a rehearsal, would be repeatable or revisable a second time around. Without death, there would be no anguish, no cherishing of the good, no commitment worthy of the name, no meaningful sacrifice, no nobility in the face of non-being. Since an immortal population would have no call to procreate, the abolition of death would also mean the abolition of youth, and with it the wonder and curiosity that rarely survives the passage to adulthood.

Kevin Warwick envisages a human being liberated from the inconvenience of sleep, and whose thoughts can be communicated without the need to struggle with miscommunicating language or speech. Virtual reality enthusiasts like Ray Kurzweil anticipate the artificial stimulation of 'neurological correlates of spiritual experiences', and the generation of pleasurable sensations and feelings independent of the objects of pleasure and the effort to attain them. What applies to the impermanence of human life is equally applicable here: we cannot be upstanding without the interruption of sleep, we cannot be creators of meaning without suffering the inadequacy of language, and we cannot find a home in the world – indeed, there can be no 'world' – unless things first exceed our grasp. Death, sleep, language, things, by removing us from ourselves, allow us to recover our given nature in a human way.

A Fatal Contradiction

The means by which human beings reason and make sense of their world, while often opposed to the forces of nature, are thus fundamentally corporeal in their significance. There is no act of human perception, thought, imagination, calculation or judgement which is not the act of a living, suffering, decaying body. The argument that humans have acquired the knowledge and competence to dictate the path of their own evolution thus founders on the fact that this capacity is itself only meaningful to a corporeal, suffering being. In the post-humanist account offered by Max More,[28] humans are biochemical machines of such sophistication and complexity that they have developed the emergent, non-mechanical properties of consciousness and reason. Yet for More the emancipatory possibilities of consciousness remain shackled by our biological–mechanical heritage, which makes us prisoners of our genes, our mortality, our uncontrollable emotions, hormones, and neural events. 'The whole appeal of seeing that we are a complex functionally interrelated collection of mechanical parts is that it opens up an appealing prospect: that technology will allow us to modify our nature, to alter ourselves, to augment and shape ourselves according to our values.'[29]

Yet on the basis of what values, what self-evidences, what certainties, is this project of 'transbiomorphosis' to be grounded? Does not the post-humanist programme simply reintroduce the dualism of mind and body which More otherwise takes to be 'scientifically and philosophically indefensible', in which the capacity to reason and choose is uprooted from the world of sentient experience? It is unsurprising that More draws confidence for his vision from Nietzche's *Übermensch*, proposing that humans become 'artists of the self',[30] and thus making the sheer will to self-transformation, the restless defiance of everything that endures, the constant and uncompromising renunciation of all mores, habits, limitations and conventions, the quintessential expression of freedom.[31]

The contradiction in this post-humanist perspective is noted by Hans Jonas, who points out that the capacity to even contemplate intervening in our evolution is a product of evolution – and a perishable product at that. On what premises, therefore, do today's genetic and cyborg revolutionaries claim the authority to make the human species an object of instrumental manipulation? Unlike Marx's proletarians, they certainly have something to lose other than their chains, for they assume a degree of power and depth of knowledge that is historically unprecedented. Yet if their judgement is that human nature is flawed and inadequate, then so must be their judgement of that inadequacy, as well as the remedy they propose for its failings, for these too are the contingent products of evolution. If, on the other hand, they proclaim the superiority of their wisdom, then in asserting their qualification to change humanity they reaffirm its premise – the adequacy of their inherited natural constitution, which should not therefore be endangered.

> Some kind of authority must be asserted for the determination of models, and unless we subscribe to dualism and say that the cognitive subject is from above the world, this authority can only base itself on an essential sufficiency of our nature such as it has evolved within this world . . . [T]his innate sufficiency of human nature . . . we must posit as the enabling premise for any creative steering of

> destiny . . . Most evidently, the authority which it imparts can never include the disfiguring, endangering, or refashioning of itself. No gain is worth this price, no hope of gain justifies this risk. And yet, today, this very share of transcendence too is in danger of being thrown into the crucible of bio-technological alchemy – as if the enabling condition for all our freedom to revise the given were itself among the revisables.[32]

Humankind is not, Jonas points out, the object of a programmed schedule of completion running, as in the growth and maturation of the individual, from the unfinished to the finished, from the provisional to the definitive, from the not yet to the true and full being. The human person is an incomplete finality, an imperfect perfection. But humanity, in its incompleteness and imperfection, is always already there, fully and without qualification, and it must therefore be preserved as such. As Paul Virilio cites St. Hildergarde of Bingen: *Homo Est Clausura Mirabilium Dei* – 'Man is the closing point of the marvels of the universe'.[33]

> The basic error of the ontology of 'not yet' and its eschatological hope is repudiated by the plain truth – ground for neither jubilation nor dejection – that genuine man is always already there and was there throughout known history: in his heights and his depths, his greatness and wretchedness, his bliss and torment, his justice and his guilt – in short, in all the *ambiguity* that is inseparable from his humanity. Wishing to abolish this constitutive ambiguity is wishing to abolish man in his unfathomable freedom . . . The really unambiguous man of utopia can only be the flattened, behaviourally conditioned homunculus of futuristic psychological engineering.[34]

The constitutive ambiguity of the human essence is precisely why the much-lauded 'dignity of man' is always only a potential – a natural capacity for uprightness which must be constantly exercised and reaffirmed – and why the dignified being of Man can only be declared with unpardonable vanity. But the fulfilment of this possibility is every person's responsibility, and one which must, Jonas asserts, never be put at risk: 'the possibility of there being responsibility in the world, which is bound to the existence of men, is of all objects of responsibility the first'.[35]

Needs and Imperatives

So the improvement of the human species is a possibility which no human being is qualified to fulfil, and its pursuit risks the very survival of that possibility itself. But what if the re-engineering of humanity were a *necessity*, rather than a mere possibility, as some of today's cyber-enthusiasts maintain?

In the account offered by Kevin Warwick, humans will in the future be living in a world dominated by intelligent and self-replicating machines, a world 'in which humans, if they still exist, will be subservient'.[36] Theodore Kaczynski, the American 'Unabomber', similarly predicted in his manifesto, 'Industrial Society and its Future', that our tendency to delegate tasks to ever-more sophisticated machines may eventually reach a point 'at which the decisions necessary to keep the system running will be so complex that human beings will be incapable of making them intelligently. At that stage the machines will be in effective control.'[37] In his heretical essay, 'Why the Future Doesn't Need Us', the respected computer scientist and co-founder of Sun Microsystems, Bill Joy, became the first prominent figure in the world of information technology to share these concerns, acknowledging 'that I may be working to create tools which will enable the construction of the technology that may replace our species'.[38]

In the view of Hans Moravec, the mismatch between humans' natural and culturally acquired competencies and their technological environments is already apparent. 'The world we inhabit is radically different, culturally and physically, from the one to which we adapted biologically.'

> Today, as our machines approach human competence across the board, our stone-age biology and our information-age lives grow ever more mismatched . . . As societal roles become yet more complex, specialised, and far removed from our inborn predispositions, they require increasing years of rehearsal to master, while providing fewer visceral rewards. The essential functioning of a technical society eludes the understanding of an increasing fraction of the population . . . The mismatch between instinct and necessity induces alienation in the midst of unprecedented physical plenty.[39]

For cyborg theorists such as Warwick, Kurzweil and Moravec, the solution to humans' dwindling control over their environments and the prospect of eventual extinction is to use the technology we have spawned to enhance our powers of thought, reasoning, memory, longevity and action. As Alvin and Heidi Toffler counsel against the Kaczynski-inspired pessimism of Joy:

'The very technologies they regard as most dangerous – robotics, genetics and nanotech – may very well help us expand the human brain's capabilities and make it possible for us to use those technologies in completely new ways.'[40]

What is desired by the cyborg enthusiasts is, in effect, the elimination or revision of those features of human existence which, being censored, repressed, harmed and alienated by modern social conditions, lead to disempowerment, frustration and suffering. The elimination of our capacity to suffer is not, however, a satisfactory answer to suffering – or at least not a *human* one. The human answer to suffering is *relief from suffering*, which must also suffer the precariousness of this relief and the knowledge of the burden that has been lightened. Because the elimination of human's capacity to suffer would mean the creation of a post-human being, the goal and beneficiary of this solution *cannot be humanity itself*. The need for the cyborg, in other words, is not a *human* need, a need whose satisfaction would reaffirm the essence of humanity. It is, rather, a *technological imperative*, for the true purpose of the re-engineering of the human being is the abolition of the obstacles presented by people to the reproduction of machines. As George Dyson candidly describes it, humans have today become 'bottlenecks' in the circulation and processing of knowledge and information:

> Most of the time, despite the perception of being inundated with information, we will remain out of the loop. Human beings have only limited time and ability to communicate: you can watch television, check your E-mail, and talk on your cellular phone at the same time, but that's the limit. We are now the bottleneck – able to absorb a limited amount of information while producing even less, from the point of view of machines.[41]

From the point of view of machines, the pedestrian individual is a hindrance, an obstructive inconvenience which must be removed, as Virilio describes it, by the total mobilisation and motorisation of the person – by the transformation of the person into a prosthesis of the machine.[42] 'In the second half of the twenty-first century', Kurzweil confidently predicts, 'you'll be able to read a book in a few seconds'.[43] Will this be an ability or an imperative, one wonders; will we be able to choose a more leisurely read (say a few minutes per book)? Kurzweil's promise is in any case meaningless. His prescription for a post-human intelligence has no other aim than that of reconciling individuals to 'the point of view of the machine': reducing the human subject to quantifiable units of memory and processing power which can be speeded up, magnified, rendered more accurate, efficient, predictable and powerful, but which are incapable of furnishing a sense of purpose or meaning to their existence.

Subjectivity and Forgetfulness

The necessity of the cyborg is nothing other than the technological imperative, sometimes resisted but increasingly treated as the acceptable price of progress, that humans be modified to function in a world, imagined, sought-after or anticipated, which is devoid of human references – a world we can no longer call 'home'. We can see this ambition, most pertinently, in the text where the term was originally coined. 'Cyborgs and Space', written by two senior American research scientists and published in *Astronautics* in 1960, was a response to an invitation from NASA to consider ways of modifying human nature 'to permit man's existence in environments which differ radically from those provided by nature as we know it'. In their account of 'self-regulating man-machine systems', Manfred Clynes and Nathan Kline rejected the path of genetic engineering in favour of 'biochemical, physiological, and electronic modifications of man's existing modus vivendi'. They proposed instead psychotropic medicine to keep astronauts awake and alert for 'weeks or even a few months', prophylactic drugs which would be automatically injected in response to detected radiation, induced hypothermia to reduce energy needs during long journeys, the modification of cardiovascular functioning (either by drugs or electrical stimulation) to suit different environments, and the 'sterilisation of the gastrointestinal tract, plus intravenous or direct intragastric feeding', to eliminate faecal waste.

'An inverse fuel cell,' they add, 'capable of reducing CO^2 to its components with removal of the carbon and recirculation of the oxygen, would eliminate the necessity for lung breathing.' 'Solving the many technological problems involved in manned space flight by adapting man to his environment, rather than vice versa,' they concluded, 'will not only mark a significant step forward in man's scientific progress, but may well provide a new and larger dimension for man's spirit as well.'[44]

Can the spirit of humanity be enlarged by its liberation from the natural needs, rhythms and magnitudes of the human body, from its intrinsic vitality, its capacity for self-animation, self-healing and survival? More importantly, what credence can we give to a scientific vision of humankind *which does not understand itself*, which denies its own rootedness in the sensible and ambiguous world, which glorifies the negation of its own conditions of possibility? To reconcile the human being with the imperatives of an inhuman environment is to make that being a functionary of its environment, and of our refusal to change that environment or resist its allure in favour of a familiar, more human world.

This refusal of scientists to return human beings, humble but uplifted, to the experienced world in which they truly dwell, is what aligns modern science with the adversaries of the human condition. The origins of this betrayal lie, as Edmund Husserl wrote in the 1930s, in a scientific project which, by failing to account for its own existence, its groundedness in the vague and fluid typifications of sensory experience, mistakes the results of its formalised method – the idealised reduction of the world to exact, measurable and universally translatable essences – for the world itself:

> In this way, the world of our experience is from the beginning interpreted by recourse to an 'idealisation' – but it is no longer seen that this idealisation, which leads to the exact space of geometry, to the exact time of physics, to exact causal laws, and which makes us see the world of our experience as being thus determined in itself, is itself the result of a function of cognitive methods, a result based on the data of our immediate experience. This experience in its immediacy knows neither exact space nor objective time and causality.[45]

This original experience, though self-evident and indubitable in its substance, is permeated by mystery and uncertainty, by the inexactitude of its object and by the impossibility of illuminating that inexactitude, of recovering lived experience, in a precise and objective way. It is from this metaphysical breach at the heart of human existence that positivist science offers an escape, promising a means of understanding the world that no longer has to understand itself – an experience, exemplified in the being of the cyborg, devoid of the unspeakable doubt, the wonder, the incommunicable convictions, that are definitive of the human subject.

> Mathematical science of nature is a technical marvel for the purpose of accomplishing inductions whose fruitfulness, probability, exactitude, and calculability could previously not even be suspected. As an accomplishment it is a triumph of the human spirit. With regard to the rationality of its methods and theories, however, it is a thoroughly relative science. It presupposes as data principles that are themselves thoroughly lacking in actual rationality. In so far as the intuitive environing world, purely subjective as it is, is forgotten in the scientific thematic, the working subject is also forgotten.[46]

The vision of the cyborg emanates from the belief, apparent as much in the physico-computational accounts of human consciousness as in the reduction of life to a universally translatable language of genes, 'that the infinite totality of what is in general is intrinsically a rational all-encompassing unity that can be mastered, without anything left over, by a corresponding universal science'.[47] Today this orthodoxy not only permits those who aspire to re-engineer the human being to forget the subject who makes that aspiration possible. It also provides a programme, a manifesto, to make this state of amnesia constitutive of the post-human condition. We should therefore heed Merleau-Ponty's warning, made before the biotech revolution had cast its fatal spell:

> Thinking 'operationally' has become a sort of absolute artificialism, such as we see in the ideology of cybernetics, where human creations are derived from a natural information process, itself conceived on the model of human machines. If this kind of thinking were to extend its reign to man and history; if, pretending to ignore what we know of them through our own situations, it were to set out to construct man and history on the basis of a few abstract indices . . . then, since man really becomes the *manipulandum* he takes himself to be, we enter into a cultural regimen where there is neither truth nor falsity concerning man and history, into a sleep, or a nightmare, from which there is no awakening.[48]

Notes

Preface

1 Hans Jonas, 'Biological Engineering – A Preview', in *Philosophical Essays: From Ancient Creed to Technological Man*, Englewood Cliffs, NJ: Prentice-Hall, 1974, p. 144.
2 Hans Jonas, *The Imperative of Responsibility: In Search of an Ethics for the Technological Age*, Chicago: University of Chicago Press, 1984, p. 7.
3 Lee M. Silver, *Remaking Eden: Cloning, Genetic Engineering and the Future of Humankind?*, London: Phoenix, 1999. The only aspect of the future of biotechnology which Silver appears to disapprove of – but which he believes will be impossible to prevent – is unequal economic access to the technologies of human genetic enhancement. He predicts this will lead to increasing genetic inequality and eventually, rather like the genetic caste system envisaged by Huxley in *Brave New World*, to the division of humanity into separate species. Were this to happen, incidentally, the only practical response of the Left, consistent with its egalitarian philosophy, would be to use state power to eugenically equalise the genetic resources of the citizenry.
4 Silver, *Remaking Eden*, p. 264.
5 Hans Jonas, 'Toward an Ontological Grounding for an Ethics for the Future', in *Mortality and Morality: A Search for the Good after Auschwitz*, Evanston, Ill.: Northwestern University Press, 1996, p. 105 (his emphasis).

Introduction

1 Gregory Stock, *Redesigning Humans: Our Inevitable Genetic Future*, New York: Houghton Mifflin, 2002, p. 151.
2 Francis Fukuyama, *Our Posthuman Future: Consequences of the Biotechnology Revolution*, New York: Farrar, Straus and Giroux, 2002, chapter 12.
3 Liza Mundy, 'A World of Their Own', *Washington Post*, 31 March 2002.
4 Stock, *Redesigning Humans*, p. 93 (my emphasis).

1 The Revolution in Molecular Biology

1 Nicholas Wade, *Life Script: The Genome and the New Medicine*, London: Simon and Schuster, 2001, p. 71. Wade gives the example of the dystrophin gene, which is activated in the brain as well as the muscles, but which produces a much smaller protein in the brain cells.

2 It is also becoming apparent that non-coding RNA – RNA which is transcribed from seemingly redundant DNA, which is not translated into protein, and which some scientists now estimate accounts for up to *ninety-eight per cent* of the transcriptional output of the human genome – fulfils a variety of indispensable regulatory functions by interacting with different RNA transcripts, with DNA and with proteins, 'to form networks that can regulate gene activity with almost infinite potential complexity'. (Carina Dennis, 'The brave new world of RNA', *Nature*, vol. 418, no. 6894, 11 July 2002, pp. 122–4.)

3 Daniel C. Williams, Richard M. Van Frank, William L. Muth, J. Paul Burnett, 'Cytoplasmic Inclusion Bodies in *Escherichia coli* Producing Biosynthetic Human Insulin Proteins', *Science*, vol. 215, no. 4533, 5 February 1982, pp. 687–8.

4 Paul Blum, Mark Velligan, Norman Lin and Abdul Matin, 'DnaK-Mediated Alterations in Human Growth Hormone Protein Inclusion Bodies', *Bio/Technology*, vol. 10, no. 3, March 1992, pp. 301–4.

5 Gary Taubes, 'Misfolding the Way to Diseases', *Science*, vol. 271, no. 5255, 15 March 1996, pp. 1493–5; R. John Ellis and Teresa J.T. Pinheiro, 'Danger – misfolding proteins', *Nature*, vol. 416, no. 6880, 4 April 2002, pp. 483–4.

6 See Calestous Juma, *The Gene Hunters: Biotechnology and the Scramble for Seeds*, London: Zed Books, 1989, pp. 117–24.

7 Cited in Liz Fletcher, 'GM crops are no panacea for poverty', *Nature Biotechnology*, vol. 19, no. 9, September 2001, pp. 797–8.

8 John Vidal, 'New Delhi opens door to GM crops', *Guardian*, 28 March 2002.

9 According to GeneWatch UK, three-quarters of the area used for genetically engineered crops in 2000 was devoted to plants modified to tolerate herbicides, with the remaining quarter given over to insect-resistant plants, or plants carrying both genetic traits. (Anon, 'Genetic Engineering: A Review of Developments in 2000', *GeneWatch UK Briefing*, no. 13, January 2001. <www.genewatch.org>)

10 The Flavr Savr tomato was engineered in such a way that the fruit produced only limited amounts of polygalacturonase (PG), an enzyme that degrades pectin and with it the walls of the fruit. This was achieved by producing a complementary copy of the gene that codes for PG, then inserting it into the tomato's genome. When both the original and the complementary gene sequence produced messenger RNA – the molecule which carries protein-synthesising instructions from the chromosomes to the ribosomes in the cell's cytoplasm – these sequences of mRNA, since they were also complementary to each other, bound together like the paired DNA strands of the original double helix. Transfer RNAs – the molecules responsible for assembling the chains of amino acids – were consequently prevented from reading and attaching to those sequences specifying the appropriate amino acids which, when joined together, would make up the PG enzyme. This method of silencing genes is now widely used.

11 Marc Lappé and Britt Bailey, *Against the Grain: The Genetic Transformation of Global Agriculture*, London: Earthscan, 1999, p. 117. The tomato was eventually withdrawn from the market.

12 Mark Harvey, 'Genetic Modification as a Bio-Socio-Economic Process: One Case of Tomato Purée', *CRIC Discussion Paper*, no. 31, November 1999; Mark Harvey, 'Cultivation and Comprehension: How Genetic Modification Irreversibly Alters the Human Engagement with Nature', *Sociological Research Online*, vol. 4, no. 3, 1999.

13 Michael Wilson, 'Genetically Modified Tomatoes: Seeking Firmer Tomatoes with Better Flavour', in Donald Bruce and Ann Bruce, eds, *Engineering Genesis: The Ethics of Genetic Engineering in Non-Human Species*, London: Earthscan, 1998, p. 51.

14 James Meek, 'Scientists create killer moth to control pests', *Guardian*, 5 March 2001.
15 James Meek, 'Scientists plan to wipe out malaria with GM mosquitoes', *Guardian*, 3 September 2001.
16 Anthoula Lazaris, Steven Arcidiacono, Yue Huang *et al*, 'Spider Silk Fibres Spun from Soluble Recombinant Silk Produced in Mammalian Cells', *Science*, vol. 295, no. 5554, 18 January 2002, pp. 472–6.
17 Tim Radford, 'Rubber trees may yield blood protein', *Guardian*, 10 February 2001.
18 G. Wright, A. Carver, D. Cottom *et al*, 'High level expression of active human alpha-1-antitrypsin in the milk of transgenic sheep', *Bio/Technology*, vol. 9, no. 9, September 1991, pp. 830–4.
19 Andrew Clark, 'PPL moves from Dolly to cows', *Guardian*, 15 December 2000. In March 2002, PPL's shares fell to an all-time low when it was announced that phase II trials of the AAT drug in the US had been abandoned after emphysema sufferers displayed exacerbated symptoms of wheezing, leading some to suggest that the transgenic 'biofacturing' process had compromised the purity of the drug, and forcing the company to postpone the product's expected release date until 2007. The City is now pressurising the company to sell its xenotransplantation and stem cell research operations and concentrate on biofactured pharmaceuticals. (Andrew Clark, 'Fresh blow for Dolly firm', *Guardian*, 19 March 2002; Andrew Clark, 'PPL to sell stem cell stake', *Guardian*, 26 March 2002.) This whole area has also raised a number of important ethical issues, including the question of ownership and consent relating to the fact that the human DNA used in many of PPL's experiments was taken from the blood sample of a Danish woman who was never consulted or informed about its use. (Antony Barnett, 'Dolly firm put woman's gene into sheep', *Observer*, 2 July 2000.)
20 See I. Wilmut, A.E. Schnieke, J. McWhir *et al*, 'Viable offspring derived from fetal and adult mammalian cells', *Nature*, vol. 385, no. 6619, 27 February 1997.
21 K.H.S. Campbell, J. McWhir, W.A. Ritchie, I. Wilmut, 'Sheep cloned by nuclear transfer from a cultured cell line', *Nature*, vol. 380, no. 6569, 7 March 1996.
22 The fact that Megan and Morag were cloned from an embryo whose cells had already begun to differentiate proved to Campbell and Wilmut that the cloning of fully differentiated somatic cells was a practical goal. The first successful cloning of domestic animals had actually been accomplished a decade earlier by Steen Willadsen at the British Agricultural Research Council's Unit on Reproductive Physiology and Biochemistry in Cambridge, and by the American Neal First and colleagues at the University of Wisconsin. First's research into cattle cloning was funded by the chemical giant W.R. Grace, and led in 1991 to its subsidary, American Breeders Service (ABS), winning the first patent ever granted for embryo cloning. These early successes depended, however, on the use of donor nuclei from early embryos whose cells had not yet differentiated. (See S.M. Willadsen, 'Nuclear transplantation in sheep embryos', *Nature*, vol. 320, no. 6057, 6 March 1986, pp. 63–5; Randall S. Prather, Frank L. Barnes, Michelle M. Sims *et al*, 'Nuclear Transplantation in the Bovine Embryo: Assessment of Donor Nuclei and Recipient Oocyte', *Biology of Reproduction*, vol. 37, no. 4, November 1987, pp. 859–66.) The cloning of somatic cells by Wilmut and his collaborators proved, on the other hand, that the organic process of developmental growth through cell differentiation, whereby identical complements of DNA are selectively activated and utilised according to the age, location and needs of the specific cell, can in fact be reversed, in effect resetting the mature cell to its primordial 'totipotent' state. Hence the meaning of the comment made by Ron James, director of PPL Therapeutics, to Gina Kolata, that the creation of Dolly represented 'a step towards immortality'. (Gina Kolata, *Clone: The Road to Dolly and the Path Ahead*, Harmondsworth: Penguin, 1997, p. 188.)
23 K.J. McCreath, J. Howcroft, K.H.S. Campbell *et al*, 'Production of gene-targeted sheep by nuclear transfer from somatic cells', *Nature*, vol. 405, no. 6790, 29 July 2000.
24 Anne Simon Moffat, 'Exploring Transgenic Plants as a New Vaccine Source', *Science*, vol. 268, no. 5211, 5 May 1995, pp. 658–60; Paul Brown, 'Vaccine in GM fruit could wipe out

hepatitis B', *Guardian*, 8 September 2000; Tim Radford, 'GM apple a day may protect teeth', *Guardian*, 9 September 2000.

25 This process of making a copy of complementary DNA from an RNA sequence – called 'reverse transcription', and first detected in 1970 in two cancer-inducing viruses – exposed another chink in the theory that Francis Crick brazenly characterised as the 'Central Dogma' of molecular biology. It demonstrated, in other words, that the strictly linear and unidirectional causal sequence of DNA-to-RNA-to-protein could be reversed, and that genomes could in fact be modified by other influences.

26 Michael Wilson, 'Vaccination Made Easy: Proteins from Plants, using Genetically Modified Plant Viruses', in Bruce and Bruce, eds, *Engineering Genesis*, p. 42–3.

27 Joyce Tait, 'The Sting in the Cabbage: Genetically Modified Insect Viruses as Pesticides', in Bruce and Bruce, eds, *Engineering Genesis*, pp. 46–9; David Bishop, 'Genetically Engineered Insecticides: The Development of Environmentally Acceptable Alternatives to Chemical Insecticides', in Peter Wheale and Ruth McNally, eds, *The Bio-Revolution: Cornucopia or Pandora's Box?*, London: Pluto, 1990, pp. 115–134.

28 Cited in Mae-Wan Ho, *Genetic Engineering: Dream or Nightmare?*, Dublin: Gateway, 1999, p. 40.

29 Ho, *Genetic Engineering*, pp. 3–4.

30 Richard Dawkins, 'Open Letter', *Observer*, 21 May 2000.

31 Cited by Michael Specter, 'How Monsanto bit off more than it could chew', *Sunday Telegraph Magazine*, 14 May 2000, p. 17.

32 Dogs are known to have bred successfully with both wolves and coyotes, and lions and tigers have been interbred to produce fertile offspring.

33 Manfred Schartl, 'Platyfish and swordtails: a genetic system for the analysis of molecular mechanisms in tumor formation', *Trends in Genetics*, vol. 8, no. 5, May 1995, pp. 185–9.

34 Martin Brookes, 'Live and Let Live', *New Scientist*, vol. 163, no. 2193, 3 July 1999.

35 Philip J. Regal, 'Scientific Principles for Ecologically Based Risk Assessment of Transgenic Organisms', *Molecular Ecology*, vol. 3, no. 1, 1994, pp. 5–13.

36 An example of genetic engineering disturbing an organism's metabolic control over normal gene expression can be seen in an experiment recording the effects of the transfer of a soybean gene to a tobacco plant. A gene coding for the expression of glutamine synthetase, which enables plants to assimilate ammonia, was transferred to the genome of tobacco plants using *Agrobacterium tumefaciens* as a vector and accompanied by a promoter from the cauliflower mosaic virus. Though the new gene successfully induced expression of glutamine synthetase in the tobacco plants' leaves, the artificial construct also activated in the leaf cells a native gene coding for an enzyme normally only expressed in the root cells of the tobacco plant. (Bertrand Hirel, Marie C. Marsolier, Anne Hoarau *et al*, 'Forcing expression of a soybean root glutamine synthetase gene in tobacco leaves induces a native gene encoding cytosolic enzyme', *Plant Molecular Biology*, vol. 20, no. 2, October 1992, pp. 207–18.) The evidence indicating that many transgenic crops – such as herbicide-resistant soybeans – may be prone to lower yields than conventional varieties also points to the inability of the transgenic plant to regulate the expression of the foreign gene which, because it is permanently switched on, consumes a disproportionate amount of the organism's molecular energy. For a comprehensive account of the potential health hazards associated with genome disturbance, see John B. Fagan, 'Assessing the Safety and Nutritional Quality of Genetically Engineered Foods', <www.psrast.org/jfassess.htm>. Fagan is an internationally renowned Professor of Molecular Biology and Biochemistry who has specialised in cancer research. PSRAST stands for Physicians and Scientists for Responsible Application of Science and Technology, a global non-governmental organisation founded in 1998 which supports the independent evaluation of modern technologies.

37 Francis Crick, *Of Molecules and Men*, Seattle: Washington University Press, 1966, pp. 10, 14.

38 Cited by Michael Specter, 'How Monsanto bit off more than it could chew', *Sunday Telegraph Magazine*, 14 May 2000, pp. 11–12.

39 Richard Dawkins, *The Blind Watchmaker*, London: Penguin, 1988, p. 112.

40 The depiction of genetically altered organisms as mechanical artefacts assembled from extraneous components not only supports the legal treatment of those organisms as patentable 'inventions'. This fiction is also used to justify their protection with patents which have a breadth of cover – we shall see some examples of these in chapter 4 – normally reserved for the products of mechanical engineering. In the absence of broad patent protection, mechanical inventions can invariably be dismantled and rebuilt by competitors to perform new, but technically predictable, functions. As legal scholar Yusing Ko points out, courts have therefore reasoned that the breadth of claims justified by a patent's disclosure 'varies inversely with the degree of unpredictability of the factors involved'. They have consequently awarded much narrower patents in the chemical field on the grounds that the effects of a change in a patented chemical's structure or environmental medium cannot be predicted by extrapolating from the information contained in the patent. This is why Ko concludes that the highly unpredictable nature of biotechnology interventions makes the awarding of broad patents in this field inconsistent with the logic of patent doctrine – and thus, one assumes, ideologically driven. (Yusing Ko, 'An Economic Analysis of Biotechnology Patent Protection', *Yale Law Review*, vol. 102, no. 3, December 1992, pp. 777–804.)

40 Richard Dawkins, *The Selfish Gene*, Oxford: Oxford University Press, 1988, p. 19.

41 Much of what follows is taken from Evelyn Fox Keller, *The Century of the Gene*, Cambridge, Mass.: Harvard University Press, 2000; Mae-Wan Ho, *Genetic Engineering* (especially Chapter 7); and Eva Jablonka and Marion J. Lamb, *Epigenetic Inheritance and Evolution: The Lamarckian Dimension*, Oxford: Oxford University Press, 1999 (especially Chapter 3). See also Stanley Shostak, *Death of Life: The Legacy of Molecular Biology*, London: Macmillan, 1998.

43 Steven Rose, *Lifelines: Biology, Freedom, Determinism*, London: Penguin, 1998, p. 221.

44 Barry Commoner, 'Deoxyribonucleic acid and the molecular basis of self-duplication', *Nature*, vol. 203, 1 August 1964, p. 490.

45 Barry Commoner, 'Failure of the Watson-Crick Theory as a Chemical Explanation of Inheritance', *Nature*, vol. 220, 26 October 1968, p. 334.

46 Keller, *The Century of the Gene*, pp. 27, 31 (my emphasis).

47 Keller, *The Century of the Gene*, pp. 70–1.

48 During the formation of frog eggs, for example, the chromosomal region containing the genes for the synthesis of ribosomal RNA is amplified 2,000-fold, the resulting excess copies then disappearing in the differentiating cells of the developing embryo. (Shostak, *Death of Life*, p. 170.)

49 Some DNA sequences, called transposons or 'jumping genes', contain the gene for an enzyme which can cut out the host sequence – sometimes along with neighbouring DNA – and then reinsert its cargo in different places in the genome. Transposons, which were first described by Barbara McClintock in the 1940s, but ignored by the male-dominated scientific establishment for a further 30 years, can also spread copies of themselves, usually via bacterial or viral infection, between species that do not interbreed. The rearrangement of DNA by transposon activity is known to be one of the principal ways that mammals increase the variety of antibodies they can produce in response to invading organisms (millions of different antibody molecules can in fact be expressed from a small number of genes). The rate of transposition in an organism has consequently been shown to be greatly increased by environmental stress, and it is thought that around 1 in every 700 human genetic mutations are a result of jumping genes, which make up about 10 per cent of the human genome. (Matt Ridley, *Genome: The Autobiography of a Species in 23 Chapters*, London: Fourth Estate, 1999, p. 129; Robert S. Schwatz, 'Jumping Genes', *New England Journal of Medicine*, vol. 332, no. 14, 6 April 1995, pp. 941–4; Barbara McClintock, 'The Significance of Responses of the Genome to Challenge', *Science*, vol. 226, no. 4676, 16 November 1984, pp. 792–801.)

50 For example, when an homologous version of a gene which, in a variety of organisms (including humans and mice) plays an indispensable role in the formation of muscle tissue,

was deleted from the genome of the *Caenorhabditis elegans* worm, normal muscle development persisted. (Lihsia Chen, Michael Krause, Bruce Draper *et al*, 'Body-Wall Muscle Formation in *Caenorhabditis elegans* Embryos that Lack the MyoD Homolog *hlh-1*', *Science*, vol. 256, no. 5054, 10 April 1992, pp. 240–3.)

51 This particular study is cited in Richard C. Strohman, 'Ancient Genomes, Wise Bodies, Unhealthy People: Limits of a Genetic Paradigm in Biology and Medicine', *Perspectives in Biology and Medicine*, vol. 37, no. 1, Autumn 1993, pp. 112–135.

52 Rose, *Lifelines*, p. 132.

53 Richard C. Strohman, 'Epigenesis and Complexity: The Coming Kuhnian Revolution in Biology', *Nature Biotechnology*, vol. 15, March 1997, p. 199.

54 Ralph J. Greenspan, 'The flexible genome'. *Nature Reviews Genetics*, vol. 2, no. 5, May 2001, pp. 383–7.

55 John Cairns, Julie Overbaugh, and Stephan Miller, 'The origin of mutants', *Nature*, vol. 335, no. 6186, 8 September 1988, pp. 142–5.

56 Miroslav Radman, 'Enzymes of evolutionary change', *Nature*, vol. 401. No. 6756, 28 October 1999, pp. 866–869.

57 Jablonka and Lamb, *Epigenetic Inheritance and Evolution*, p. 69.

58 James A. Shapiro, 'Genome organisation, natural genetic engineering and adaptive mutation', *Trends in Genetics*, vol. 13, no. 3, March 1997, pp. 98–104.

59 Andrew Gudkov and Boris Kopnin, 'Gene Amplification in Multidrug-Resistant Cells: Molecular and Karyotypic Events', *BioEssays*, vol. 3, no. 2, 1985, pp. 68–71.

60 Catherine A. Prody, Patrick Dreyfus, Ronit Zamir *et al*, '*De novo* amplification within a "silent" human cholinesterase gene in a family subjected to prolonged exposure to organophosphorous insecticides', *Proceedings of the National Academy of Sciences*, vol. 86, no. 2, January 1989, pp. 690–4.

61 John H. Campbell, 'The New Gene and Its Evolution', in K.S.W. Campbell and M.F. Day, eds, *Rates of Evolution*, London: Allen and Unwin, 1987, p. 286.

62 Pilar Cubas, Coral Vincent, Enrico Coen, 'An epigenetic mutation responsible for natural variation in floral symmetry', *Nature*, vol. 401, no. 6749, 9 September 1999, pp. 157–61.

63 Hugh D. Morgan, Heidi G.E. Sutherland, David I.K. Martin, Emma Whitelaw, 'Epigenetic inheritance at the agouti locus in the mouse', *Nature Genetics*, vol. 23, no. 3, November 1999, pp. 314–18.

64 Jablonka and Lamb, *Epigenetic Inheritance and Evolution*, pp. 134–7.

65 The following discussion derives from Jablonka and Lamb, *Epigenetic Inheritance and Evolution*, Chapter 5.

66 Jablonka and Lamb, *Epigenetic Inheritance and Evolution*, p. 188.

67 Cited in Robert Olby, *The Path to the Double Helix*, London: Macmillan, 1974, p. 432 (emphasis in original).

68 Keller, *The Century of the Gene*, pp. 100–1.

69 Barbara McClintock, 'The Significance of Responses of the Genome to Challenge', *Science*, vol. 226, no. 4676, 16 November 1984, pp. 798.

70 Barry Commoner, 'Failure of the Watson-Crick Theory as a Chemical Explanation of Inheritance', p. 340.

71 Barry Commoner, 'Deoxyribonucleic acid and the molecular basis of self-duplication', p. 491.

2 Agricultural Biotechnology and Ecological Risk

1 John B. Fagan, 'Assessing the Safety and Nutritional Quality of Genetically Engineered Foods', <www.psrast.org/jfassess.htm>

2 DNA rearrangements caused by the integration of foreign sequences have been identified, for example, in Monsanto's Roundup Ready soybean line 40–3–2. (Pieter Windels, Isabel Taverniers, Ann Depicker *et al*, 'Characterisation of Roundup Ready soybean insert', *European Food Research and Technology*, vol. 213, no. 2, 2001, pp. 107–12.)

3 Carolyn Napoli, Christine Lemieux, and Richard Jorgensen, 'Introduction of a Chimeric Chalcone Synthase Gene into Petunia Results in Reversible Co-Suppression of Homologous Genes *in trans*', *The Plant Cell*, vol. 2, no. 4, April 1990, pp. 279–89.
4 Alexander R. van der Krol, Leon A. Mur, Marcel Beld *et al*, 'Flavonoid Genes in Petunia: Addition of a Limited Number of Gene Copies May Lead to a Suppression of Gene Expression', *The Plant Cell*, vol. 2, no. 4, April 1990, pp. 291–99.
5 Peter Meyer, Felicitas Linn, Iris Heidmann *et al*, 'Endogenous and environmental factors influence 35S promoter methylation of a maize A1 gene construct in transgenic petunia and its colour phenotype', *Molecular and General Genetics*, vol. 231, no. 3, February 1992, pp. 345–52; Debora MacKenzie, 'Jumping genes confound German scientists', *New Scientist*, vol. 128, no. 1747, 15 December 1990, p. 18.
6 Ricarda A. Steinbrecher, 'Ecological Consequences of Genetic Engineering', in Brian Tokar, ed., *Redesigning Life: The Worldwide Challenge to Genetic Engineering*, London: Zed Books, 2001, p. 97.
7 Luke Anderson, *Genetic Engineering, Food, and our Environment*, Totnes: Green Books, 1999, p. 14.
8 Marc Lappé and Britt Bailey, *Against the Grain: The Genetic Transformation of Global Agriculture*, London: Earthscan, 1999, pp. 81–4.
9 Andy Coghlan, 'Splitting Headache', *New Scientist*, vol. 164, no. 2213, 20 November 1999, p. 25.
10 Lappé and Bailey, *Against the Grain*, pp. 103–6. See also Kurt Kleiner, 'Monsanto's Cotton gets the Mississippi Blues', *New Scientist*, vol. 156, no. 2106, 1 November 1997, p. 4.
11 Steinbrecher, 'Ecological Consequences of Genetic Engineering', p. 95.
12 World Wildlife Fund, *Do Genetically-Engineered (GE) Crops Reduce Pesticides? The Emerging Evidence Says Not Likely*, Toronto: WWF, 2000. <www.wwf.ca>
13 Angelika Hilbeck, Martin Baumgartner, Padruot M. Fried, and Franz Bigler, 'Effects of transgenic *Bacillus thuringiensis* corn-fed prey on mortality and development time of immature *Chrysoperla carnea* (Neuroptera: Chrysopidae)', *Environmental Entomology*, vol. 27, no. 2, April 1998, p. 485. According to the US Environmental Protection Agency, 20 million acres of Bt corn were planted in the US in 1999.
14 J. Koskella and G. Stotzky, 'Microbial Utilisation of Free and Clay-Bound Insecticidal Toxins from *Bacillus thuringiensis* and their Retention of Insecticidal Activity after Incubation with Microbes', *Applied and Environmental Microbiology*, September 1997, vol. 63, no. 9, pp. 3561–8; H. Tapp and G. Stotzky, 'Persistence of the insecticidal toxin from *Bacillus thuringiensis* subsp. *Kurstaki* in soil', *Soil Biology and Biochemistry*, vol. 30, no. 4, 1998, pp. 471–6. In one study it was found that the composted leaves of Bt cotton caused a significant transient *increase* in total bacterial and fungal populations in the soil. The suggestion that the transgenic plants decomposed faster than the parental plants was a phenomenon attributed by the researchers not to the transgenes themselves, but to unintentional changes to plant characteristics partly induced by the general modification of the genome. (K.K. Donegan, C.J. Palm, V.J. Fielan *et al*, 'Changes in levels, species and DNA fingerprints of soil microorganisms associated with cotton expressing the *Bacillus thuringiensis* var. *kurstaki* endotoxin', *Applied Soil Ecology*, vol. 2, no. 2, June 1995, pp. 111–24.) Another recent study has shown that the toxin may persist in the soil for 180 days after being released from the roots of Bt plants, but that it is not subsequently taken up by other crops later grown in the soil. (Deepak Saxena and G. Stotzky, '*Bt* toxin uptake from soil by plants', *Nature Biotechnology*, vol. 19, no. 3, March 2001, p. 199.)
15 Angelika Hilbeck, Martin Baumgartner, Padruot M. Fried, and Franz Bigler, 'Effects of transgenic *Bacillus thuringiensis* corn-fed prey on mortality and development time of immature *Chrysoperla carnea* (Neuroptera: Chrysopidae)', *Environmental Entomology*, vol. 27, no. 2, April 1998, pp. 480–7; Angelika Hilbeck, William J. Moar, Marianne Pusztai-Carey *et al*, 'Toxicity of *Bacillus thuringiensis* Cry1Ab toxin to the predator *Chrysoperla carnea* (Neuroptera: Chrysopidae)', *Environmental Entomology*, vol. 27, no. 5, October 1998, pp. 1255–63.

16 A. Nicholas, E. Birch, Irene E. Geoghegan, Michael E.N. Majerus *et al*, 'Tri-trophic interactions involving pest aphids, predatory 2–spot ladybirds and transgenic potatoes expressing snowdrop lectin for aphid resistance', *Molecular Breeding*, vol. 5, no. 1, 1999, pp. 75–83.

17 John E. Losey, Linda S. Rayor, Maureen E. Carter, 'Transgenic Pollen Harms Monarch Larvae', *Nature*, vol. 399, no. 6733, 20 May 1999, p. 214.

18 Laura C. Hansen Jesse, John J. Obrycki, 'Field deposition of Bt transgenic corn pollen: lethal effects on the monarch butterfly', *Oecologia*, vol. 125, no. 2, 2000, pp. 241–8.

19 There is already evidence that some insects (such as the diamondback moth) have established Bt resistance. See, for example, H. Hama, K. Suzuki, and H. Tanaka, 'Inheritance and stability of resistance to *Bacillus thuringiensis* formulations of the diamondback moth, *Plutella xylostella* (Linnaeus) (Lepidoptera: Yponomeutidae)', *Applied Entomology and Zoology*, vol. 27, no. 3, 1992, pp. 355–62; and Bruce E. Tabashnik, Yong-Biao Liu, Naomi Finson *et al*, 'One gene in diamondback moth confers resistance to four *Bacillus thuringiensis* toxins', *Proceedings of the National Academy of Sciences*, vol. 94, no. 5, March 1997, pp. 1640–4. Eight species of insect are known to have already developed resistance to *non*-transgenic *B. thuringiensis*, and indeed more than 500 insects and mites have acquired resistance to some form of pesticide (William H. McGaughey and Mark E. Whalon, 'Managing Insect Resistance to *Bacillus thuringiensis* Toxins', *Science*, vol. 258, no. 5087, 27 November 1992, p. 1451–1455). Since 1945 the total volume of pesticides used in the US has grown more than eight-fold, but the proportion of crops lost to insects remains thirteen per cent higher (Lappé and Bailey, *Against the Grain*, p. 102).

20 Jocelyn Kaiser, 'Pests Overwhelm *Bt* Cotton Crop', *Science*, vol. 273, no. 5274, 26 July 1996, p. 423.

21 Anderson, *Genetic Engineering, Food, and Our Environment*, p. 28.

22 If it is to be successful, this strategy really requires that the gene that codes for resistance is fully recessive (in other words, that it has to be inherited from both parents before its effects are expressed in the organism). Research indicates, however, that resistance to the Bt toxin can be encoded, as it appears to be with the European corn borer, in an incompletely dominant autosomal gene, and can thus be passed on to progeny regardless of the other parent's DNA. See F. Huang, L.L. Buschman, R.A. Higgins, W.H. McGaughey, 'Inheritance of resistance to *Bacillus thuringiensis* toxin (Dipel ES) in the European corn borer', *Science* vol. 284, no. 5416, 7 May 1999, p. 965–7.

23 Steinbrecher, 'Ecological Consequences of Genetic Engineering', p. 87.

24 By 1998 in Australia, for example, there were at least two reported cases of resistance to glyphosate found in the rigid ryegrass weed. See Stanley Robert and Ute Baumann, 'Resistance to the Herbicide Glyphosate', *Nature*, vol. 395, no. 6697, 3 September 1998, pp. 25–6.

25 See Rikke B. Jørgensen and Bente Andersen, 'Spontaneous Hybridisation Between Oilseed Rape (*Brassica napus*) and Weedy *B. capestris* (Brassicaceae): A Risk of Growing Genetically Modified Oilseed Rape', *American Journal of Botany*, vol. 81, no. 12, December 1994, pp. 1620–26; Thomas R. Mikkelsen, Bent Andersen and Rikke Bagger Jørgensen, 'The Risk of Crop Transgene Spread', *Nature*, vol. 380, no. 6569, 7 March 1996, p. 31.

26 Andy Coghlan, 'Gone with the Wind', *New Scientist*, vol. 162, no. 2182, 17 April 1999, p. 25. It should be pointed out that for the conventional crop the scientists used sterile male rape plants. Because they do not produce pollen themselves, these plants are much more susceptible to fertilisation by airborne pollen than non-sterile ones. Hence the researchers described their findings as a 'worst case scenario'.

27 Anne-Marie Chèvre, Frédérique Eber, Alain Baranger, Michel Renard, 'Gene Flow from Transgenic Crops', *Nature*, vol. 389, no. 6654, 30 October 1997, p. 924.

28 Linda Hall, Keith Topinka, John Huffman *et al*, 'Pollen flow between herbicide-resistant *Brassica napus* is the cause of multiple-resistant *B. napus* volunteers', *Weed Science*, vol. 48, no. 6, 2000, pp. 688–94.

29 Joy Bergelson, Colin B. Purrington and Gale Wichmann, 'Promiscuity in Transgenic Plants', *Nature*, vol. 395, no. 6697, 3 September 1998, p. 25.

30 Steinbrecher, 'Ecological Consequences of Genetic Engineering', p. 92.
31 Paul Brown, 'Mexico's vital gene reservoir polluted by modified maize', *Guardian*, 19 April 2002.
32 David Quist and Ignacio H. Chapela, 'Transgenic DNA introgressed into traditional maize landraces in Oaxaca, Mexico', *Nature*, vol. 414, no. 6863, 29 November 2001, pp. 541–3. In an unprecedented response to industry criticism of this report, the editor of *Nature* subsequently published an admission that flaws in the study, allegedly identified by the scientific authors of two accompanying letters, meant it should never have been published in the first place (see *Nature*, vol. 416, no. 6881, 11 April 2002, pp. 600–602). Quist and Chapela were also given the right to a reply, in which they argued that the central findings of their study were not in dispute. The impenetrable technical content of the ensuing exchange excluded all but the most specialist of scientists, thus leading thoughtful members of a befuddled public to suspend their hunger for truth and ask: what are the political stakes in this debate, and whose side am I on? George Monbiot's meticulous exposure of the relationship between the Bivings Group, a PR company contracted to Monsanto, and an internet smear campaign launched to discredit the original *Nature* article, was a welcome clarification of the interests at stake. (George Monbiot, 'The fake persuaders', *Guardian*, 14 May 2002, and 'Corporate phantoms', *Guardian*, 29 May 2002.)
33 Researchers in the US recently said they were 'shocked' to discover, for example, that wild sunflower weeds become hardier and produce fifty per cent more seed when crossed with a cultivated sunflower genetically engineered to be toxic to moth larvae. (Andy Coghlan, 'Weeds do well out of modified crops', *New Scientist*, vol. 175, no. 2356, 17 August 2002, p. 11.)
34 Jane Rissler and Margaret Mellon, *The Ecological Risks of Engineered Crops*, Cambridge, Mass.: MIT, 1996, pp. 51–2.
35 Mae-Wan Ho, *Genetic Engineering: Dream or Nightmare?*, Dublin: Gateway 1999, pp. 35–6.
36 Debora MacKenzie, 'Stray genes highlight superweed danger', *New Scientist*, vol. 168, no. 2261, 21 October 2000, p. 6.
37 These concerns were raised by Jane Rissler and Margaret Mellon of the US National Wildlife Federation in their submission to the US Department of Agriculture Animal and Plant Health Inspection Service (USDA APHIS) in 1992. The submission, 'On an Interpretive Ruling Concerning Transgenic Tomatoes', is available at <www.ucsusa.org/agriculture/aphis.aug92.html>
38 The Massachusetts company, Aqua Bounty Farms, hopes to be the first cultivator of transgenic salmon to gain regulatory approval to market their product – 'AcquAdvantage salmon' – in the US. The company holds the licence for a patent issued by the European Patent Office in 2001 to the Canadian Seabright Corporation (now Genesis Group Inc.), a company that supports the commercial application of research conducted at Newfoundland's Memorial University. The patent covers Atlantic salmon 'and all other fish species carrying an additional gene for faster growth'. (Ted Warren, 'Newfoundland group patents genetically modified salmon', *The Navigator*, vol. 4, no. 10, October 2001.) It should be pointed out that the growth-enhanced salmon, like many transgenic plants, can be bred to be sterile, but the risk of producing fertile fish cannot be completely eliminated (nor can the danger of sterile transgenic fish mating unsuccessfully with their wild cousins, displacing their non-transgenic fertile competitors in the process, and thus driving the species to extinction). Research into genetically engineered salmon conducted in Scotland was abandoned in 1992 because of fears of the consequences of escaping fish. To give a measure of the risks involved, note that in the first five months of the year 2000 the number of Scottish farmed salmon which escaped their cages and fled into the wild totalled 395,000, according to a report cited by Matthew Fort, 'Swimming Against the Tide', *Guardian*, 9 June 2000. See also Tony Reichhardt, 'Will souped up salmon sink or swim?', *Nature*, vol. 406, no. 6791, 6 July 2000, pp. 10–12.
39 See M. Syvanen, 'Horizontal Gene Transfer: Evidence and Possible Consequences', *Annual Review of Genetics*, vol. 28, 1994, pp. 237–61.

40 Frank Gebhard and Kornelia Smalla, 'Transformation of *Acinetobacter* sp. strain BD413 by transgenic sugar beet DNA', *Applied and Environmental Microbiology*, vol. 64, no. 4, April 1998, pp. 1550–3.

41 See Michael G. Lorenz and Wilfried Wackernagel, 'Bacterial Gene Transfer by Natural Genetic Transformation in the Environment', *Microbiological Reviews*, vol. 58, no. 3, September 1994, pp. 563–602.

42 Ho, *Genetic Engineering*, p. 20.

43 See Jaan Suurküla, 'Horizontal Transfer: An Introduction'. <www.psrast.org/hrtrintr.htm>

44 Mae-Wan Ho and Joe Cummins, 'Stop Release of GM Insects!', *ISIS News*, no. 9/10, July 2001. (www.i-sis.org)

45 Mae-Wan Ho, Terje Traavik, Orjan Olsvik *et al*, 'Gene Technology and Gene Ecology of Infectious Diseases', *Microbial Ecology in Health and Disease*, vol. 10, no. 1, 1998, pp. 33–59.

46 Ann E. Greene and Richard F. Allison, 'Recombination Between Viral RNA and Transgenic Plant Transcripts', *Science*, vol. 263, no. 5152, 11 March 1994, pp. 1423–5; James E. Scoelz and William M. Wintermantel, 'Expansion of Viral Host Range through Complementation and Recombination in Transgenic Plants', *The Plant Cell*, vol. 5, no. 11, November 1993, pp. 1669–79. For a more extensive discussion of the risks of viral recombination and transcapsidation (the encapsulation and 'disguising' of a viral genome with the protein coat of a different virus) see Rissler and Mellon, *The Ecological Risks of Engineered Crops*, pp. 62–70. The ability of viruses to recombine was acknowledged by Monsanto's director of regulatory science, Roy Fuchs, in an article by Stan Grossfeld, 'Genetic Engineering Debate Shifting to US', *The Boston Globe*, 23 September 1998.

47 Mae-Wan Ho, Angela Ryan and Joe Cummins, 'Cauliflower Mosaic Viral Promoter – A Recipe for Disaster?', *Microbial Ecology in Health and Disease*, vol. 11, no. 4, 1999, pp. 194–7; and by the same authors, 'Hazardous CaMV Promoter?', *Nature Biotechnology*, vol. 18, no. 4, April 2000, p. 363; and by Ho and Cummins, 'Hazards of Transgenic Plants Containing the Cauliflower Mosaic Viral Promoter', *Microbial Ecology in Health and Disease*, vol. 12, no. 1, 2000, pp. 6–11

48 Douwe Zuidema, Alexander Schouten, Magda Usmany *et al*, 'Expression of cauliflower mosaic virus gene I in insect cells using a novel polyhedrin-based baculovirus expression vector', *Journal of General Virology*, vol. 71, part 10, October 1990, pp. 2201–09.

49 Draft minutes of the committee, cited in Antony Barnett, 'Health fear over GM cattle feed', *Observer*, 15 October 2000.

50 John Vidal, 'Scientists question safety of GM maize', *Guardian*, 4 November 2000.

51 Walter Doerfler, Rainer Schubbert, Hilde Heller *et al*, 'Integration of foreign DNA and its consequences in mammalian systems', *Trends in Biotechnology*, vol. 15, no. 8, August 1997, pp. 297–301.

52 At the time of writing, producers in the US are not required to label their products as genetically modified, and indeed a scientific review by the government's Food and Drug Administration (FDA) concluded in May 2000 that labelling was unnecessary. The European Union, on the other hand, introduced legislation passed in 1998 which requires all genetically engineered food solids to be labelled under the Novel Foods Act. An enormous range of products seem to have escaped the remit of this ruling, however, including by-products like vegetable oil from genetically engineered corn or soya; meat, eggs, and dairy products from livestock fed on genetically engineered animal feed; lecithin, which derives from soya oil and is widely used in confectionary, yoghurts, and sauces; and numerous enzymes and additives which are extracted from genetically engineered bacteria and used to stabilise, sweeten, or enhance the flavour of processed foods. Some member states, including Italy, Austria and Luxembourg, have taken further action and banned the importing of specific transgenic products (in this case maize engineered by Novartis to express both the Bt toxin and resistance to the herbicide glufosinate). As public concern about the genetic engineering of food began to be registered by politicians and the media, the biotech industry in Europe has seen its output and profits slide. In the space of four years, the value

of US corn exports to the EU fell from $360 million to near zero, soya exports fell from $2.6 billion to less than $1 billion, and Canada's Europe-bound exports of canola dropped from $500 million to virtually nothing (see *Biodemocracy News*, no. 27 <www.purefood.org>). In spite of fierce lobbying from US biotech companies, the European parliament voted in July 2002 to introduce strict labelling and contamination rules for all genetically modified food and animal feed sold in the EU. Though eggs, meat and milk from animals reared on genetically modified feed narrowly escaped the ruling, the labelling requirement does extend from genetically modified products to the thousands of foodstuffs produced from genetically modified organisms, even when they do not contain detectable amounts of foreign DNA or protein. The ruling also reduced the maximum level of permitted contamination of non-transgenic products from 1 per cent to 0.5 per cent. If the bill is approved, as is expected, by EU member states, the stigmatising impact of labels, combined with the diseconomies of compulsory segregation and traceability paper trails, could, according to biotech lobbyists, cost US industry $4 billion worth of trade a year – good reason, they argue, for the US government to take the issue to the WTO.

53 James Meikle, 'Beekeepers seek GM halt after honey contamination', *Guardian*, 17 May 2000.

54 James Meikle, 'Rogue GM seeds taint UK crops', *Guardian*, 18 May 2000.

55 Andy Coghlan, 'Sowing Dissent', *New Scientist*, vol. 166, no. 2240, 27 May 2000, p. 4.

56 Anne Simon Moffat, 'Exploring Transgenic Plants as a New Vaccine Source', *Science*, vol. 268, no. 5211, 5 May 1995, pp. 658–60.

57 Rissler and Mellon, *The Ecological Risks of Engineered Crops*, p. 6.

58 Jeremy Rifkin, *The Biotech Century: How Genetic Commerce Will Change the World*, London: Phoenix, 1999, pp. 110–11.

59 Ho, *Genetic Engineering*, p. 141.

60 It is true that many plants – notably tubers, bulbs and runners – reproduce asexually, in effect producing natural clones of themselves by dividing or sending out runners or roots. But the difference here is that while each new plant in this case develops in a specific environment to which it adapts its fluid genome (like twins in a family), commercially produced clones are grown in the sterile and standardised environment of the petri dish and laboratory.

61 Donna J. Haraway, 'A Cyborg Manifesto: Science, Technology, and Socialist-Feminism in the Late Twentieth Century', in *Simians, Cyborgs, and Women: The Reinvention of Nature*, London: Free Association Books, 1991, pp. 176, 177, 155.

62 According to Rissler and Mellon, this has happened to at least six wild relatives of cultivated crops in the US. (Rissler and Mellon, *The Ecological Risks of Engineered Crops*, p. 56.) An example of this in the animal kingdom is the plight of Spain's white-headed duck, which is threatened with extinction by the aggressive sexual behaviour of the ruddy duck. The latter was introduced to Britain from North America by the conservationist Sir Peter Scott in the 1940s. Escaping from captivity it quickly established a self-sustaining feral population which migrates to Spain and mates with its white-headed Mediterranean cousin. (Paul Brown, '£1,000 to Shoot a Ruddy Randy Duck', *Guardian*, 3 June 2000.)

63 Intercropping is not only a way of hedging your bets, a kind of 'defensive diversification' aimed at avoiding outright ruination when predators strike. As researchers in China studying the effects of genetic diversity on the control of rice blast showed, planting disease-susceptible rice varieties in mixtures with resistant varieties improved the yield and reduced blast severity amongst the *former* by 89 per cent and 94 per cent respectively, when compared with a monocultured crop. (Youyong Zhu, Hairu Chen, Jinghua Fan *et al*, 'Genetic diversity and disease control in rice', *Nature*, vol. 406, no. 6797, 17 August 2000, pp. 718–22.)

64 Lappé and Bailey, *Against the Grain*, p. 99.

65 David Suzuki and Peter Knudtson, *Genethics: The Ethics of Engineering Life*, London: Unwin Hyman, 1989, pp. 296–8.

66 Material deprivation may also be an overriding factor, as appears to have been the case when, in the midst of an acute hunger crisis in August 2002, Mozambique reluctantly accepted a shipment of genetically engineered maize from the US. (Associated Press, 'Mozambique takes GM maize', *Guardian*, 15 August 2002.)

67 Calestous Juma, *The Gene Hunters: Biotechnology and the Scramble for Seeds*, London: Zed Books, 1989, p. 131.

68 It is also foolhardy to assume the permanent inability of some crops to evolve into wild weeds. It took two centuries of cultivating proso millet in Canada, for example, before weedy varieties began to appear. Examples in the animal kingdom are also evident: horses were domesticated for thousands of years before they became feral when the Spaniards reintroduced them to North America, and domesticated bees in Brazil became a lethal menace when they were interbred with African honey bees in the 1950s, creating a highly aggressive and venomous hybrid which had spread to Mexico by the 1980s.

69 David Pimentel, Lori Lach, Rodolfo Zuniga, Doug Morrison, 'Environmental and Economic Costs of Non-Indigenous Species in the United States', *Bioscience*, vol. 50, no. 1, 2000, pp. 53–65.

70 Rissler and Mellon, *The Ecological Risks of Engineered Crops*, pp. 23, 34, 55, 30, 52.

71 Vandan Shiva, *Biopiracy: The Plunder of Nature and Knowledge*, Boston, Mass.: South End Press, 1997, pp. 93–4.

72 Press Association, 'Experts Try to Untangle Knotweed', *Guardian*, 17 May 2000.

73 Allison A. Snow and Pedro Morán Palma, 'Commercialisation of Transgenic Plants: Potential Ecological Risks', *BioScience*, vol. 47, no. 2, February 1997, p. 94.

74 See Rifkin, *The Biotech Century*, pp. 75–7; Pat Spallone, *Generation Games: Genetic Engineering and the Future for Our Lives*, Philadelphia: Temple University Press, 1992, pp. 79–81.

75 Moyra Bremner, *GE: Genetic Engineering and You*, London: HarperCollins, 1999, p. 34. Researchers have also found that metabolic changes in herbicide-resistant soya crops cause the plants to produce significantly reduced levels of phytoestrogen – a plant substance similar to the human hormone oestrogen. The health implications of this are unclear. (Marc A. Lappé, E. Britt Bailey, Chandra Childress and Kenneth D.R. Setchell, 'Alterations in Clinically Important Phytoestrogens in Genetically Modified, Herbicide-Tolerant Soybeans', *Journal of Medicinal Food*, vol. 1, no. 4, 1998/1999, pp. 241–5.)

76 Tomoko Inose and Kousaku Murata, 'Enhanced accumulation of toxic compound in yeast cells having high glycolytic activity: a case study in the safety of genetically engineered yeast', *International Journal of Food Science and Technology*, vol. 30, no. 2, April 1995, pp. 141–6. A mutagen is a substance which causes changes to the genetic constitution of cells. This includes carcinogens and teratogens (agents which cause congenital abnormalities in developing foetuses).

77 Julie A. Nordlee, Steve L. Taylor, Jeffrey A. Townsend *et al*, 'Identification of Brazil-Nut Allergen in Transgenic Soybeans', *The New England Journal of Medicine*, vol. 334, no. 11, 14 March 1996, pp. 688–92; and Marion Nestle, 'Allergies to Transgenic Foods – Questions of Policy', *The New England Journal of Medicine*, vol. 334, no. 11, 14 March 1996, pp. 726–7.

78 Antony Barnett, 'GM Medicine "Risks the Lives of Diabetics"', *Observer*, 7 May 2000. It has also been reported that loss of hypoglycaemic awareness and other symptoms are shared by diabetics using enzyme-modified porcine insulin, and that the real cause is 'the *switch* from one chemical form of insulin (porcine or bovine) to another (human insulin), irrespective of the way in which the latter is synthesised' (Michael J. Reiss and Roger Straughan, *Improving Nature? The Science and Ethics of Genetic Engineering*, Cambridge: Cambridge University Press, 1996, p. 100).

79 Arthur N. Mayeno and Gerald J. Gleich, 'Eosinophilia-myalgia syndrome and tryptophan production: a cautionary tale', *Trends in Biotechnology*, vol. 12, no. 9, 1994, pp. 346–52.

80 Curiously, an alternative explanation for the poisonings has been proposed by the investigative journalist Bob Woffinden. In his view, the symptoms observed in the US in 1989 were identical to those of the infamous 'cooking oil disaster' in Spain in 1981. In the

official government account, the cause of the 1,000 deaths and 25,000 or more cases of serious illness was the illegal dilution of Spanish olive oil with imported industrial-grade rapeseed oil that had been made deliberately unfit for consumption by the addition of the toxic chemical aniline. Widespread scientific and medical unease about the cooking oil theory, however, especially among epidemiologists, suggested to Woffinden that both 'toxic oil syndrome' and the L-tryptophan mystery were actually deliberately concealed cases of organophosphate poisoning, deriving from vegetables that had been consumed with copious amounts of chemical pesticides. (Bob Woffinden, 'Cover-up', *Guardian Weekend*, 25 August 2001.)

81 Andrew Osborn, 'Euro Vote "Lets GM Companies off the Hook"', *Guardian*, 13 April 2000. One positive accomplishment of the European Parliament and Council of Ministers is the latest (March 2001) Directive 2001/18/EC on the deliberate release of genetically modified organisms (GMOs). Most important in this directive is the requirement that producers of GMOs provide proof that the organisms have 'genetic and phenotypic stability'. With the evidence mounting that genetic engineering radically undermines the stability of plant genomes, many ecologists believe that the new Directive, if properly implemented, could bring about the end of agricultural genetic engineering. (See Mae-Wan Ho and Angela Ryan, 'Europe's New Rules Could Sink All GMOs', *ISIS News*, no. 11/12, October 2001.)

82 Research has in fact already discovered that farm workers regularly exposed to topical Bt pesticides may develop allergic skin sensitisation and raised levels of specific antibodies. (I. Leonard Bernstein, Jonathan A. Bernstein, Maureen Miller *et al*, 'Immune Responses in Farm Workers after Exposure to *Bacillus Thuringiensis* Pesticides', *Environmental Health Perspectives*, vol. 107, no. 7, July 1999, pp. 575–82.)

83 Stanley W.B. Ewen and Arpad Pusztai, 'Effects of diets containing genetically modified potatoes expressing *Galanthus nivalis* lectin on rat small intestine', *The Lancet*, vol. 354, no. 9187, 16 October 1999, pp. 1353–4.

84 Scott Kilman and Sarah Lueck, 'Kraft Taco Shell Puts Focus On Biotechnology Oversight', *The Wall Street Journal*, 25 September 2000. Though Aventis responded by sacking three of its executives and setting aside $100 million (£65 million) to cover compensation claims, the US government announced in January 2001 that it would spend $20 million in taxpayers money to buy from Aventis its unsaleable maize seed.

85 Cited by Stuart Laidlaw, 'Starlink fallout could cost billions', *Toronto Star*, 9 January 2001.

86 Rachel Nowak, 'Disaster in the making', *New Scientist*, no. 2273, 13 January 2001, pp. 4–5.

87 Dutch researchers have shown that bacterial DNA can linger in the human gut long enough to be taken up by compatible resident bacteria, and that the rate of gene flow increases tenfold when the normal bacterial inhabitants are reduced – as they are in people (or animals) being treated with antibiotics. See Debora MacKenzie, 'Gut Reaction', *New Scientist*, vol. 161, no. 2171, 30 January 1999, p. 4.

88 Though genetic scientists maintain that the antibiotics they use are selected because they lack medical application, Joe Cummins reports that kanamycin 'is used prior to endoscopy of colon and rectum and to treat ocular infections. It is used in blunt trauma emergency treatment and has been found to be effective against *E. coli 0157* without causing release of verotoxin'. (Joe Cummins, 'Kanamycin Still Used and Cross-Reacts with New Antibiotics', *ISIS News*, no. 9/10, July 2001.)

89 Mae-Wan Ho, 'Monsanto's GM Cottons and Gonorrhoea', *ISIS News*, no. 7/8, February 2001.

90 Ho, *Genetic Engineering*, p. 191.

91 According to the World Health Organisation, despite the much-vaunted 'epidemiological transition' (from communicable to non-communicable diseases), of the 54 million deaths from all causes in the world in 1998, a quarter were due to infectious and parasitic diseases (including respiratory infections), the majority of which have evolved some degree of resistance to anti-microbial drugs. Of these 13 million deaths from infectious diseases,

ninety-seven per cent occurred in low- and middle-income countries. (WHO, *The World Health Report 1999*, Geneva 1999, p. 98.)

92 Anderson, *Genetic Engineering, Food, and our Environment*, p. 55.

93 According to some estimates, as much as half the food consumed by rural communities in the developing world may consist of wild resources (Ho, *Genetic Engineering*, p. 139).

94 Charles Benbrook, 'Evidence of the Magnitude and Consequences of the Roundup Ready Soybean Yield Drag from University-Based Varietal Trials in 1998', *Ag BioTech InfoNet Technical Paper*, no 1, 13 July 1999, <www.biotech-info.net> Benbrook served as Executive Director on the Board on Agriculture for the US National Academy of Sciences, and is an expert on Integrated Pest Management (IPM).

95 Ho, *Genetic Engineering*, p. 142.

96 Roger Highfield, 'Monsanto weed killer "wipes out beneficial insects"', *Independent*, 12 October 1999.

97 Lappé and Bailey, *Against the Grain*, p. 41.

98 Lennart Hardell and Mikael Eriksson, 'A Case-Control Study of Non-Hodgkin Lymphoma and Exposure to Pesticides', *Cancer*, vol. 85, no. 6, 15 march 1999, pp. 1353–60.

99 Caroline Cox, 'Glyphosate (Roundup)', *Journal of Pesticide Reform*, vol. 18, no. 3, Fall 1998, p. 3.

100 Cox, 'Glyphosate (Roundup)', p. 4.

101 Lappé and Bailey, *Against the Grain*, pp. 75–6, 125.

102 George Monbiot, *The Captive State: The Corporate Takeover of Britain*, London: Macmillan, 2000, pp. 264–5.

103 Ulrich Beck, *World Risk Society*, Cambridge: Polity, 1999.

3 The Life Science Industry

1 Action Group on Erosion, Technology and Concentration (ETC), 'Biopiracy +10: Captain Hook Awards 2002', *ETC Communiqué*, no. 75, March/April 2002. ETC, which was formerly the Rural Advancement Foundation International (RAFI), is an international non-governmental organisation headquartered in Winnipeg, Canada, which is dedicated to conservation of biodiversity and sustainable agriculture. It is well-regarded in the field of rural development, and has performed studies commissioned by the United Nations Development Programme (UNDP). See: <www.etcgroup.org>

2 Brian Tokar, 'Monsanto: A Checkered History', *The Ecologist*, vol. 28, no. 5, September/October 1998, pp. 254–61.

3 Michael Grunwald, 'Monsanto Hid Decades of Pollution', *Washington Post*, 1 January 2002; David Teather, 'Monsanto found guilty of polluting', *Guardian*, 25 February 2002.

4 Aspartame is produced by the Monsanto subsidiary G.D. Searle and sold under the brand names Nutrasweet and Equal. Adverse reactions to the product among sensitive consumers have been reported to medical scientists for some time, while the association between aspartame and brain cancer is made by J.W. Olney, N.B. Farber, E. Spitznagel, and L.N. Robins, 'Increasing brain tumor rates: is there a link to aspartame?', *Journal of Neuropathology and Experimental Neurology*, vol. 55, no. 11, November 1996, pp. 1115–23.

5 Monsanto sells BST under the trade name Posilac. Sales of Posilac netted Monsanto around $200 million in 1998, which is roughly the same amount the tax payer gives the US government to buy the surplus milk produced by the nation's farmers. (George Monbiot, *Captive State: The Corporate Takeover of Britain*, London: Macmillan, 2000, p. 234.) On the label of the Posilac bottle farmers are warned of the possible danger of 'reduced pregnancy rates in injected cows', 'increases in cystic ovaries and disorders of the uterus', 'decreases in gestation length and birth weight of calves', 'increased twinning ratios', 'increased risk for clinical mastitis', and 'increases in somatic cell counts'. (Cited by Jeremy Rifkin, *The Biotech Century: How Genetic Commerce Will Change the World*, London: Phoenix, 1999, p. 99.) Milk from cows treated with recombinant bovine growth hormone has also been found to

contain elevated levels of insulin-like growth factor 1 (IGF-1), a natural chemical which stimulates milk production in both cows and humans. Because the chemical can survive digestion, it may enter the blood stream and, by raising IGF-1 levels in humans, cause a range of metabolic disorders. There are, according to US cancer specialist Samuel Epstein, 'converging lines of evidence implicating IGF-1 in rBGH milk as a potential risk factor for breast and gastrointestinal cancers' (Samuel S. Epstein, 'Unlabeled Milk from Cows Treated with Biosynthetic Growth Hormones: A Case of Regulatory Abdication', *International Journal of Health Services*, vol. 26, no. 1, 1996, pp. 173–85). See also Brewster Kneen, *Farmageddon: Food and the Culture of Biotechnology*, Gabriola Island, British Columbia: New Society Publishers, 1999, pp. 63–93.

6 Lindsay Toub, 'Monsanto's Empire', *New Internationalist*, no. 293, August 1997, p. 16.

7 Richard Wray, 'US drugs group drops Monsanto', *Guardian*, 29 November 2001.

8 Tokar, 'Monsanto: A Checkered History', p. 257.

9 In the Summer of 2001, Monsanto announced plans to drop its technology use fees on transgenic corn and soybean seeds in time for the 2002 growing season, and instead to collect funds directly through royalty payments from the licensed seed supplier.

10 Cited by Seth Shulman, *Owning the Future*, Boston: Houghton Mifflin, 1999, p. 103.

11 Marvin Hayenga, 'Structural Change in the Biotech Seed and Chemical Industrial Complex', *AgBioForum*, vol. 1, no. 2, 1998, pp. 43–55. <www.agbioforum.missouri.edu>

12 The Saskatchewan farmer, Percy Schmeiser, lost his defence against the Monsanto suit and was ordered by a Canadian judge in March 2001 to pay $10,000 licensing fees and up to $75,000 in illegitimate profits from his 1998 crop.

13 Rick Weiss, 'Seeds of discord: Monsanto's gene police raise alarm on farmers' rights, rural tradition', *Washington Post*, 3 February 1999.

14 Brian Tokar, 'Challenging Biotechnology', in Brian Tokar, ed., *Redesigning Life? The Worldwide Challenge to Genetic Engineering*, London: Zed Books, 2001, p. 9.

15 Bill Heffernan, 'Study on Concentration in US Agriculture: Report to the National Farmers Union', 5 February 1999. <www.greens.org/s-r/gga/heffernan.html>

16 Jeremy Rifkin, *The Age of Access*, New York: Tarcher/Putnam, 2000, pp. 6, 67.

17 Heffernan, 'Study on Concentration in US Agriculture'.

18 Hugh Warwick, *Syngenta: Switching off Farmers' Rights*, GeneWatch UK, 2000, p. 11.

19 Hope Shand, 'Gene Giants: Understanding the "Life Industry"', Brian Tokar, ed., *Redesigning Life: The Worldwide Challenge to Genetic Engineering*, London: Zed Books, 2001, p. 228.

20 Terry Macalister, 'Bayer doubles stake in Dolly parent PPL', *Guardian*, 13 July 2001.

21 Pamela Sherrid, 'Please pass the bioengineered butter: DuPont's new push into agricultural biotech', *U.S. News Online*, 2 March 1998. <www.usnews.com>

22 Warwick, 'Syngenta: Switching off Farmers' Rights', p. 11.

23 ETC, 'Concentration in Corporate Power: The Unmentioned Agenda', *ETC Communiqué*, no. 71, July/August 2001.

24 ETC, 'The Five Gene Giants are Becoming Four', *ETC News Release*, 9 April 2002.

25 Anon, 'Business: Hold My Hand', *The Economist*, vol. 351, no. 8119, 15 May 1999, pp. 73–4.

26 Frances Moore Lappé, Joseph Collins, Peter Rosset and Luis Esparza, *World Hunger: Twelve Myths*, London: Earthscan, 1998, pp. 69–70.

27 RAFI, 'Earmarked for Extinction? Seminis Eliminates 2,000 Varieties', *RAFI Communiqué*, 17 July 2000.

28 Sarah Anderson and John Cavanagh, *Top 200: The Rise of Corporate Global Power*, Washington: Institute for Policy Studies, 2000, p. 5.

29 Shand, 'Gene Giants', p. 231.

30 Shand, 'Gene Giants', p. 231.

31 Heffernan, 'Study on Concentration in US Agriculture'.

32 Transfer pricing involves manipulating the prices at which the various subsidiaries of a single company buy and sell their products to one another so as to ensure that the highest

profits are registered in the countries with the lowest tax rates. Transfer pricing, the use of tax havens, and aggressive demands for tax rebates, enabled seven US corporations listed amongst the biggest 200 in the world – including General Motors, Texaco and PepsiCo – to pay less than zero in federal corporate taxes (normally levied at thirty-five per cent) during 1998. (Anderson and Cavanagh, *Top 200*, p. 6.)

33 Sally Lehrman, 'Foundations funding biomedical bodies "should shift focus"', *Nature*, vol. 383, no. 6596, 12 September 1996, p. 112.

34 Lori Andrews and Dorothy Nelkin, *Body Bazaar: The Market for Human Tissue in the Biotechnology Age*, New York: Crown, 2001, pp. 47–8.

35 Sheldon Krimsky, James G. Ennis, and Robert Weissman, 'Academic-Corporate Ties in Biotechnology: A Quantitative Study', *Science, Technology, and Human Values*, vol. 16, no. 3, Summer 1991, pp. 275–287.

36 S. Krimsky, L.S. Rothenberg, P. Stott, and G. Kyle, 'Financial Interests of Authors in Scientific Journals: A Pilot Study of 14 Publications', *Science and Engineering News*, vol. 2, no. 4, October 1996, pp. 395–410.

37 Will Woodward, 'Universities "need a Filkin" to police deals with firms', *Guardian*, 27 December 2001.

38 Monbiot, *Captive State*, pp. 287–8.

39 Kevin Maguire, 'University accepts tobacco "blood money"', *Guardian*, 5 December 2000; Richard Smith, 'A tainted university', *Guardian*, 21 May 2001.

40 Monbiot, *Captive State*, p. 291.

41 Lousie Jury, 'Crop experts linked to biotech firms', *Independent*, 9 July 1998.

42 Julian Borger, '"Conflict of interest" in GM food report', *Guardian*, 6 April 2000.

43 See <www.edmonds-institute.org/door.html>. See also Jennifer Ferrara, 'Revolving Doors: Monsanto and the Regulators', *The Ecologist*, vol. 28, no. 5, September/October 1998, pp. 280–6. The list includes Michael Kantor, former secretary of commerce and US trade representative, who was recruited to the board of directors at Monsanto; Marcia Hale, a former assistant to the president and director for intergovernmental affairs, who became Monsanto's director of international government affairs; Linda J. Fisher, former assistant administrator of the EPA's Office of Pollution Prevention, Pesticides and Toxic Substances, who is now Monsanto's vice-president of government and public affairs; Josh King, former director of production for White House events, who became Monsanto's director of global communication; Michael A. Friedman, former acting commissioner of the Department of Health and Human Services at the FDA, who is now senior vice-president for clinical affairs at G. D. Searle & Co., a pharmaceutical division of Monsanto; William D. Ruckelshaus, former chief administrator of the EPA who joined (in the 1980s) the board of directors at Monsanto; Jack Watson, formerly chief of staff to the US president Jimmy Carter, who is now a staff lawyer with Monsanto; Michael Taylor, who as the FDA's deputy commissioner for policy in 1994 wrote the labelling guidelines for milk containing rBGH, but who was later exposed as a former Monsanto lawyer and is now head of Monsanto's Washington office; Charles W. Burson, who was assistant to the president and chief of staff to vice-president Al Gore before he became general counsel for Monsanto in 2001; Terry Medley, who held several senior posts in the US Department of Agriculture and sat on the FDA's food advisory committee, but is now director of regulatory and external affairs of DuPont's Agricultural Enterprise; Clayton K. Yeutter, former secretary of the Department of Agriculture and a US government trade representative, who joined the board of directors of Mycogen Corporation; Larry Zeph, formerly a biologist working for the EPA, who became regulatory science manager at Pioneer Hi-Bred International; L. Val Giddings, former biotechnology regulator and biosafety negotiator for the Department of Agriculture, who is now vice-president for food and agriculture of the Biotechnology Industry Organisation.

44 Robert Cohen, 'FDA regulation meant to promote rBGH milk resulted in antibiotic resistance'. <www.psrast.org/bghsalmonella.htm>

45 John Vidal, 'GM lobby takes root in Bush's cabinet', *Guardian*, 1 February 2001.

46 Marcia Angell, 'Is Academic Medicine for Sale?', *New England Journal of Medicine*, vol. 342, no. 20, 18 May 2000, pp. 1516–8.

47 Jeanne Lenzer, 'Alteplase for stroke: money and optimistic claims buttress the "brain attack" campaign', *British Medical Journal*, vol. 324, no. 7339, 23 March 2002, pp. 723–6.

48 Henry Thomas Stelfox, Grace Chua, Keith O'Rourke, and Allan Detsky, 'Conflict of Interest in the Debate over Calcium-Channel Antagonists', *New England Journal of Medicine*, vol. 338, no. 2, 8 January 1998, pp. 101–6.

49 Deborah E. Barnes and Lisa A. Bero, 'Why Review Articles on the Health Effects of Passive Smoking Reach Different Conclusions', *Journal of the American Medical Association*, vol. 279, no. 19, 20 May 1998, pp. 1566–70.

50 Mark Frieberg, Bernard Saffran, Tammy J. Stinson *et al*, 'Evaluation of Conflict of Interest in Economic Analyses of New Drugs Used in Oncology', *Journal of the American Medical Association*, vol. 282, no. 15, 20 October 1999, pp. 1453–7.

51 David Blumenthal, Michael Gluck, Karen Seashore Louis *et al*, 'University–Industry Research Relationships in Biotechnology: Implications for the Industry', *Science*, vol. 232, no. 4756, 13 June 1986, pp. 1361–6. A trade secret is generally information which would give a commercial advantage to competitors were it to fall into their hands. It is normally enforced by serving injunctions against unauthorised disclosure, and may be favoured over patent protection because it is cheaper, quicker, and does not require the public disclosure of technical information. Blumenthal and his colleagues repeated this study a decade later, and concluded that 'the involvement of industry with academic institutions has increased, but the characteristics of the relationship have remained remarkably stable'. (David Blumenthal, NancyAnne Causino, Eric G. Campbell and Karen Seashore Louis, 'Relationships Between Academic Institutions and Industry in the Life Sciences – An Industry Survey', *New England Journal of Medicine*, vol. 334, no. 6, 8 February 1996, pp. 368–73; David Blumenthal, Eric G. Campbell, NancyAnne Causino, Karen Seashore Louis, 'Participation of Life-Science Faculty in Research Relations with Industry', *New England Journal of Medicine*, vol. 335, no. 23, 5 December 1996, pp. 1734–9.)

52 Eric G. Campbell, Karen Seashore Louis, David Blumenthal, 'Looking a Gift Horse in the Mouth', *Journal of the American Medical Association*, vol. 279, no. 13, 1 April 1998, pp. 995–999.

53 David Blumenthal, Eric G. Campbell, Melissa S. Anderson *et al*, 'Withholding Research Results in Academic Life Science', *Journal of the American Medical Association*, vol. 277, no. 15, 16 April 1997, pp. 1224–8.

54 Laura Bonetta, 'Inquiry into clinical scandal at Canadian research hospital', *Nature Medicine*, vol. 4, no. 10, October 1998, p. 1095; Nancy Olivieri, 'Letter to the Editor', *Nature Medicine*, vol. 5, no. 1, January 1999, p. 3; Laura Bonetta, 'Canadian fight over thalassemia drug worsens', *Nature Medicine*, vol. 5, no. 11, November 1999, p. 1223. Olivieri eventually decided to publish her findings in Nancy F. Olivieri, Gary M. Brittenham, Christine E. McLaren *et al*, 'Long-Term Safety and Effectiveness of Iron-Chelation Therapy with Deferiprone for Thalassemia Major', *New England Journal of Medicine*, vol. 339, no. 7, 13 August 1998, pp. 417–23.

55 Gretchen Vogel, 'Long-Suppressed Study Finally Sees Light of Day', *Science*, vol. 276, no. 5312, 25 April 1997, pp. 525–6.

56 Andrews and Nelkin, *Body Bazaar*, p. 57.

57 7 K. Kleiner, 'Milk hormone dispute boils over into court', *New Scientist*, vol. 142, no. 1923, 30 April 1994, pp. 6–7.

58 Jane Akre, 'Got Milk? Get Fired: The Inside Story of Censorship at Fox News', *In These Times*, vol. 25, no. 13, 28 May 2001.

59 Mitchel Cohen, 'Biotechnology and the New World Order', in Brian Tokar, ed., *Redesigning Life: The Worldwide Challenge to Genetic Engineering*, London: Zed Books, 2001, pp. 306–7.

60 Frank Davidoff, Catherine D. DeAngelis, Jeffrey M. Drazen *et al*, 'Sponsorship, Authorship, and Accountability', *New England Journal of Medicine*, vol. 345, no. 11, 13 September 2001, pp. 825–6.

61 Sarah Boseley, 'Scandal of scientists who take money for papers ghostwritten by drug companies', *Guardian*, 7 February 2002.
62 See *Science*, vol. 291, no. 5507, 16 February 2001; *Science*, vol. 296, no. 5565, 5 April 2002.
63 Stanley W.B. Ewen and Arpad Pusztai, 'Effects of diets containing genetically modified potatoes expressing *Galanthus nivalis* lectin on rat small intestine', *The Lancet*, vol. 354, no. 9187, 16 October 1999, pp. 1353–4.
64 One of several criticisms made was that the rats used in Pusztai's experiment were fed on a protein-deficient diet, which may alone have caused the impaired organ growth which he documented.
65 Pusztai's own web site is at < www.freenetpages.co.uk/hp/a.pusztai/ >.
66 Eric Schiff, *Industrialisation without National Patents*, Princeton: Princeton University Press, 1971, p. 122.
67 As Merges and Nelson point out, drawing on the work of the industry historian Arthur Bright, the validation of Edison's broad patent also considerably slowed the pace of technological innovations in the incandescent lighting industry in the US and Great Britain, as both the Edison Company and its competitors lost the economic incentive to pursue improvements in carbon filament lamp design. (Robert P. Merges and Richard R. Nelson, 'On the Complex Economics of Patent Scope', *Columbia Law Review*, vol. 90, no. 4, May 1990, pp. 839–916.)
68 Schiff, *Industrialisation without National Patents*, p. 104.
69 With the merger of Ciba and Sandoz to form Novartis in 1996, Ciba Speciality Chemicals was spun off as a separate company.
70 Eric G. Campbell, Brian R. Clarridge, Manjusha Gokhale *et al*, 'Data Withholding in Academic Genetics: Evidence from a National Survey', *Journal of the American Medical Association*, vol. 287, no. 4, 23/30 January 2002, pp. 473–80.
71 'At a certain stage of development, the material productive forces of society come into conflict with the existing relations of production or – this merely expresses the same thing in legal terms – with the property relations within the framework of which they have operated hitherto. From forms of development of the productive forces these relations turn into their fetters.' Karl Marx, *A Contribution to the Critique of Political Economy*, Moscow: Progress Publishers, 1970, p. 21.

4 Manufacturing Scarcity

1 Frederick H. Buttel and Jill Belsky, 'Biotechnology, Plant Breeding, and Intellectual Property: Social and Ethical Dimensions', *Science, Technology and Human Values*, vol. 12, no. 1, Winter 1987.
2 See Richard C. Lewontin and Jean-Pierre Berlan, 'The Political Economy of Agricultural Research: The Case of Hybrid Corn', in C. Ronald Carroll, John H. Vandermeer and Peter Rosset, eds, *Agroecology*, New York: McGraw-Hill, 1990, p. 619.
3 William Lesser, 'Intellectual Property Rights and Concentration in Agricultural Biotechnology', *AgBioForum*, vol. 1, no. 1, 1998, pp. 56–61. <www.agbioforum.missouri.edu>
4 Lawrence Busch, William B. Lacy, Jeffrey Burkhardt, and Laura R. Lacy, *Plants, Power, and Profit: Social, Economic, and Ethical Consequences of the New Biotechnologies*, Oxford: Blackwell, 1991, pp. 124–9.
5 See Jack Ralph Kloppenburg Jr., *First the Seed: The Political Economy of Plant Biotechnology, 1492–2000*, Cambridge: Cambridge University Press, 1988, p. 93.
6 Lewontin and Berlan, 'The Political Economy of Agricultural Research', p. 622.
7 Jean-Pierre Berlan and R.C. Lewontin, 'The Political Economy of Hybrid Corn', *Monthly Review*, vol. 38. No. 3, July-August 1986, pp. 35–47.
8 See <www.upov.org>
9 The 1930 Plant Patent Act, which covers asexually reproducing plant varieties (but excludes virtually every variety of staple food crop, including potatoes), was until 1980 considered

by the US Patent Office to be a special dispensation granted to a narrow range of (mainly ornamental) plants.

10 Naomi Roht-Arriaza, 'Of Seeds and Shamans: The Appropriation of the Scientific and Technical Knowledge of Indigenous and Local Communities', *Michigan Journal of International Law*, vol. 17, Summer 1996, p. 942.

11 Roht-Arriaza, 'Of Seeds and Shamans', p. 945.

12 For a description of these changes to the corporate scene, see Calestous Juma, *The Gene Hunters: Biotechnology and the Scramble for Seeds*, London: Zed Books, 1989, pp. 81–4.

13 It should be noted that the trend towards corporate concentration in the biotechnology industry in the 1990s mirrored developments in the global economy as a whole. Between 1990 and 2000 the worldwide value of corporate mergers and acquisitions increased from $462 billion to over $3.5 *trillion*, according to ETC (formerly RAFI). (Action Group on Erosion, Technology, and Concentration, 'Concentration in Corporate Power: The Unmentioned Agenda', *ETC Communiqué*, no. 71, July/August 2001.) Between 1994 and 1997 alone, some 27,600 companies merged – more than during the entire merger-mad decade of the 1980s. (Seth Shulman, *Owning the Future*, New York: Houghton Mifflin, 1999, p. 163.)

14 George Monbiot, *Captive State: The Corporate Takeover of Britain*, London: Macmillan, 2000, p. 253.

15 Jeremy Rifkin, *The Biotech Century: How Genetic Commerce Will Change the World*, London: Pheonix, 1998, p. 41.

16 Vandana Shiva, *Biopiracy: The Plunder of Nature and Knowledge*, Boston: South End Press, 1997, p. 5.

17 Cited by Andrew Kimbrell, *The Human Body Shop: The Cloning, Engineering, and Marketing of Life*, Washington: Gateway, 1997, pp. 230–1.

18 Rifkin, *The Biotech Century*, p. 43; Kimbrell, *The Human Body Shop*, p. 234.

19 Marjorie Sun, 'Scientists Settle Cell Line Dispute', *Science*, vol. 220, no. 4595, 22 April 1983, pp. 393–4.

20 Sandra Blakeslee, 'Patient sues for title to own cells', *Nature*, vo. 311, no. 5983, 20 September 1984, p. 198.

21 Kimbrell, *The Human Body Shop*, pp. 246–52.

22 The EU 'Directive on the Legal Protection of Biotechnological Inventions' (98/44/EC), which was conceived with the aim of bringing European patent laws into line with those in the US, was finally passed after fierce lobbying by representatives of the biotech industry. While purporting to uphold 'the difference between inventions and discoveries', the directive goes on to rule (Article 5) that 'An element isolated from the human body or otherwise produced by means of a technical process, including the sequence or partial sequence of a gene, may constitute a patentable invention, even if the structure of that element is identical to that of a natural element.'

23 Shulman, *Owning the Future*, p. 180.

24 James Meek, 'Patenting our genes', *Guardian*, 26 June 2000.

25 Angela Ryan, 'Human Gene Patenting Roundup', *ISIS News*, no. 6, September 2000, p. 17.

26 The appeal against the patent was defeated on the grounds that the scientists who submitted the claim had excised an intron – one of the non-coding sequences normally edited-out by the cell during transcription – from the gene, and that the patented gene was therefore not a product of nature. (Andy Coghlan, 'DNA "not life" rules patent office', *New Scientist*, vol. 145, no. 1962, 28 January 1995.)

27 Genewatch UK, 'Genetic Engineering: A Review of Developments in 2000', *GeneWatch Briefing*, no. 13, January 2001. <www.genewatch.org>

28 Cited in John Sulston and Georgina Ferry, *The Common Thread: A Story of Science, Ethics and the Human Genome*, London: Bantam, 2002, p. 88.

29 Declan Butler, 'Patent on umbilical-cord cells rejected in Europe', *Nature*, vol. 399, no. 6737, 17 June 1999, p. 626.

30 Rifkin, *The Biotech Century*, pp. 61–2.
31 Vicki Glaser, 'Geron issued UK Dolly patents', *Nature Biotechnology*, vol. 18, no. 3, March 2000, pp. 256–7.
32 Shulman, *Owning the Future*, pp. 54–6.
33 Shulman, *Owning the Future*, pp. 91–102.
34 According to the American Intellectual Property Law Association, the average legal costs of a run-of-the-mill patent infringement case that goes to trial is around $1.2 million. Patent litigation in the biotechnology industry increased sixty-nine per cent between 1995 and 1996 alone. (Shulman, *Owning the Future*, pp. 171–3.)
35 RAFI, 'Monsanto's "Submarine Patent" Torpedoes Ag Biotech', *RAFI News Release*, 26 April 2001.
36 To strengthen proprietory control over transgenic species one US biotech company, AviGenetics, is working to create strains of poultry which have been engineered to carry a genetic 'trademark' or 'copyright' sequence. (James Meek, 'Genetic chickens get DNA copyright tag', *Guardian*, 21 September 2000.)
37 One can also see the same contradiction appearing in inverted form in the views of some ecologists, who argue both that genetic engineering is 'unnatural' and therefore ridden with risk, *and* that the deliberate manipulation of genomes is insufficient to make the natural organism a human 'invention'.
38 See Vandana Shiva, 'The Threat to Third World Farmers', *The Ecologist*, vol. 30, no. 6, September 2000, pp. 40–3.
39 In the outrage that followed, officials in India – where turmeric is widely used to treat the cuts and grazes of children – eventually unearthed an article describing its medicinal properties in a 1953 issue of the *Journal of the Indian Medical Association*. As a result, the award of the US patent was overturned.
40 Of the world's twenty most important food crops, none are in fact indigenous either to North America or to Australia (see Jack R. Kloppenburg, Jr., and Daniel Lee Kleinman, 'Seeds of Controversy: National Property Versus Common Heritage', in Jack R. Kloppenburg, Jr., ed., *Seeds and Sovereignty: The Use and Control of Plant Genetic Resources*, Durham: Duke University Press, 1988, pp. 179–184). It should be noted that not all of North America's 'foreign' food crops were imported by European settlers. Many were already being cultivated by the indigenous population (including maize, although the varieties grown today are thought to be descended from Mexican seeds transported north over a thousand years ago). When it comes to industrially processed crops, as Kloppenburg shows, genetic interdependence is of course much greater, with many non-indigenous crops grown in the Third World for export.
41 UNDP, *UN Development Report 2001: Making New Technologies Work for Human Development*, New York: Oxford University Press, 2001, p. 39.
42 Shulman, *Owning the Future*, p. 133.
43 Rifkin, *The Biotech Century*, p. 49.
44 Shulman, *Owning the Future*, p. 133.
45 Roht-Arriaza, 'Of Seeds and Shamans', p. 928.
46 Antony Barnett, 'This plant keeps him alive. Now its secret is "stolen" to make us thin', *Observer*, 17 June 2001; Andrew Clark, 'Phytopharm hails bushmen's fatbuster', *Guardian*, 6 December 2001.
47 Antony Barnett, 'Bushmen victory over drug firms', *Observer*, 31 March 2002.
48 Luke Harding, 'India outraged as US company wins patents on rice', *Guardian*, 23 August 2001.
49 Antony Barnett, 'Thai fury at US "piracy" of rice gene', *Observer*, 28 November 2001. Further details of this project are available on the University of Florida's website: <www.napa.ufl.edu/2001news/jasmine.htm>
50 Following protests by a global coalition of environmental groups, and the evidence that an Indian agrochemical company had been producing neem-based fungicide since 1980, this

particular patent was revoked in May 2000. Ten other neem-related patents have been granted by the European Patent Office, and W.R. Grace has filed similar patents in the US. (Ulrike Hellerer, K.S. Jarayaman, 'Greens persuade Europe to revoke patent on neem tree', *Nature*, vol. 405, no. 6784, 18 May 2000, pp. 266–7.)

51 ETC, 'Proctor's Gamble: Yellow Bean Patent Owner Sues Sixteen Farmers and Processors in US', *ETC News Release*, 17 December 2001.

52 RAFI, 'Bracing for "El Nuna": Andean Groups Hopping Mad About Popping-Bean Patent', *RAFI News Release*, 20 March 2001.

53 Leslie Roberts, 'Scientific Split Over Sampling Strategy', *Science*, vol. 252, no. 5013, 21 June 1999, p. 1615.

54 Denis J. Murphy, 'Production of novel oils in plants', *Current Opinion in Biotechnology*, vol. 10, no. 2, 1999, pp. 175–80.

55 Lawrence Busch *et al*, *Plants, Power, and Profit*, p. 172.

56 Lawrence Busch *et al*, *Plants, Power, and Profit*, p. 173.

57 GeneWatch UK, 'Privatising Knowledge, Patenting Genes: The Race to Control Genetic Information', *GeneWatch Briefing*, no. 11, June 2000.

58 RAFI, 'Updates: Vanilla and Biotechnology', *RAFI Communiqué*, 30 July 1991. Madagascar meets around three-quarters of the global vanilla market, and is dependent on the trade for up to a tenth of its export earnings. US consumption accounts for fifty-eight per cent of global production. (Juma, *The Gene Hunters*, p. 141.)

59 Hope Shand, 'Gene Giants: Understanding the "Life Industry"', in Brian Tokar, ed., *Redesigning Life: The Worldwide Challenge to Genetic Engineering*, London: Zed Books, 2001, p. 228.

60 In this process the enzyme glucose isomerase, which is produced using a cultured microorganism, is immobilised or 'fixed' in a stable, non-soluble and therefore re-usable form. It is then added to liquefied corn starch, where it acts to catalyse the conversion (the 'isomerisation') of glucose into fructose. Commercialisation of this technology in the US has cut sugar cane imports by eighty per cent since 1975, causing Philippines exports to the US, for example, to decline from 1.75 million tons in 1980 to 0.29 million tons in 1991. The predicament of the estimated fifty million people in Third World countries who are thought to depend on cane sugar exports for their livelihoods is not helped by the fiercely protected European beet sugar industry, political support for which makes the EU the world's largest producer of sugar, and the second largest exporter of the commodity. (See Robin Jenkins, 'Keeping the Sugar Barons Sweet', *Grain*, March 2001. www.grain.org/publications/mar013–en-p.htm)

61 Frances Moore Lappé, Joseph Collins and Peter Rosset, *World Hunger: Twelve Myths*, London: Earthscan, 1998, p. 2.

62 John Vidal, 'World obese catch up with the underfed', *Guardian*, 6 March 2000; John Vidal, 'Scandal of the food Britain throws away', *Guardian*, 4 April 2000.

63 Lappé, Collins, and Rosset, *World Hunger*, pp. 8–9.

64 Lappé, Collins, and Rosset, *World Hunger*, p. 9; Andrew Simms, *Selling Suicide: Farming, False Promises, and Genetic Engineering in Developing Countries*, London: Christian Aid, 1999, p. 4; Jeremy Rifkin, *Beyond Beef: The Rise and Fall of the Cattle Culture*, New York: Plume, 1993, p. 163.

65 Frederick H. Buttel, Martin Kenney, Jack Kloppenburg, Jr., 'From Green Revolution to Biorevolution: Some Observations on the Changing Technological Bases of Economic Transformation in the Third World', *Economic Development and Cultural Change*, vol. 34, no. 1, October 1985, pp. 31–55.

66 Lappé, Collins, and Rosset, *World Hunger*, p. 59.

67 Lappé, Collins, and Rosset, *World Hunger*, p. 61.

68 Donald K. Freebairn, 'Did the Green Revolution Concentrate Incomes? A Quantitative Study of Research Reports', *World Development*, vol. 23, no. 2, 1995, pp. 265–79.

69 Lappé, Collins, and Rosset, *World Hunger*, p. 68.

70 Vandana Shiva, *Stolen Harvest: The Hijacking of the Global Food Supply*, London: Zed Books, 2000, pp. 12–14, 58–9.
71 Lappé, Collins, and Rosset, *World Hunger*, pp. 70–1.
72 Richard Lewontin, 'Genes in the Food!', *New York Review of Books*, vol. 48, no. 10, 21 June 2001, p. 84.
73 Jules Pretty, 'Against the grain', *Guardian*, 17 January 2001.
74 Antje Lorch, 'Push and Pull: Biological Control of Stemborer and Striga', *Biotechnology and Development Monitor*, no. 43, 2000, p. 22; Fred Pearce, 'An ordinary miracle', *New Scientist*, vol. 169, no. 2276, 3 February 2001, pp. 16–17.
75 Peter Rosset and Medea Benjamin, eds, *The Greening of the Revolution: Cuba's Experiment with Organic Agriculture*, Melbourne: Ocean Press, 1994. For a wider review of the various organic and agroecological approaches and projects being developed across the South, see Nicholas Parrott and Terry Marsden, *The Real Green Revolution: Organic and Agroecological Farming in the South*, London: Greenpeace Environmental Trust, 2002.
76 Anuradha Mittal, 'Land Loss, Poverty and Hunger: How World Bank, IMF and WTO Policies Undermine Small Farmers and Food Security', *Ecologist*, vol. 30, no. 6, September 2000, p. 44.
77 Kevin Watkins, *Rigged Rules and Double Standards: Trade, Globalisation, and the Fight Against Poverty*, Oxford: Oxfam International, 2002, p. 141.
78 Mittal, 'Land Loss, Poverty and Hunger', p. 44.
79 Simms, *Selling Suicide*, p. 7. In the UK the corresponding figure is around thirty per cent. See J. van Wijk, 'Farm seed saving in Europe under pressure', *Biotechnology and Development Monitor*, vol. 17, pp. 13–14.
80 RAFI, 'The Gene Giants: Update on Consolidation in the Life Industry', *RAFI Communiqué*, 30 March 1999.
81 Sarah Sexton and Nicholas Hildyard, 'Genetic Engineering and World Hunger', *Synthesis/Regeneration*, no. 19, Spring 1999. With a quarter of farmed land in the US under soybean cultivation – the crop being the country's biggest export commodity – human consumers, particularly in the developing world, are now becoming a commercial target. As well as trying to penetrate local oilseed markets in India, the American Soybean Association is, according to Shiva, 'promoting "analogue" dals – soybean extrusions shaped into pellets that look like black gram, green gram, pigeon pea, lentil, and kidney bean. The diet they envision would be a monoculture of soybean; only its appearance would be diverse.' (Shiva, *Stolen Harvest*, p. 31.)
82 Rifkin, *Beyond Beef*, p. 161. Rifkin also cites Frances Moore Lappé: 'An acre of cereal can produce five times more protein than an acre devoted to meat production; legumes (beans, peas, lentils) can produce ten times more; and leafy vegetables, fifteen times more . . . spinach can produce up to twenty-six times more protein per acre than beef' (p. 162). Elsewhere Lappé points out that 'For every 16 pounds of grain and soy fed to beef cattle in the United States we only get 1 pound back in meat on our plates.' (Frances Moore Lappé, *Diet for a Small Planet*, New York: Ballantine, 1991, p. 69.)
83 R.E. Webb and W.M. Bruce, cited in Kloppenburg, *First the Seed*, p. 126.
84 GeneWatch UK, 'The Next Generation of GM Foods: Good for Whose Health?', *GeneWatch Briefing*, no. 10, April 2000. Included in this category is the likely development of an 'eat fat stay thin' pill by SmithKline Beecham. Researchers for the company successfully transferred to mice a human gene which codes for the protein UCP-3. When the gene was overexpressed, the mice were able to significantly increase their food intake without weight gain. The researchers conclude that, despite the lack of evidence linking UCP-3 with the aetiology of obesity, 'enhancement of UCP-3 expression, or stimulation of its activity, is a promising approach to the treatment of this disease'. (John C. Clapham, Jonathan R.S. Arch, Helen Chapman *et al*, 'Mice overexpressing human uncoupling protein-3 in skeletal muscle are hyperphagic and lean', *Nature*, vol. 406, no. 6794, 27 July 2000, pp. 415–8.)

85 Dylan Jackson, *Lifestyle Drugs Outlook to 2005: The New Blockbusters*, London: Datamonitor, 1999. The author of this report estimates that lifestyle drugs could make up ten per cent of the total pharmaceutical market by 2005.

86 The Western cosmetics market – worth around $40 billion to US retailers alone – is also a key target for the biotech industry. The New Jersey company GeneLink, for example, recently announced in a press release that it had filed a patent for a mail order genetic test kit that predicts an individual's susceptibility to the oxidative stresses that cause skin ageing. The company has entered into partnership with a major cosmetics firm with the aim of producing 'skin care products that block oxidative stress . . . reduce expression of this variant gene and retard skin ageing'. (Anon, 'GeneLink, Inc. Announces the Filing of US Patent Application for SNPS Based Method to Detect Susceptibility to Connective Tissue Breakdown and Skin Aging', GeneLink Press Release, 30 October 2001.)

87 GeneWatch UK, 'Biotech Deals Put Lung Cancer Vaccine Hopes in the Hands of Japan Tobacco', *GeneWatch Press Release*, November 2001.

88 Gary H. Toenniessen, 'Vitamin A Deficiency and Golden Rice: The Role of the Rockefeller Foundation', Rockefeller Foundation, 14 November 2000.

89 Paul Brown, 'GM rice promoters "have gone too far"', *Guardian*, 10 February 2001.

90 Brown, 'GM rice promoters "have gone too far"'.

91 Vandana Shiva, 'Poverty and Globalisation', *BBC Reith Lectures*, May 2000. As well as providing havens for beneficial insects and other wildlife, serving medicinal uses, and being an important source of animal fodder, in many marginal lands nutritious 'weeds' are often the first line of defence against famine when crops fail due to environmental stresses that weeds can by definition withstand.

92 Kathryn G. Dewey, 'Nutrition and Agricultural Change', in C. Ronald Carroll, John H. Vandermeer and Peter Rosset, eds, *Agroecology*, New York: McGraw-Hill, 1990, pp. 459–80.

93 Mae-Wan Ho, 'The "Golden Rice" – An Exercise in How Not to Do Science', *ISIS Sustainable Audit*, no. 1. <www.i-sis.org> As Ho points out, the technique used to make the golden rice is classic first generation genetic engineering, which includes the insertion of an antibiotic-resistance marker gene as well as the powerful cauliflower mosaic virus promoter.

94 RAFI, 'Update on Trojan Trade Reps, Golden Rice, and the Search for Higher Ground', *RAFI News Release*, 12 October 2000.

95 Martha L. Crouch, 'How the Terminator Terminates', *Occasional Papers Series*, The Edmonds Institute, 1998; Mae-Wan Ho, 'Killing Fields Near You: Terminator Crops at Large', *ISIS News*, no. 7/8, February 2001. Because 100 per cent reliability in gene control systems cannot be guaranteed (some fertile pollen and seeds are likely to be produced), because pollen-sterile plants can still be cross-fertilised by wild pollen, and because horizontal gene transfer remains a risk, ecologists have not reacted favourably to the recent claim by biotech companies and both UK and US advisory bodies that this technology could have conservationist value in preventing unintended gene flow from transgenic organisms. One major worry is that the system will actually increase the instability of plant genomes, a concern which is corroborated by the discovery that 'site-specific' recombinase enzymes can act unfaithfully by cutting and joining DNA at inappropriate sites, thus 'scrambling' the host genome. This has been observed in transgenic mice, where high levels of Cre expression in the testes of the animals led to the rearrangement of chromosomes and the inducement of complete male sterility. The scientists who discovered this wisely recommend against using Cre-mediated recombination in human gene therapy, and suggest that in other cases, once recombination has occurred, 'it would be prudent to remove or inactivate the *Cre* gene as rapidly as possible'. (Edward E. Schidt, Deborah S. Taylor, Justin R. Prigge *et al*, 'Illegitimate Cre-dependent chromosome rearrangement in transgenic mouse spermatids', *Proceedings of the National Academy of Sciences*, vol. 97, no. 25, 5 December 2000, pp. 13702–7.) It should also be pointed out that the terminator toxin 'barnase', which is used in US Patent 5,750,867 for Maintenance of

Male-Sterile Plants now owned by Aventis, has been found to cause pathological alteration of renal function in the kidneys of rats, and the use of the barnase gene may therefore present a significant risk to wildlife. (Olga N. Ilinskaya and Spiros Vamvakas, 'Nephrotoxic effects of bacterial ribonucleases in the isolated perfused rat kidney', *Toxicology*, vol. 120, no. 1, 1997, pp. 55–63.)

96 Gloria J. Hutchinson, 'Genetic Seeds'. <www.esotericworldnews.com/genetic.htm>

97 Richard A. Steinbrecher and Pat Roy Mooney, 'Terminator technology: the threat to world food security', *The Ecologist*, vol. 28, no. 5, September-October 1998, p. 277.

98 Using genetic engineering to sterilise plant gametes is also being treated as a functional means of creating more uniform hybrids (by interplanting male-sterile plants with female-sterile partners) and avoiding the labour-intensive task of detassling self-pollinating crops such as wheat (M. De Block, D. Debrouwer, T. Moens, 'The development of a nuclear male sterility system in wheat. Expression of the *barnase* gene under the control of tapetum specific promoters', *Theoretical and Applied Genetics*, vol. 95, no. 1/2, 1997, pp. 125–131). The latest patent awarded to Syngenta in May 2001 for terminator technology (US Patent 6,228,643) is described as a method for preventing unwanted gene flow from transgenic plants. Promoting the genetic sterilisation of plants as a biosafety tool now seems to be the main strategy for gaining public acceptance of this technology.

99 Mae-Wan Ho, 'Killing Fields Near You: Terminator Crops at Large', *ISIS News*, no. 7/8, February 2001.

100 RAFI, 'Terminator on Trial', *RAFI News Release*, 12 May 2000; Antony Barnett, '"Junkie" GM gene threat to Third World farmers', *Observer*, 2 April 2000.

101 Hugh Warwick, *Syngenta: Switching Off Farmers' Rights?*, Genewatch UK, 2000, p. 15.

102 Antony Barnett, 'Gene scientists disable plants' immune system', *Observer*, 8 October 2000. See also Warwick, *Syngenta: Switching Off Farmers' Rights?*.

5 Animal Biotechnology and Ethics

1 The animal, named Noah, was born live and seemingly healthy, but died two days later, apparently from a bacterial infection. (Philip Cohen, 'Bad copies', *New Scientist*, vol. 169, no. 2276, 3 February 2001, p. 7.) Greater success seems to have been had with the cloning of the endangered wild animal the 'mouflon', which is found on the Mediterranean islands of Sardinia, Corsica and Cyprus, and is thought to be an ancestor of the modern domesticated sheep. Cells from the ovaries of two deceased mouflons were fused with enucleated sheep eggs, and the resulting blastocysts implanted in the wombs of sheep. One viable offspring was born. (Pasqualino Loi, Grazyna Ptak, Barbara Barboni *et al*, 'Genetic rescue of an endangered mammal by cross-species nuclear transfer using post-mortem somatic cells', *Nature Biotechnology*, vol. 19, no. 10, October 2001, pp. 962–4.)

2 James Meek, 'The cow, her cloned bison calf and a new era in saving endangered species', *Guardian*, 9 October 2000; Sanjida O'Connell, 'Black and white, not bred all over', *Guardian*, 21 December 2000; Robert McKie, 'How Noah could clone a new ark', *Observer*, 7 January 2001; James Meek, 'Scientists pledge to clone extinct Tasmanian tiger', *Guardian*, 29 May 2002.

3 Bernard E. Rollin, 'On *telos* and genetic engineering', in Alan Holland and Andrew Johnson, eds, *Animal Biotechnology and Ethics*, London: Chapman and Hall, 1998, p. 169.

4 Much was made of the threat to the rare Herdwick sheep, for example, posed by the UK government's slaughter programme in the wake of the 2001 foot and mouth outbreak (which originated, as with BSE, in the feeding of waste meat back to animals). The Herdwick flocks graze the Lakeland fells, but have no need of fences or shepherds due to their inherited knowledge of their landscape, its dangers, refuges and rewards.

5 Since, for example, the cloned bucardo mentioned above will be female, it will be the responsibility of scientists to manufacture a male as a mate. 'Once we have cloned burcados we will have to look at ways of creating males', says ACT scientist Phil Damiani. 'That

means finding a method of adding Y chromosomes to embryos'. (Robert McKie, 'How Noah could clone a new ark', *Observer*, 7 January 2001.)

6 Vandana Shiva, 'Reductionism and Regeneration: A Crisis in Science', in Maria Mies and Vandana Shiva, *Ecofeminism*, London: Zed Books, 1993, pp. 23–4.

7 Michael Banner, 'Ethics, society and policy: a way forward', in Holland and Johnson, eds, *Animal Biotechnology and Ethics*, pp. 326–7.

8 The claim made by Ian Wilmut and others that cloning is not strictly speaking a form of genetic engineering, since it merely reproduces the genetic material of a given cell's nucleus, is rather disingenuous. Research into the cloning of animals would have had little financial backing if it did not offer a means of mass producing identical genetically altered mammals, such as research mice or genetically enhanced cattle. The managing director of PPL Therapuetics, Ron James, told *New York Times* journalist Gina Kolata that Wilmut had indeed persuaded him to finance the Roslin experiments by arguing that cloning promised to be a more efficient route towards the creation of drug-producing animals. Hence instead of injecting naked DNA directly into the nuclei of fertilised eggs, then waiting in hope for surviving animals to be born with the foreign gene, scientists can now achieve satisfactory manipulation of a cell's DNA before they fuse that DNA with an enucleated egg and enable the embryo to develop. 'Instead of making the animal and then working out which animal is best', PPL's research director told Kolata, 'we wanted to do our selection at the cell level'. (Gina Kolata, *Clone: The Road to Dolly and the Path Ahead*, Harmondsworth: Penguin, 1997, pp. 183, 197.) And Wilmut has himself admitted that 'making precise changes in the DNA of animals . . . was the original, and continues to be the main, aim of nuclear transfer research'. (Ian Wilmut, 'Dolly: The Age of Biological Control', in Justine Burley, ed., *The Genetic Revolution and Human Rights*, Oxford: Oxford University Press, 1999, p. 24.) It should also be pointed out that what is currently defined as a 'genetic clone' is in fact something less than the 'true clones' (identical twins) that result from the natural splitting of a fertilised egg. This is because, in the case of nuclear transfer, a small amount of genetic material – contained in the circular chromosomes of mitochondrial DNA located in the cell's cytoplasm – remains in the enucleated egg cell and is incorporated into the cells of the developing animal. (The mitochondria of mother and child are always identical, because in the normal fertilisation of the ovum the sperm cell contributes only its nucleus while its mitochondrial DNA is degraded. Mitochondrial DNA is thus used to trace female lineage in the same way that the Y chromosome is used to identify male ancestry.) This also means that, when the egg is taken from a different species (as in the cloning of the gaur, the bucardo and the panda), the cloned animal will contain important genetic material (mitochondria are now thought to play a significant role in the metabolisation and oxidisation of chemicals) from a species with which it would never sexually reproduce. Even a clone which has not been directly genetically engineered can in this sense be described as 'transgenic'. (The exception to this would of course occur if a cell nucleus were fused with an enucleated egg taken from the same female animal.)

9 See K. Gordon, E. Lee, J.A. Vitale *et al*, 'Production of human-tissue plasminogen-activator in transgenic mouse milk', *Bio/Technology*, vol. 5, no. 11, November 1987, pp. 1183–7; Harry Meade and Carol Ziomek, 'Urine as a substitute for milk?', *Nature Biotechnology*, vol. 16, January 1998, pp. 21–2; Michael K. Dyck, Dominic Gagné, Mariette Ouellet *et al*, 'Seminal vesicle production and secretion of growth hormone into seminal fluid', *Nature Biotechnology*, vol. 17, November 1999, pp. 1087–90.

10 Jay Rutovitz and Sue Mayer, *Genetically Modified and Cloned Animals. All in a Good Cause?*, GeneWatch UK: 2002, p. 39. <www.genewatch.org>

11 Karl M. Ebert, James P. Selgrath, Paul DiTullio *et al*, 'Transgenic production of a variant of human tissue-type plasminogen activator in goat milk: generation of transgenic goats and analysis of expression', *Bio/Technology*, vol. 9, no. 9, September 1991, pp. 835–8.

12 The first transgenic sheep created for this purpose were produced by Wilmut and his

colleagues through the microinjection of foreign DNA into newly fertilised sheep eggs. (See A.J. Clark, H. Bessos, J.O. Bishop *et al*, 'Expression of human anti-hemophilic factor IX in the milk of transgenic sheep', *Bio/Technology*, vol. 7, no. 5, May 1989, pp. 487–92.) The researchers subsequently achieved the same results by cloning genetically engineered foetal cells. (See Angelika E. Schnieke, Alexander J. Kind, William A. Ritchie *et al*, 'Human Factor IX Transgenic Sheep Produced by Transfer of Nuclei from Transfected Fetal Fibroblasts', *Science*, vol. 278, no. 5346, 19 December 1997.) One of the reasons why animals rather than bacteria are being used to synthesise medically useful proteins like factor IX is that the length of bacterial plasmids is insufficient to incorporate the long strips of DNA necessary to synthesise complex proteins. The DNA sequence that expresses factor VIII, the long protein in which the majority of haemophiliacs are deficient, is apparently too lengthy even for mammalian chromosomes. (See 'Discussion I' in Peter Wheale and Ruth McNally, eds, *The Bio-Revolution: Cornucopia or Pandora's Box?*, London: Pluto, 1990, p. 47.)

13 G. Wright, A. Carver, D. Cottom *et al*, 'High level expression of active human alpha-1-antitrypsin in the milk of transgenic sheep', *Bio/Technology*, vol. 9, no. 9, September 1991, pp. 830–4; Ian Wilmut, 'Pharmaceuticals from Milk: Producing Pharmaceuticals in Sheep Milk', in Bruce and Bruce, eds, *Engineering Genesis*.

14 Paul Krimpenfort, Adriana Rademakers, Will Eyestone *et al*, 'Generation of transgenic dairy cattle using *"in vitro"* embryo production', *Bio/Technology*, vol. 9, no. 9, September 1991, pp. 844–7.

15 Ajay Sharma, Mike J. Martin, Jeannine F. Okabe *et al*, 'An Isologous Porcine Promoter Permits High Level Expression of Human Hemoglobin in Transgenic Swine', *Biotechnology*, vol. 12, no. 1, January 1994, pp. 55–9.

16 Helen Sang, 'Transgenic chickens – methods and potential applications', *Trends in Biotechnology*, vol. 12, no. 10, September 1994, pp. 415–20; James Meek, 'Genetic chickens get DNA copyright tag', *Guardian*, 21 September 2000.

17 Cited in Rutovitz and Mayer, *Genetically Modified and Cloned Animals*, pp. 31, 37.

18 T. Ben Mepham, Robert D. Combes, Michael Balls *et al*, 'The use of Transgenic Animals in the European Union: The Report and Recommendations of ECVAM Workshop 28', *ATLA*, vol. 26, no. 1, 1998, p. 27.

19 The normal method for 'deleting' genes involves introducing into the targeted cells a non-functional genetic sequence which is flanked at both ends by DNA that is identical in sequence to the DNA that flanks the targeted gene. In a process that is still poorly understood, but which is similar to the alignment and recombination of homologous sections of the paired chromosomes during meiosis, the transgenic construct finds the targeted gene and displaces it, along with an undetermined length of its adjacent sequences, when the foreign but identical flanking sequences smoothly integrate themselves into the host DNA. The success of this technique – called 'homologous recombination' – varies from species to species (it does not work efficiently in human cells). It is thought to play a role in the horizontal transfer of genes, and is one natural mechanism by which 'disarmed' viral vectors may reacquire their infectiousness through contact with other viruses.

20 James Meek, 'Cloned pigs give vital boost to future of transplants', *Guardian*, 3 January 2002; Yifan Dai, Todd D. Vaught, Jeremy Boone *et al*, 'Targeted disruption of the α1,3-galactosyltransferase gene in cloned pigs', *Nature Biotechnology*, vol. 20, no. 3, March 2002, pp. 251–5. There are of course widespread concerns amongst ecologists and medical researchers that xenotransplantation will allow new and unknown microorganisms, harmless to their natural hosts, to cross the species barrier, causing infectious disease, spreading cancer-causing retroviruses, and potentially creating mutant viruses as deadly as HIV, Ebola, or BSE. Pigs are already known to harbour endogenous retroviruses which have been found to infect human cells *in vitro*. The Ebola and Marburg monkey viruses have caused large disease outbreaks in humans, HIV is widely believed to have derived from a monkey retrovirus, and millions of people in the 1950s were infected with the non-virulent monkey virus SV40 after vaccines were contaminated by the monkey kidney cells in which they were produced. (Declan Butler,

'Last chance to stop and think on risks of xenotransplants', *Nature*, vol. 391, no. 6665, 22 January 1998, pp. 320–4.).

21 Rutovitz and Mayer, *Genetically Modified and Cloned Animals*, pp. 49–59.

22 Home Office, *Statistics of Scientific Procedures on Living Animals. Great Britain 1999*, Cm 4841, Norwich: TSO, 2000; Home Office, *Statistics of Scientific Procedures on Living Animals. Great Britain 1996*, Cm 3722, Norwich: TSO, 1997.

23 Kathleen A. Mahon, Ana B. Chepelinsky, Jaspal S. Khillan *et al*, 'Oncogenesis of the Lens in Transgenic Mice', *Science*, vol 235, no. 4796, 27 March 1987, pp. 1622–8.

24 Rutovitz and Mayer, *Genetically Modified and Cloned Animals*, pp. 19–26.

25 Paul Krimpenfort, Adriana Rademakers, Will Eyestone *et al*, 'Generation of transgenic dairy cattle using "*in vitro*" embryo production', *Bio/Technology*, vol. 9, no. 9, September 1991, pp. 844–7.

26 Jocelyn Kaiser, 'Cloned Pigs May Help Overcome Rejection', *Science*, vol. 295, no. 5552, 4 January 2002, pp. 25–7.

27 I. Wilmut, A.E. Schnieke, J. McWhir *et al*, 'Viable offspring derived from fetal and adult mammalian cells', *Nature*, vol. 385, no. 6619, 27 February 1997.

28 James Meek and Kirsty Scott, 'Dolly hobbles back into the limelight', *Guardian*, 5 January 2002.

29 Wendy Dean, Fatima Santos, Miodrag Stojkovic *et al*, 'Conservation of methylation reprogramming in mammalian development: Aberrant reprogramming in cloned embryos', *Proceedings of the National Academy of Sciences*, vol. 98, no. 24, 20 November 2001, pp. 13734–8.

30 Davor Solter, 'Mammalian Cloning: Advances and Limitations', *Nature Reviews Genetics*, vol. 1, no. 3, December 2000, pp. 199–207.

31 Vernon G. Pursel, Carl A. Pinkert, Kurt F. Miller *et al*, 'Genetic Engineering of Livestock', *Science*, vol. 244, no. 4910, 16 June 1989, pp. 1281–8.

32 Cited in Rutovitz and Mayer, *Genetically Modified and Cloned Animals*, p. 26.

33 Peter Singer, *Animal Liberation*, 2nd edition, London: Jonathan Cape, 1990.

34 Jeremy Bentham, *An Introduction to the Principles of Morals and Legislation*, London: Methuen, 1982, p. 183 (his emphasis).

35 John Stuart Mill, 'Whewell on Moral Philosophy', in *Collected Works of John Stuart Mill. Vol. X: Essays on Ethics, Religion and Society*, London: Routledge and Kegan Paul, 1969, p. 187.

36 BMA, *Our Genetic Future: The Science and Ethics of Genetic Technology*, Oxford: Oxford University Press, 1992, p. 4.

37 C. Ray Greek and Jean Swingle Greek, *Sacred Cows and Golden Geese: The Human Cost of Experiments on Animals*, New York: Continuum, 2000, p. 66.

38 Trisha Gura, 'Systems for Identifying New Drugs Are Often Faulty', *Science*, vol. 278, no. 5340, 7 November 1997, pp. 1041–2.

39 An oncogene is a gene which, though normally involved in assisting healthy cell division, has mutated to a form which promotes uncontrolled cell growth (cancer). (Non-mutated oncogenes are properly designated 'proto-oncogenes'.) A tumour-suppressor gene, on the other hand, normally functions to *prevent* uncontrolled cell growth, and its deletion or mutation can also therefore lead to cancer.

40 The genetic engineering of animals for medical research is partly premised on the discovery that humans share a large part of their genome with other animals. Genes from different species which are significantly similar in their nucleotide sequences, and which are assumed to derive from a common ancestral gene, are called 'homologous genes'. Many human oncogenes and tumour-suppressor genes have 'homologues' in other animals such as rodents.

41 Tyler Jacks, Amin Fazeli, Earlene M. Schmitt *et al*, 'Effects of an *Rb* mutation in the mouse', *Nature*, vol. 359, no. 6393, 24 September 1992, pp. 295–200.

42 Gura, 'Systems for Identifying New Drugs Are Often Faulty', pp. 1041–2

43 Mary A. Bedell, David A. Largaespada, Nancy A. Jenkins, Neal G. Copeland, 'Mouse

models of human disease. Part II: Recent progress and future directions', *Genes and Development*, vol. 11, no. 1, 1 January 1997, pp. 11–43; Rick Weiss, 'Creation of flawed animals raises new ethics issues', *Washington Post*, 7 June 1998.

44 Jianglin Fan, Mireille Challah, Teruo Watanabe, 'Transgenic rabbit models for biomedical research: Current status, basic methods and future perspectives', *Pathology International*, vol. 49, no. 7, 1999, pp. 583–94.

45 There are generally two different senses in which the term 'intrinsic value' is used in environmental philosophy, though often the two are conflated. In the first sense, the value of something is regarded as 'intrinsic' when it is recognised as an end in itself, rather than a source of or instrument for a greater or more important value. In the second sense, the term applies to something which has value independent of its relationship to other sources of value. Though some environmentalists – such as Holmes Rolston – contend that the second definition applies to all natural organisms, it is more commonly argued that only humans and animals, because their own existence matters to them, have intrinsic value in the second sense. The fact that nature may be a source of intrinsic value – aesthetic contemplation, for example – in the first definition, does not invalidate the observation that both nature and the experience of intrinsic value it delivers can be and are commonly 'instrumentalised' (maximised, preserved, regulated) in a utilitarian manner – a manner which most ethical thinkers believe is not appropriate for humans and animals.

46 Tom Regan, *The Case for Animal Rights*, Berkeley: The University of California Press, 1983.

47 Joyce D'Silva, 'Campaigning against transgenic technology', in Holland and Johnson, eds, *Animal Biotechnology and Ethics*.

48 Andrew George, 'Animal biotechnology in medicine', in Holland and Johnson, eds, *Animal Biotechnology and Ethics*, pp. 44–5. Though it seems reasonable to characterise the artificial selection of desirable phenotypes for breeding as a primitive form of 'biotechnology', to refer to this practice – as George does – as 'genetic engineering' is, for reasons outlined in the first chapter, disingenuous at best, and largely ideological in its function.

49 Stephen R.L. Clark, 'Natural Integrity and Biotechnology', in David S. Oderberg and Jacqueline A. Laing, eds, *Human Lives: Critical Essays on Consequentialist Bioethics*, Basingstoke: Macmillan, 1997, p. 65.

50 David E. Cooper, 'Intervention, humility and animal integrity', in Holland and Johnson, eds, *Animal Biotechnology and Ethics*, p. 146.

51 Henk Verhoog, 'The Concept of Intrinsic Value and Transgenic Animals', *Journal of Agricultural and Environmental Ethics*, vol. 5 no. 2, 1992, p. 158.

52 Hans Jonas, 'Seventeenth Century and After: The Meaning of the Scientific and Technological Revolution', in *Philosophical Essays: From Ancient Creed to Technological Man*, Englewood Cliffs, Prentice-Hall, 1974, p. 70.

53 See Leon R. Kass, 'Teleology, Darwinism, and the Place of Man', in *Toward a More Natural Science: Biology and Human Affairs*, New York: Free Press, 1985. The quote from Darwin is taken from the Introduction to *The Origin of Species*, and is cited by Kass (p. 259).

54 Bernard E. Rollin, 'On *telos* and genetic engineering', in Holland and Johnson, eds, *Animal Biotechnology and Ethics*, p. 162.

55 Alan Holland, 'The Biotic Community: A Philosophical Critique of Genetic Engineering', in Wheale and McNally, eds, *The Bio-Revolution*, p. 170.

56 Stephen R.L. Clark, 'Natural Integrity and Biotechnology', in Oderberg and Laing, eds, *Human Lives*, p. 67.

57 See Mark Ridley, *The Problems of Evolution*, Oxford: Oxford University Press, 1985, pp. 89–92.

58 R.K. Colwell, cited by Verhoog, 'The Concept of Intrinsic Value and Transgenic Animals', *Journal of Agricultural and Environmental Ethics*, p. 156.

59 Bernard E. Rollin, *The Frankenstein Syndrome: Ethical and Social Issues in the Genetic Engineering of Animals*, New York: Cambridge University Press, 1995, p. 193. See also, by the same author, 'On *telos* and genetic engineering', in Holland and Johnson, eds, *Animal Biotechnology and Ethics*.

60 Cited by Rick Weiss, 'Creation of flawed animals raises new ethics issues', *Washington Post*, 7 June 1998.
61 Rollin, *The Frankenstein Syndrome*, p. 179.
62 'Feathers are a waste', declared Israeli geneticist Avigdor Cahaner, noting the inconvenience of feathers to the manufacturing of chicken meat. Featherless male chickens are also incapable of mating, since without feathers on their wings they cannot raise themselves above the hen. (Robert Uhlig, 'The little red rooster, oven ready and low in calories', *Daily Telegraph*, 21 May 2002.)
63 Andy Coghlan, 'Pressure group broods over altered turkeys', *New Scientist*, vol. 138, no. 1875, 29 May 1993, p. 9. The technique used here utilises the same 'antisense' technology employed by Calgene in its slow-softening tomato.
64 Clark, 'Natural Integrity and Biotechnology', in Oderberg and Laing, eds, *Human Lives*, p. 69.
65 Holland, 'The Biotic Community: A Philosophical Critique of Genetic Engineering', in Wheale and McNally, eds, *The Bio-Revolution*, p. 172.
66 Cooper, 'Intervention, humility and animal integrity', in Holland and Johnson, eds, *Animal Biotechnology and Ethics*.
67 See Kate Soper, *What is Nature?*, Oxford: Blackwell, 1995.
68 Niels Bohr, 'Light and Life', in *Atomic Physics and Human Knowledge*, New York: John Wiley and Sons, 1958, p. 9.
69 I have already noted that the idea that nature has independent or intrinsic value (the value of nature's 'autonomy' being one example of this) carries little credibility amongst philosophers.
70 Ted Benton argues for a critical interpretation of Marx's *Economic and Philosophical Manuscripts* along these lines in *Natural Relations: Ecology, Animal Rights and Social Justice*, London: Verso, 1993, Ch. 2.
71 John O'Neill, *Ecology, Policy and Politics: Human Well-Being and the Natural World*, London: Routledge, 1993, p. 151.
72 O'Neill, *Ecology, Policy and Politics*, p. 151.
73 See D.R. Crocker, 'Anthropomorphism: Bad Practice, Honest Prejudice?', in Georgina Ferry, ed., *The Understanding of Animals*, Oxford: Blackwell, 1984.
74 See Carolyn Merchant, *The Death of Nature: Women, Ecology and the Scientific Revolution*, New York: HarperCollins, 1990, chapter 7.
75 David F. Noble, *The Religion of Technology: The Divinity of Man and the Spirit of Invention*, Harmondsworth: Penguin, 1999, p. 181.

6 Health and Disease

1 Tim Radford, 'Scientists finish first draft of DNA blueprint', *Guardian*, 26 June 2000.
2 Tim Radford, 'Scientists revel in a day of glory', *Guardian*, 27 June 2000; James Meek, 'US wins the hype war', *Guardian*, 27 June 2000.
3 In the four months between October 2000 and February 2001, patent applications for human gene sequences in the US grew by thirty-eight per cent to over 175,000 (Madeleine Bunting, 'The profits that kill', *Guardian*, 12 February 2001). Applications for patents on several thousand DNA sequences were launched by the National Institutes of Health (NIH) in the early 1990s, though the US government agency was sensible enough to withdraw them subsequently. The private sequencing company Celera, which competed with the public consortium to map the human genome, for the most part avoided the aggressive patenting strategy adopted by companies like Incyte and Human Genome Sciences, choosing instead to pursue its commercial goals by licensing its private DNA database and related programmes for analysing the data to paying subscribers. While key figures in the Human Genome Project, partly inspired by the success of the 'open source' movement in the computer software industry, ensured that the results of the Project's own sequencing

efforts were released for free public use on the internet every evening, Celera was using this same material to improve its own commercial database. This strategy was recently rewarded with a three-year deal, involving an undisclosed sum of money, brokered with the UK government's main medical research funding agency, the Medical Research Council, allowing British scientists access to Celera's genetic data. For a UK insider's view of the Human Genome Project, and an earnest critique of the commercialisation of genetic information, see John Sulston and Georgina Ferry, *The Common Thread: The Story of Science, Ethics and the Human Genome*, London: Bantam Press, 2002.

4 Francis Collins, cited by Tim Radford, '30 years on . . . a brave new genetic world', *Guardian*, 9 February 2001.

5 Charles Murray, 'Genetics of the Right', *Prospect*, no. 51, April 2000, p. 29.

6 See 'The Human Genome', *Nature*, vol. 409, no. 6822, 15 February 2001. Celera predicts that humans have around 39,000 genes, but concedes that the evidence for some 12,000 of these is weak.

7 Tom Abate, 'Genome discovery shocks scientists: genetic blueprint contains far fewer genes than thought', *San Fancisco Chronicle*, 11 February 2001.

8 This, of course, is the contradictory argument of the animal experimentation industry: that non-human animals are sufficiently similar to *homo sapiens* to justify the use of the former as 'models' to investigate human disease and its treatment, but not similar enough to command the respect due to humans themselves. This argument, as already mentioned, has been demolished by Ray and Jean Greek's exposure of the tenuous relationship between animal experimentation and the advancement of medical knowledge and clinical treatment. (C. Ray Greek and Jean Swingle Greek, *Sacred Cows and Golden Geese: The Human Costs of Experiments on Animals*, New York: Continuum, 2000.)

9 Robin McKie, 'Revealed: the secret of human behaviour', *Observer*, 11 February 2001.

10 Aravinda Chakravarti, '. . . to a future of genetic medicine', *Nature*, vol. 409, no. 6822, 15 February 2001, p. 823.

11 Craig Venter later revealed that, unknown to his colleagues, most of the DNA used by Celera in its sequencing effort had actually come from his own cells, thus giving added meaning to his ownership claim over the human genome. (Robin McKie, 'Human genome "is mine"', *Guardian*, 28 March 2002.)

12 As Keller astutely observes, the obsession with 'disease genes' and the evasion of a genetic definition of normality also allows geneticists to preach 'freedom' (from suffering and disease) whilst clinging fiercely to its antithesis: the paradigm of genetic determinism. (Evelyn Fox Keller, 'Nature, Nurture, and the Human Genome Project', in Daniel J. Kevles and Leroy Hood, eds, *The Code of Codes: Scientific and Social Issues in the Human Genome Project*, Cambridge, Mass.: Harvard University Press, 1993.)

13 L.L. Cavalli-Sforza, A.C. Wilson, C.R. Cantor *et al*, 'Call for a Worldwide Survey of Human Genetic Diversity: A Vanishing Opportunity for the Human Genome Project', *Genomics*, vol. 11, 1991, pp. 490–1.

14 Leslie Roberts, 'Anthropologists Climb (Gingerly) on Board', *Science*, vol. 258, no. 5086, 20 November 1992, pp. 1300–1301.

15 R.C. Lewontin, 'The Apportionment of Human Diversity', *Evolutionary Biology*, vol. 6, 1972, pp. 381–98; Guido Barbujani, Arianna Magagni, Eric Minch, and L. Luca Cavalli-Sforza, 'An apportionment of human DNA diversity', *Proceedings of the National Academy of Sciences*, vol. 94, no. 9, April 1997, pp. 4516–19.

16 Robert W. Wallace, 'The Human Genome Diversity Project: Medical Benefits Versus Ethical Concerns', *Molecular Medicine Today*, vol. 4, no. 2, February 1998, pp. 59–62.

17 Alex Mauron and Jean-Marie Thévoz, 'Germ-Line Engineering: A Few European Voices', *Journal of Medicine and Philosophy*, vol. 16, no. 6, 1991, p. 656.

18 Matt Ridley, *Genome: The Autobiography of a Species in 23 Chapters*, London: Fourth Estate, 1999, pp. 192–4.

19 Kenan Malik, 'Darwinian Fallacy', *Prospect*, no. 36, December 1998.

20 David J. Weatherall, 'Single gene disorders or complex traits: lessons from the thalassaemias and other monogenic diseases', *British Medical Journal*, vol. 321, no. 7269, 4 November 2000, pp. 1117–20.
21 Matt Ridley, *Genome: The Autobiography of a Species in 23 Chapters*, pp. 142, 191.
22 Richard C. Strohman, 'Ancient Genomes, Wise Bodies, Unhealthy People: Limits of a Genetic Paradigm in Biology and Medicine', *Perspectives in Biology and Medicine*, vol. 37, no. 1, Autumn 1993, pp. 112–45
23 An 'homozygote' is an organism or cell in which the two copies ('alleles') of a specific gene are identical. When the alleles are different, the organism or cell is said to be 'heterozygous' for that gene, or is described as a 'heterozygote'. A recessive allele will only manifest its effects in a homozygote. In a heterozygote, the recessive allele is silenced by the effects of the other, 'dominant' allele.
24 Julian Zielenski and Lap-Chee Tsui, 'Cystic Fibrosis: Genotypic and Phenotypic Variations', *Annual Review of Genetics*, vol. 29, 1995, pp. 777–807.
25 <http: / / genet.sickkids.on.ca>
26 Philippe M. Frossard, 'Identification of cystic fibrosis mutations in Oman', *Clinical Genetics*, vol. 57, 2000, pp. 235–6.
27 Julian Zielenski and Lap-Chee Tsui, 'Cystic Fibrosis: Genotypic and Phenotypic Variations', *Annual Review of Genetics*, vol. 29, 1995, pp. 794–6; Eitan Kerem and Malka Nissim-Rafinia, Zvi Argaman *et al*, 'A Missense Cystic Fibrosis Transmembrane Conductance Regulator Mutation With Variable Phenotype', *Pediatrics*, vol. 100, no. 3, September 1997, p. e5.
28 Katrina M. Dipple and Edward R.B. McCabe, 'Genes Convert "Simple" Mendelian Disorders to Complex Traits', *Molecular Genetics and Metabolism*, vol. 71, no. 1/2, September/October 2000, pp. 43–50.
29 David J. Weatherall, 'Single gene disorders or complex traits: lessons from the thalassaemias and other monogenic diseases', *British Medical Journal*, vol. 321, no. 7269, 4 November 2000, pp. 1117–20.
30 Ruth Hubbard and Elijah Wald, *Exploding the Gene Myth*, Boston: Beacon Press, 1999, pp. 64–5.
31 Ernest Beutler, Vincent J. Felitti, James A. Koziol *et al*, 'Penetrance of 845G→A (C282Y) *HFE* hereditary haemochromatosis mutation in the USA', *Lancet*, vol. 359, no. 9302, 19 January 2002, pp. 211–8.
32 See <www.pkunews.org>
33 Jaakko Kaprio, 'Commentary: Role of other genes and environment should not be overlooked in monogenic disease', *British Medical Journal*, vol. 322, no. 7293, 28 April 2001, p. 1023.
34 Joseph X. DiMario, Akif Uzman, and R.C. Strohman, 'Fiber Regeneration is not Persistent in Dystrophic (MDX) Mouse Skeletal Muscle', *Developmental Biology* vol. 148, no. 1, 1991, pp. 314–21.
35 S.B. England, L.V. Nicholson, M.A. Johnson *et al*, 'Very Mild Muscular Dystrophy Associated with the Deletion of 46% of Dystrophin', *Nature*, vol. 343, no. 6254, 11 January 1990, pp. 180–2.
36 Q.L. Lu, G. E. Morris, S.D. Wilton, T. Ly, O.V. Artem'yeva, P. Strong, and T.A. Partridge, 'Massive Idiosyncratic Exon Skipping Corrects the Nonsense Mutation in Dystrophic Mouse Muscle and Produces Functional Revertant Fibers by Clonal Expansion', *Journal of Cell Biology*, vol. 148, no. 5, 6 March 2000, pp. 985–95.
37 Hugh C. Hendrie, Adesola Ogunniyi, Kathleen S. Hall *et al*, 'Incidence of Dementia and Alzheimer Disease in 2 Communities: Yoruba Residing in Ibadan, Nigeria, and African Americans Residing in Indianapolis, Indiana', *Journal of the American Medical Association*, vol. 285, no. 6, 14 February 2001, pp. 739–47.
38 Lon White, Helen Petrovitch, G. Webster Ross *et al*, 'Prevalence of Dementia in Older Japanese-American Men in Hawaii', *Journal of the American Medical Association*, vol. 276, no. 12, 25 September 1996, pp. 955–60.

39 Linsay A. Farrer, 'Intercontinental Epidemiology of Alzheimer Disease: A Global Approach to Bad Gene Hunting', *Journal of the American Medical Association*, vol. 285, no. 6, 14 February 2001, pp. 796–8.
40 This case is cited by Ruth Hubbard and R.C. Lewontin, 'Pitfalls of Genetic Testing', *New England Journal of Medicine*, vol. 334, no. 18, May 2, 1996, pp. 1192–3.
41 John J. Mulvihill, 'Craniofacial syndromes: no such thing as a single gene disease', *Nature Genetics*, vol. 9, no. 2, February 1995, pp.101–3.
42 Katrina M. Dipple and Edward R.B. McCabe, 'Genes Convert "Simple" Mendelian Disorders to Complex Traits', *Molecular Genetics and Metabolism*, vol. 71, no. 1/2, September/October 2000, pp. 43–50.
43 Between 1969 and 1996, for example, 14 different chromosomal regions were identified as possible genetic causes of manic depression, yet none of the studies have been consistently replicated. See Neil Risch and David Botstein, 'A manic depressive history', *Nature Genetics*, vol. 12, no. 4, April 1996, pp. 351–3.
44 Neil A. Holtzman and Teresa M. Marteau, 'Will Genetics Revolutionise Medicine?', *New England Journal of Medicine*, vol. 343, no. 2, 13 July 2000, p. 141.
45 See André Gorz, 'Medicine, Health and Society', in Gorz, *Ecology as Politics*, Boston: South End Press, 1980; Thomas McKeown, *The Role of Medicine: Dream, Mirage or Nemesis*, Oxford: Blackwell, 1979.
46 Since the main infectious diseases were airborne rather than waterborne, Lewontin argues that improvements in sanitation were incidental to the decline in mortality rates during the nineteenth century, which instead reflected 'a general trend of increase in the real wage, an increase in the state of nutrition of European populations, and a decrease in the number of hours worked'. (Richard Lewontin, *The Triple Helix: Gene, Organism and Environment*, Cambridge, Mass.: Harvard University Press, 2001, p. 104.)
47 Matt Ridley, *Genome: The Autobiography of a Species in 23 Chapters*, p. 236.
48 Peter G. Gosselin and Paul Jacobs, 'Patent Office Now at Heart of Gene Debate', *Los Angeles Times*, 7 February 2000. In March 2000 Myriad Genetics announced a licensing agreement with a British company, Rosgen, to sell cancer screening in the UK (James Meek, 'US genetics firm in cancer check deal', *Guardian*, 9 March 2000). The world-renowned French cancer research centre, the Institut Curie in Paris, is currently contesting the European Patent Office's decision to grant these patent rights to Myriad, arguing that it is preventing other scientists from developing faster, cheaper and more reliable breast cancer tests. (Jon Henley, 'Cancer unit fights US gene patent', *Guardian*, 8 September 2001.) The European Parliament has also passed a resolution committing its members to challenging the patent.
49 Fergus J. Crouch, Michelle L. DeShano, M. Anne Blackwood *et al*, 'BRCA1 Mutations in Women Attending Clinics that Evaluate the Risk of Breast Cancer', *New England Journal of Medicine*, vol. 336, no. 20, 15 May 1997, pp. 1409–15.
50 E. B. Claus, J. Schildkraut, E.S. Iversen Jr., D. Berry, and G. Parmigiani, 'Effect of BRCA1 and BRCA2 in the association between breast cancer risk and family history', *Journal of the National Cancer Institute*, vol. 90, no. 23, 2 December 1998, pp. 1824–9.
51 Jeffery P. Struewing, Patricia Hartge, Sholom Wacholder *et al*, 'The Risk of Cancer Associated with Specific Mutations of BRCA1 and BRCA2 among Ashkenazi Jews', *New England Journal of Medicine*, vol. 336, no. 20, 15 May 1997, pp. 1401–8.
52 Linda D. Voss and Jean Mulligan, 'Bullying in school: are short pupils at risk? Questionnaire study in a cohort', *British Medical Journal*, vol. 320, no. 7235, 4 March 2000, pp. 612–13.
53 G.A. Colditz, B.A. Rosner, F.E. Speizer, 'Risk factors for breast cancer according to family history of breast cancer', *Journal of the National Cancer Institute*, vol. 88, no. 6, 20 March 1996, pp. 365–71; Kenneth van Golen, Kara Milliron, Seena Davies, and Sofia D. Merajver, '*BRCA*-Associated Cancer Risk: Molecular Biology and Clinical Practice', *Journal of Laboratory and Clinical Medicine*, vol. 134, no. 1, July 1999, pp. 11–18.
54 <www.medinfo.cam.ac.uk/phgu/info_database/Diseases/Cancer/cancer.asp>

55 Eric R. Fearon, 'Human Cancer Syndromes: Clues to the Origin and Nature of Cancer', *Science*, vol. 278, no. 5340, 7 November 1997, pp. 1043–50.

56 Robert N. Hoover, 'Cancer – nature, nurture, or both', *New England Journal of Medicine*, vol. 343, no. 2, 13 July 2000, pp. 135–6.

57 Paul Lichtenstein, Niels V. Holm, Pia K. Verkasalo, 'Environmental and Heritable Factors in the Causation of Cancer: Analysis of Cohorts of Twins from Sweden, Denmark, and Finland', *New England Journal of Medicine*, vol. 343, no. 2, 13 July 2000, pp. 78–85.

58 R.G. Ziegler, R.N. Hoover, M.C. Pike *et al*, 'Migration patterns and breast cancer risk in Asian-American women', *Journal of the National Cancer Institute*, vol. 85, no. 22, 17 November 1993, pp. 1819–27.

59 Edward Goldsmith, 'Are the Experts Lying?', *The Ecologist*, vol. 28, no. 2, March/April 1998, pp. 51–2. Contrary to the claim that this increase can be attributed to the effects of smoking, 'about 75% of the increased cancer incidence since 1950 has been in sites other than the lung' (Samuel Epstein, 'Winning the War Against Cancer? . . . Are They Even Fighting it?', *The Ecologist*, vol. 28, no. 2, March/April 1998, p. 71).

60 George Monbiot, 'Purporting to be beating cancer', *Guardian*, 4 January 2001. A famous study in 1990, reporting on a dramatic 30 per cent drop in age-specific cancer mortality rates among young Israeli women between 1976 and 1986, correlated this sudden and impressive improvement with the banning in 1978 of several carcinogenic pesticides from dairy cow feed. (Jerome B. Westin and Elihu Richter, 'The Israeli Breast-Cancer Anomaly', *Annals for the New York Academy of Sciences*, vol. 609, 1990, pp. 269–79.)

61 Until it de-merged its pharmaceuticals, agrochemicals, and 'specialities' businesses to form the new company Zeneca in 1993, ICI exclusively funded Breast Cancer Awareness Month. Zeneca merged with Astra AB in 1999. The 'non-operating private foundation', AstraZeneca Healthcare Foundation, remains the 'principal sponsor' of the US National Breast Cancer Awareness Month (NBCAM) programme. <www.nbcam.org/board.cfm?section=board>

62 Joann G. Elmore, Mary B. Barton, Victoria M. Moceri *et al*, 'Ten-year risk of false positive screening mammograms and clinical breast examinations', *New England Journal of Medicine*, vol. 338, no. 16, 16 April 1998, pp. 1089–96.

63 Peter C. Gøtzsche and Ole Olsen, 'Is screening for breast cancer with mammography justifiable?', *Lancet*, vol. 355, no. 9198, 8 January 2000, pp. 129–34; Ole Olsen and Peter C. Gøtzsche, 'Cochrane review on screening for breast cancer with mammography', *Lancet*, vol. 358, no. 9290, 20 October 2001, pp. 1340–2.

64 Lennarth Nyström, Ingvar Andersson, Nils Bjurstam *et al*, 'Long-term effects of mammography screening: updated overview of the Swedish randomised trials', *Lancet*, vol. 359, no. 9310, 16 March 2002, pp. 909–19.

65 According to a recent Canadian study, the risk of developing a tumour of the womb lining increases four-fold in people who take tamoxifen, while the risk of deep vein thrombosis doubles. (B.P. Will, K.M. Nobrega, J-M. Berthelot, 'First do no harm: extending the debate on the provision of preventive tamoxifen', *British Journal of Cancer*, vol. 85, no. 9, November 2001, pp. 1280–8.)

66 Monte Paulsen, 'The politics of cancer: why the medical establishment blames victims instead of carcinogens', *Utne Reader*, no. 60, November/December 1993, pp. 81–9.

67 J.K. Cruickshank, J.C. Mbanya, R. Wilks *et al*, 'Sick genes, sick individuals or sick populations with chronic disease? The emergence of diabetes and high blood pressure in African-origin populations', *International Journal of Epidemiology*, vol. 30, 2001, pp. 111–17.

68 M. Vrijheid, H. Dolk, B. Armstrong *et al*, 'Chromosomal congenital abnormalities and residence near hazardous-waste landfill sites', *Lancet*, vol. 359, no. 9303, 26 January 2002, pp. 320–22. The same researchers have reported an increased risk of comparable magnitude of non-chromosomal congenital abnormalities in pregnancy cases within three kilometres of hazardous-waste landfill sites. (H. Dolk, M. Vrijheid, B. Armstrong *et al*, 'Risk of congenital abnormalities near hazardous-waste landfill sites in Europe: the EUROHAZCON study', *Lancet*, vol. 352, no. 9126, 8 August 1998, pp. 423–7.) See also: Paul Elliott, David Briggs,

Sara Morris *et al*, 'Risk of adverse birth outcomes in populations living near landfill sites', *British Medical Journal*, vol. 323, no. 7309, 18 August 2001, pp. 363–8.

69 Mihael H. Polymeropoulos, Christian Lavedan, Elisabeth Leroy *et al*, 'Mutation in the alpha-Synuclein Gene Identified in Families with Parkinson's Disease', *Science*, vol. 276, no. 5321, 27 June 1997, pp. 2045–7.

70 Tohru Kitada, Shuichi Asakawa, Nobutaka Hattori *et al*, 'Mutations in the *parkin* gene cause autosomal recessive juvenile parkinsonism', *Nature*, vol. 392, no. 6676, 9 April 1998, pp. 605–8; Udall Parkinson's Disease Research Center of Excellence, 'Duke Researchers Find Specific Genetic Link to Broad Spectrum of Parkinson's Diseases Cases', Parkin Gene Press Release: Duke University Medical Centre, 5 October 2000.

71 P. Chan, C.M. Tanner, X. Jiang and J.W. Lagston, 'Failure to find the alpha-synuclein gene missense mutation (G209A) in 100 patients with younger onset Parkinson's disease', *Neurology*, vol. 50, no. 2, 1998, pp. 513–4.

72 <www.parkinsons-foundation.org>; <www.parkinsonsinstitute.org>

73 J. William Langston, Philip Ballard, James W. Tetrud, Ian Irwin, 'Chronic parkinsonism in humans due to a product of meridine-analog synthesis', *Science*, vol. 219, no. 4587, 1983, pp. 979–80.

74 Mona Thiruchelvam, Eric K. Richfield, Raymond B. Baggs *et al*, 'The Nigrostriatal Dopaminergic System as a Preferential Target of Repeated Exposures to Combined Paraquat and Maneb: Implications for Parkinson's Disease', *Journal of Neuroscience*, vol. 20, no. 24, 15 December 2000, pp. 9207–14.

75 Ranjita Betarbet, Todd B. Sherer, Gillian MacKenzie *et al*, 'Chronic systemic pesticide exposure reproduces features of Parkinson's disease', *Nature Neuroscience*, vol. 3, no. 12, December 2000, pp. 1301–6.

76 R.C. Lewontin, *The Doctrine of DNA: Biology as Ideology*, Harmondsworth: Penguin, 1993, pp. 67–8.

77 WHO, *The World Health Report 1999*, Geneva: WHO, 1999, pp. 104–108, 24.

78 Robin McKie, 'Gene cures "will not help Third World"', *Observer*, 11 February 2001.

79 The 'normal' lifetime risk of breast cancer for women is calculated as 12.6 per cent (Neil A. Holtzman and Theresa M. Marteau, 'Will Genetics Revolutionise Medicine?', *New England Journal of Medicine*, vol. 343, no. 2, 13 July 2000, pp. 141–4.)

80 Lewontin, *The Doctrine of DNA*, p. 51.

81 Lewontin, *The Doctrine of DNA*, p. 46.

82 Richard G. Wilkinson, *Unhealthy Societies: The Afflictions of Inequality*, London: Routledge, 1996.

83 Richard C. Strohman, 'Ancient Genomes, Wise Bodies, Unhealthy People: Limits of a Genetic Paradigm in Biology and Medicine', *Perspectives in Biology and Medicine*, vol. 37, no. 1, Autumn 1993, p. 139.

84 Strohman, 'Ancient Genomes, Wise Bodies, Unhealthy People', pp. 129, 140.

85 Erich Fromm, *The Sane Society*, London: Routledge and Kegan Paul, 1963, p. 133.

7 From Pharmacogenetics to Gene Therapy

1 Dorothy Nelkin, 'The Social Power of Genetic Information', in Daniel J. Kevles and Leroy Hood, eds, *The Code of Codes: The Scientific and Social Issues in the Human Genome Project*, Cambridge, Mass.: Harvard University Press, 1993, p. 188.

2 C. Roland Wolf and Gillian Smith, 'Pharmacogenetics', *British Medical Bulletin*, vol. 55, no. 2, 1999, pp. 366–86.

3 C. Roland Wolf, Gillian Smith, Robert L. Smith, 'Pharmacogenetics', *British Medical Journal*, vol. 320, no. 7240, 8 April 2000, pp. 987–90.

4 Wolfgang Sadée, 'Pharmacogenomics', *British Medical Journal*, vol. 319, no. 7220, 13 November 1999, pp. 1286–9.

5 E.J. Calabrese, 'Genetic predisposition to occupationally-related diseases: current status and

future directions', in Philippe Grandjean, ed., *Ecogenetics: Genetic Predisposition to the Toxic Effects of Chemicals*, London: Chapman and Hall, 1991, p. 22.

6 Jason Lazarou, Bruce H. Pomeranz, Paul N. Corey, 'Incidence of Adverse Drug Reactions in Hospitalized Patients: A Meta-analysis of Prospective Studies', *Journal of the American Medical Association*, vol. 279, no. 15, 15 April 1998, pp. 1200–1205.

7 Robert Snedden, 'The Challenge of Pharmacogenetics and Pharmacogenomics', *New Genetics and Society*, vol. 19, no. 2, August 2000, p. 149.

8 Ruth Hubbard and Elijah Wald, *Exploding the Gene Myth*, Boston: Beacon, 1999, p. 110.

9 When an egg and sperm cell fuse, the result is a single cell with two nuclei, each containing half the genetic material found in the diploid nucleus of a somatic cell. These two 'pronuclei' do not immediately fuse into one, but rather duplicate themselves first and then combine with their mate after the cell has divided into two.

10 This latter technique was successfully applied to mice recently by Ralph Brinster and his colleagues at the University of Pennsylvania. They used a retrovirus to carry a non-invasive marker or 'reporter' gene into the nuclei of mouse spermatogonial stem cells, then transplanted the genetically modified cells into the mice. Of the progeny of these male mice, 4.5 per cent were found to be transgenic, and the transgene was transmitted to and expressed in subsequent generations. (Makoto Nagano, Clayton J. Brinster, Kyle E. Orwig *et al*, 'Transgenic mice produced by retroviral transduction of male germ-line stem cells', *Proceedings of the National Academy of Sciences*, vol. 98, no. 23, 6 November 2001, pp. 13090–5.)

11 Paul A. Martin, 'Genes as drugs: the social shaping of gene therapy and the reconstruction of genetic disease', *Sociology of Health and Illness*, vol. 21, no. 5, 1999, pp. 517–38.

12 W. French Anderson, 'Human Gene Therapy', *Science*, vol. 256, no. 5058, 8 May 1992, pp. 808–13.

13 Marina Cavazzana-Calvo, Salima Hacein-Bey, Geneviève de Saint Basile *et al*, 'Gene Therapy of Human Severe Combined Immunodeficiency (SCID)-X1 Disease', *Science*, vol. 288, no. 5466, 28 April 2000, pp. 669–72.

14 James Meikle, 'Pioneering gene treatment gives frail toddler a new lease of life', *Guardian*, 4 April 2002.

15 R. Michael Blaese, Kenneth W. Culver, A. Dusty Miller *et al*, 'T Lyphocyte-Directed Gene Therapy for ADA$^-$ SCID: Initial Trial Results After 4 Years', *Science*, vol. 270, no. 5235, 20 October 1995, pp. 475–80.

16 Donald B. Kohn, Michael S. Hershfield, Denise Carbonaro *et al*, 'T lymphocytes with a normal ADA gene accumulate after transplantation of transduced autologous umbilical cord blood CD34+ cells in ADA-deficient SCID neonates', *Nature Medicine*, vol. 4, no. 7, July 1998, pp. 775–80.

17 Nicholas Wade, 'Gene experiment comes close to crossing ethicists' line', *New York Times*, 23 December 2001.

18 Michael R. Knowles, Kathy W. Hohneker, R.N. Zhaoqing Zhou *et al*, 'A Controlled Study of Adenoviral-Vector-Mediated Gene Transfer in the Nasal Epithelium of Patients with Cystic Fibrosis', *New England Journal of Medicine*, vol. 333, no. 13, 18 September 1995, pp. 823–31; Ronald G. Crystal, Noel G. McElvaney, Melissa A. Rosenfeld *et al*, 'Administration of an adenovirus containing the human CFTR cDNA to the respiratory tract of individuals with cystic fibrosis', *Nature Genetics*, vol. 8, no. 1, September 1994, pp. 42–51.

19 Sally Lehrman, 'Virus Treatment Questioned After Gene Therapy Death', *Nature*, vol. 401, no. 6753, 7 October 1999, pp. 517–8; Paul Smaglik, 'Tighter watch urged on adenoviral vectors', *Nature*, vol. 402, no. 6763, 16 December 1999, p. 707; Eliot Marshall, 'Gene Therapy on Trial', *Science*, vol. 288, no. 5468, 12 May 2000, pp. 951–6.

20 Rick Weiss and Deborah Nelson, 'Victim's Dad Faults Gene Therapy Team', *Washington Post*, 3 February 2000.

21 A. Donsante, C. Vogler, N. Muzyczka *et al*, 'Observed incidence of tumorigenesis in long-term rodent studies of rAAV vectors', *Gene Therapy*, vol. 8, no. 17, September 2001, pp. 1343–6.

22 Daniel G. Miller, Elizabeth A. Rutledge and David W. Russell, 'Chromosomal effects of adeno-associated virus vector integration', *Nature Genetics*, vol. 30, no. 2, February 2002, pp. 147–8.

23 In March 2001, an international group of scientists reported that they had successfully enhanced the cognitive plasticity and learning and memory performance of mice by the insertion of a gene whose product inhibits the action of the neural enzyme calcineurin. Using a technique that enabled them to turn off the expression of the transgene when required, the geneticists believe their research 'may provide a clear target for potential treatment of learning and memory disorders'. (Gaël Malleret, Ursula Haditsch, David Genoux *et al*, 'Inducible and Reversible Enhancement of Learning, Memory, and Long-Term Potentiation by Genetic Inhibition of Calcineurin', *Cell*, vol. 104, no. 5, 9 March 2001, pp. 675–86.)

24 Rick Weiss, 'Cosmetic gene therapy's thorny traits', *Washington Post*, 12 October 1997.

25 Robin McKie, 'Coming soon: a pill to cure forgetfulness', *Observer*, 11 November 2001.

26 Peter Kramer, *Listening to Prozac*, New York: Pantheon, 1993. See also Francis Fukuyama, *Our Posthuman Future: Consequences of the Biotechnology Revolution*, New York: Farrar, Straus and Giroux, 2002, chapter 3.

27 Christie Aschwanden, 'Gene Cheats', *New Scientist*, vol. 165, no. 2221, 15 January 2000, pp. 24–9.

28 Anon, 'Patenting Genes – Stifling Research and Jeopardising Healthcare', Econexus and GeneWatch UK, April 2001. <www.genewatch.org>

29 Andrew Kimbrell, *The Human Body Shop: The Cloning, Engineering, and Marketing of Life*, Washington: Gateway, 1997, p. 170–6.

30 David Galton, *In Our Own Image: Eugenics and the Genetic Modification of People*, London: Little, Brown, 2001, p. 53.

31 Gina Kolata, 'Selling growth drug for children: the legal and ethical questions', *New York Times*, 15 August 1994.

32 Hubbard and Wald, *Exploding the Gene Myth*, pp. 69–71.

33 Kimbrell, *The Human Body Shop*, p. 175.

34 Daniel Rudman, Axel G. Feller, Hoskote S. Nagraj, 'Effects of Human Growth Hormone in Men Over 60 Years Old', *New England Journal of Medicine*, vol. 323, no. 1, 5 July 1990, pp. 1–6.

35 Tania Branigan, 'Girl "stretched" in £12,000 operation', *Guardian*, 14 November 2000.

36 Philip Cohen, 'Dolly helps the infertile', *New Scientist*, vol. 158, no. 2133, 9 May 1998, p. 6.

37 Rachel Nowak, 'Sex . . . who needs it?', *New Scientist*, vol. 173, no. 2326, 19 January 2002, p. 5.

38 Kimbrell, *The Human Body Shop*, pp. 95–6.

39 J.J. Eppig and M.J. O'Brien, 'Development in vitro of mouse oocytes from primordial follicles', *Biology of Reproduction*, vol. 54, no. 1, January 1996, pp. 197–207.

40 Ralph L. Brinster and James W. Zimmermann, 'Spermatogenesis following male germ-cell transplantation', *Proceedings of the National Academy of Sciences*, vol. 91, no. 24, November 1994, pp. 11298–302; David E. Clouthier, Mary R. Avarbock, Shanna D. Maika *et al*, 'Rat spermatogenesis in mouse testis', *Nature*, vol. 381, no. 6581, 30 May 1996, pp. 418–21.

41 Richard Dawkins, 'What's Wrong with Cloning', in Martha C. Nussbaum and Cass R. Sunstein, eds, *Clones and Clones: Facts and Fantasies about Human Cloning*, New York: W.W. Norton, 1999, pp. 57–8.

42 Leon R. Kass, 'The Wisdom of Repugnance', in Leon R. Kass and James Q. Wilson, *The Ethics of Human Cloning*, Washington, D.C.: AEI Press, 1998, p. 19.

43 The conception of the human as an industrial process and product is of course well entrenched in our technocratic culture, where it is nourished and encouraged by the logic of economic rationality. Consider, amongst other well-documented cases, the perverted logic of a legal decision made in the US in 1976. Here, a woman who had for several years earned a regular income from the sale of her rare form of blood, was granted tax exemption for the

money she spent on a requisite diet and on travelling to the clinic. Satisfied by the arguments of her lawyer, the court ruled that, since the woman was essentially functioning as both a 'factory' and a 'container' for her commercial product, the cost of raw materials (food) and transport (travel) were tax deductible. (Kimbrell, *The Human Body Shop*, pp. 19–21.) She did fail, however, in her claim for a ten per cent deduction owing to 'depreciation' (the blood donor's plasma eventually loses its ability to regenerate). (Lori Andrews and Dorothy Nelkin, *Body Bazaar: The Market for Human Tissue in the Biotechnology Age*, New York: Crown, 2001, p. 196, n58.)

44 Sarah Boseley, 'One birth in 80 from a test tube', *Guardian*, 28 June 2001.

45 One should not forget that average sperm counts amongst men in developed countries have fallen precipitously to around a half of what they were three decades ago, and that the most likely cause of declining male fertility is the build up of oestrogen-mimicking industrial chemicals, as well as the synthetic oestrogen (ethanol oestradiol) used in the female contraceptive pill, in the environment and water supply. (Paul Brown, 'Fish clue to human fertility decline', *Guardian*, 18 March 2002.)

46 An independent commission recommended the destruction of the frozen embryos, but the state legislature voted against the recommendations, and the embryos remain in a Melbourne medical centre. (Lee M. Silver, *Remaking Eden: Cloning, Genetic Engineering and the Future of Humankind*, London: Pheonix, 1999, pp. 96–98.)

47 After a series of rulings, appeals and counter-appeals, the man was granted a constitutional right to veto any proposed implantation of the embryos, leaving the frozen zygotes in biological and legal limbo. (Kimbrell, *The Human Body Shop*, pp. 106–114.)

48 Rick Weiss, 'Babies in limbo: laws outpaced by fertility advances', *Washington Post*, 8 February 1998.

49 In this case the birth mother eventually filed to 'adopt' both babies, and in the face of an embarrassing court case the client-parents relinquished the girl. (Kimbrell, *The Human Body Shop*, Washington: Gateway, 1997, pp. 127–9.)

50 Jamie Wilson, 'Surrogate twins find new parents', *Guardian*, 14 August 2001.

51 Combined with a practice called embryo fusion, in which cells from different embryos are brought together and persuaded to develop into a chimeric organism (as very occasionally happens in nature), cloning could be used to enable partners in a same-sex relationship to have a child that is genetically related to *both* of them. Portentously, the Chicago scientists who helped save the life of Fanconi anaemia sufferer Molly Nash (see below) have made significant progress developing another relevant reproductive technique, initially with the aim of allowing men who cannot produce sperm to have their own genetic offspring. By forcing the somatic cells of a man or woman to divide into *haploid* progeny cells (each with only one set of chromosomes), they claim to be able to use such 'artificial sperm' to fertilise another woman's egg – a procedure they believe may be commercially available in the US to lesbian couples in less than two years. (Jonathan Leake, 'Artificial sperm may let lesbians father babies', *Sunday Times*, 20 January 2002.)

52 According to a prominent American bioethicist, there are '100,000 frozen embryos left over after IVF [which could] be used to produce new stem cell lines'. Arlene Judith Klotzko, 'Embryonic victory', *Guardian*, 20 August 2001.

53 James Meek, 'Foetus designed to bar inherited illness', *Guardian*, 17 October 2000.

8 The Road to Human Cloning

1 'Declaration in Defence of Cloning and the Integrity of Scientific Research', *Free Inquiry*, vol. 17, no. 3, Summer 1997, pp. 11–12.

2 Richard Dawkins, 'Foreword', in Justine Burley, ed., *The Genetic Revolution and Human Rights*, Oxford: Oxford University Press, 1999, pp. x-xi.

3 See John Harris, 'Clones, Genes, and Human Rights', in Justine Burley, ed., *The Genetic Revolution and Human Rights*.

4 See R.C. Lewontin, 'The Confusion over Cloning', *New York Review of Books*, vol. 44, no. 16, 23 October 1997, pp. 18–23.
5 Sarah Boseley, 'Caesarean births soar to one in five', *Guardian*, 26 October 2001.
6 The increased rate and improved techniques of caesareans since the 1950s has certainly saved lives, and indeed over half of all caesareans performed in Britain are categorised as 'emergency' interventions. But the growth in caesareans also reflects the way childbirth is increasingly perceived in purely functional terms, as a medical process to be accomplished with maximum efficiency and minimum time and effort, and in such a way that potentially litigious outcomes can be eliminated. The common belief that caesarean deliveries reduce the likelihood of mothers suffering future pelvic floor disorders has not been corroborated by research. (See Alastair H. MacLennan, Anne W. Taylor, David H. Wilson, Don Wilson, 'The prevalence of pelvic floor disorders and their relationship to gender, age, parity and mode of delivery', *British Journal of Obstetrics and Gynaecology*, vol. 107, no. 12, December 2000, pp. 1460–70.) A recent study of 165 emergency caesareans performed in the UK also found that five years after the operations thirty per cent of the women who had undergone the procedure had been unable to conceive any further children. (Amelia Hill, 'Caesareans linked to risk of infertility', *Observer*, 21 April 2002.)
7 Anne Karpf, 'A pox on vaccines', *Guardian*, 16 January 2002.
8 Alan Stockdale, 'Waiting for the cure: mapping the social relations of human gene therapy research', *Sociology of Health and Illness*, vol. 21, no. 5, 1999, pp. 579–96.
9 Until December 2001, PGD in the UK was only permitted for couples at risk of transmitting life-threatening conditions – such as cystic fibrosis, sickle cell anaemia, or Duchenne muscular dystrophy – to their intended child, and had only been used in around fifty cases since its introduction in 1990. Following a legal challenge and licencing review, it can now be made available in licensed fertility clinics both as a means of improving implantation rates by screening out embryos with minor chromosomal imperfections, and as a way of identifying embryos which, when born, will provide compatible umbilical cord blood cells for transplantation to a seriously ill sibling. (Beezy Marsh, '"Perfect baby soon" as genetic test is approved', *Daily Mail*, 12 November 2001; Sarah Boseley, 'Rules eased on designer babies', *Guardian*, 13 December 2001.)
10 Ronald Dworkin, *Sovereign Virtue: The Theory and Practice of Equality*, Cambridge, Mass.: Harvard University Press, 2000, p. 433.
11 Marc Lappé, 'Ethical Issues in Manipulating the Human Germ Line', *Journal of Medicine and Philosophy*, vol. 16, no. 6, 1991, p. 627.
12 Ray Moseley, 'Maintaining the Somatic/Germ-Line Distinction: Some Ethical Drawbacks', *Journal of Medicine and Philosophy*, vol. 16, no. 6, 1991, p. 646.
13 Burke K. Zimmerman, 'Human Germ-Line Therapy: The Case for its Development and Use', *Journal of Medicine and Philosophy*, vol. 16, no. 6, 1991, pp. 597–8.
14 Gregory Stock, *Redesigning Humans: Our Inevitable Genetic Future*, New York: Houghton Mifflin, 2002, p. 13. This slippery slope may also resemble the road to moral self-entrapment which Stanley Milgram's famous study of obedience to authority revealed, where people continually acquiesce to acts of steadily increasing cruelty or moral dubiousness because a change of heart would signal admission of past wrong doing and a recognition that the road they had already travelled down was the wrong one. (See Bauman's discussion of Milgram's experiments in Zygmunt Bauman, *Modernity and the Holocaust*, Cambridge: Polity Press, 1989, pp. 157–9.)
15 Moseley, 'Maintaining the Somatic/Germ-Line Distinction', p. 644.
16 This has now become a source of considerable concern to many scientists working in the fertility industry. The problem is that intracytoplasmic sperm injection (ICSI) essentially works by circumnavigating the evolutionary barriers to conception facing genetically deficient sperm, and also that it involves an invasive mechanical assault on the integrity of the egg. Research already indicates both that the technical aspects of the injection procedure can have a detrimental effect on the healthy development of the early embryo, and that

there are higher rates of chromosomal abnormalities in eggs that have been fertilised using ICSI. (See, for example: John C.M. Dumoulin, Edith Coonen, Marijke Bras *et al*, 'Embryo development and chromosomal abnormalities after ICSI: effect of the injection procedure', *Human Reproduction*, vol. 16, no. 2, February 2001, pp. 306–12; Ervin Macas, Bruno Imthurn and Paul J. Keller, 'Increased incidence of numerical chromosome abnormalities in spermatozoa injected into human oocytes by ICSI', *Human Reproduction*, vol. 16, no. 1, January 2001, pp. 115–20.)

17 Johnjoe McFadden, 'Our genes are doomed', *Guardian*, 5 February 2001.

18 Alex Mauron and Jean-Marie Thévoz, 'Germ-Line Engineering: A Few European Voices', *Journal of Medicine and Philosophy*, vo. 16, no. 6, 1991, p. 653.

19 W. French Anderson, 'Human Gene Therapy: Why Draw a Line?', *Journal of Medicine and Philosophy*, vol. 14, no. 6, 1989, pp. 681–93.

20 Esmail D. Zanjani and W. French Anderson, 'Prospects for in Utero Human Gene Therapy', *Science*, vol. 285, no. 5436, 24 September 1999, pp. 2084–8. The genetic manipulation of embryos or foetuses may also be the only way of effectively treating genetic diseases whose biochemical agents strike or establish themselves before birth.

21 David King, Tom Shakespeare, Richard Nicholson *et al*, 'Correspondence', *Nature*, vol. 397, no. 6718, 4 February 1999, p. 383.

22 Jerry Hall and Yan-Ling Feng, 'Parthenogenetically Derived Stem Cells Form Nerve Cells', Press Release from the 57th Annual Meeting of the American Society for Reproductive Medicine, 22 October 2001.

23 Michelle Pountney, 'Making babies without men', *Herald Sun*, 23 October 2001.

24 In August 2001 President Bush announced a further compromise in his support of the opinions of religious conservatives, allowing federal funding for experiments using existing cell cultures originating from human embryos, but refusing to fund research on embryos themselves. There are seventy-eight stem cell batches sanctioned by the US government for private research, most of which are held in other countries. The vast majority of these cell lines, incidentally – indeed, *all* the human embryonic stem cells lines reported in the scientific literature to date – have been cultured using growth factors released by embryonic mouse cells with which the human cells were mixed. The risk that these cells may consequently carry mouse viruses, or that mouse cells may themselves end up in the bodies of human transplant recipients, is such that the use of these cells will almost certainly be covered by restrictive FDA xenotransplantation regulations. The task facing the leading researchers in this area is thus to generate, with the backing of private capital, new human stem cell lines using human rather than mouse feeder cells. (Justin Gillis and Ceci Connolly, 'Stem cell research faces FDA hurdle', *Washington Post*, 24 August 2001.)

25 William M. Rideout, Teruhiko Wakayama, Anton Wutz *et al*, 'Generation of mice from wild-type and targeted ES cells by nuclear cloning', *Nature Genetics*, vol. 24, no. 2, February 2000, pp. 109–110.

26 I. Wilmut, A.E. Schnieke, J. McWhir *et al*, 'Viable offspring derived from fetal and adult mammalian cells', *Nature*, vol. 385, no. 6619, 27 February 1997, p. 812.

27 James A. Thomson, Joseph Itskovitz-Eldor, Sander S. Shapiro *et al*, 'Embryonic Stem Cell Lines Derived from Human Blastocysts', *Science*, vol. 282, no. 5391, 6 November 1998, pp. 1145–7.

28 Izhak Kehat, Dorit Kenyagin-Karsenti, Mirit Snir *et al*, 'Human embryonic stem cells can differentiate into myocytes with structural and functional properties of cardiomyocytes', *Journal of Clinical Investigation*, vol. 108, no. 3, August 2001, pp. 407–14.

29 Suheir Assady, Gila Maor, Michal Amit *et al*, 'Insulin Production by Human Embronic Stem Cells', *Diabetes*, vol. 50, no. 8, August 2001, pp. 1691–7.

30 Dan S. Kaufman, Eric T. Hanson, Rachel L. Lewis *et al*, 'Hematopoietic colony-forming cells derived from human embryonic stem cells', *Proceedings of the National Academy of Sciences*, vol. 98, no. 19, 11 September 2001, pp. 10716–21.

31 Curt R. Freed, Paul E. Greene, Robert E. Breeze, 'Transplantation of Embryonic

Dopamine Neurons for Severe Parkinson's Disease', *New England Journal of Medicine*, vol. 344, no. 10, 8 March 2001, pp. 710–19; Sarah Boseley, 'Parkinson's miracle cure turns into a catastrophe', *Guardian*, 13 March 2001.

32 Lars M. Björklund, Rosario Sánchez-Pernaute, Sangini Chung *et al*, 'Embryonic stem cells develop into functional dopaminergic neurons after transplantation in a Parkinson rat model', *Proceedings of the National Academy of Sciences*, vol. 99, no. 4, 19 February 2002, pp. 2344–9.

33 David Humpherys, Kevin Eggan, Hidenori Akutsu *et al*, 'Epigenetic Instability in ES Cells and Cloned Mice', *Science*, vol. 293, no. 5527, 6 July 2001, pp. 95–97. Attempts to clone sheep as part of a commercial breeding programme at the South Australian Research and Development Institute have been hindered by rates of post-natal mortality among cloned lambs as high as nine out of ten. (Penelope Debelle, 'Death rate of cloned sheep a blow to wool industry', *Sidney Morning Herald*, 28 November 2001.) Recent attempts to clone mice from adult somatic cells have also revealed high rates of premature death associated with impaired immune systems and liver failure, indicating that the deleterious effects of cloning may be long-term or delayed. (Narumi Ogonuki, Kimiko Inoue, Yoshie Yamamoto *et al*, 'Early death of mice cloned from somatic cells', *Nature Genetics*, vol. 30, no. 3, March 2002, pp. 253–4.) While much has been made of the inefficiencies and low success rates of cloning, Lee Silver insists that these should be placed in their proper context. He argues that in the early years of IVF the ratio of implanted embryos brought successfully to term was lower than the one in thirteen ratio evident in the Dolly experiment (actually it was one in twenty-nine – with thirteen ewes used as surrogates). Though Silver concludes that 'there is no scientific basis for the belief that cloned children will be any more prone to genetic problems than naturally conceived children', a recent study showing that infants conceived with ICSI and IVF have twice as high a risk of a major birth defect as naturally conceived children suggests that proponents of cloning may in any case be entitled to lower their standards. It remains unclear whether this increased risk derives from the manipulation (and often freezing and thawing) of gametes and embryos, or from genetic anomolies which are also responsible for the adults' fertility problems. (Lee M. Silver, *Remaking Eden: Cloning, Genetic Engineering and the Future of Humankind?*, London: Pheonix, 1999, pp. 120–22; Michèle Hansen, Jennifer J. Kurinczuk, Carol Bower, Sandra Webb, 'The risk of major birth defects after intracytoplasmic sperm injection and in vitro fertilisation', *New England Journal of Medicine*, vol. 346, no. 10, 7 March, 2002, pp. 725–30.)

34 Ian Wilmut, 'Are there any normal cloned mammals?', *Nature Medicine*, vol. 8, no. 3, March 2002, pp. 215–6; Kellie L.K. Tamashiro, Teruhiko Wakayama, Hidenori Akutsu *et al*, 'Cloned mice have an obese phenotype not transmitted to their offspring', *Nature Medicine*, vol. 8, no. 3, March 2002, pp. 262–7.

35 Tim Radford, 'Stem cell "repair kit" for stroke victims', *Guardian*, 8 September 2000; James Meek, 'Stroke tests go ahead despite US warning', *Guardian*, 16 April 2001.

36 James Meek, 'Scientist urges law to boost embryo supply', *Guardian*, 8 May 2000.

37 James Meek and Martin Kettle, 'Bush under pressure over embryos made for research', *Guardian*, 12 July 2001. At the time of writing, US Senators are battling over a proposal to allow clients of fertility clinics to donate their surplus embryos to scientists conducting stem cell research.

38 Anthony Browne and Gaby Hinsliff, 'Wanted: women's eggs for research', *Observer*, 17 December 2000.

39 Of course, there may be other, less challenging methods of producing foetuses for medical use. There is, for example, the report of a woman who proposed being inseminated with her father's sperm so as to produce a foetus with genetically compatible brain cells that could be used to treat her father's Alzheimer's disease. In another case a woman with severe diabetes wanted to conceive and then abort a foetus in order to use its pancreatic cells to help her condition. (Andre Kimbrell, *The Human Body Shop: The Cloning, Engineering, and Marketing of Life*, Washington: Gateway, 1997, pp. 54–5.) There have now been several

cases in the US where parents have conceived with the express intention of producing a compatible donor of bone marrow or umbilical cord-derived blood stem cells for their fatally stricken child. If one questions the wisdom of allowing people to create a child for a purpose for which it may turn out to be *no good*, thus leaving the well-meaning parents with the unenviable task of disentangling their love for the surviving child with their disappointment over its sibling's death, then one will probably be reassured by the practice – legal in both the US and now in Britain – of using pre-implantation genetic diagnosis (PGD) to pre-select compatible embryos for this same purpose. Yet even here the success rates are estimated at eighty per cent, meaning one out of every five couples who consent to this procedure will conceive a child who fails the task for which it was chosen, and who may subsequently be expected to donate other tissue such as bone marrow or evens organs to make its existence fully worthwhile. The cloning of a seriously ill child may eventually be heralded as a more efficient solution.

40 See <www.advancedcell.com/testimony_7–18–2001_1.asp>

41 Laurence H. Tribe, 'Second Thoughts On Cloning', *New York Times*, 5 December 1997.

42 In August 2001 it was reported that a German surgeon had significantly reduced heart muscle damage in heart attack patients by transplanting their own bone marrow stem cells into arteries near the heart. (Hannah Cleaver and David Derbyshire, 'Stem cell therapy repairs a heart', *Daily Telegraph*, 25 August 2001.) Published research has also recorded the successful transformation of human adult bone marrow stem cells into mouse heart muscle tissue when the human cells were transplanted into mice with disabled immune systems, suggesting a surprising level of versatility in adult stem cells. (Catalin Toma, Mark F. Pittenger, Kevin S. Cahill *et al*, 'Human Mesenchymal Stem Cells Differentiate to a Cardiomyocyte Phenotype in the Adult Murine Heart', *Circulation*, vol. 105, no. 1, 1 January 2002, pp. 93–8.) American researchers have traced the fate of hematopoietic stem cells – which are isolated from the blood or bone marrow – when injected into mice, and shown that the stem cells can differentiate into cells of the liver, intestines, lung and skin. (Diane S. Krause, Neil D. Theise, Michael I. Collector *et al*, 'Multi-Organ, Multi-Lineage Engraftment by a Single Bone Marrow-Derived Stem Cell', *Cell*, vol. 105, no. 3, 4 May 2001, pp. 369–77.) In September 2001 Montreal scientists announced they had isolated, from the scalp of adult rodents, stem cells with the ability to differentiate into a variety of cell types, including neurones, muscle cells and fat cells. Analysis of human scalp tissue has revealed cells with similar potential. (J.G. Toma, M. Akhavan, K.J.L. Fernandes *et al*, 'Isolation of Multipotent Adult Stem Cells from the Dermis of Mammalian Skin', *Nature Cell Biology*, vol. 3, no. 9, September 2001, pp. 778–84.) Earlier in the same year, scientists in America were able to isolate stem cells from the fat deposits left over from liposuction cosmetic surgery, and to induce these multipotent cells to differentiate into cartilage, muscle fibre and bone cells. (Patricia A. Zuk, Min Zhu, Hiroshi Mizuno *et al*, 'Multilineage Cells from Human Adipose Tissue: Implications for Cell-Based Therapies', *Tissue Engineering*, vol. 7, no. 2, April 2001, pp. 211–28.) In March 2002 researchers at the University of Texas announced that when cancer patients were given transplants of stem cells extracted from the circulating blood of an adult donor, the stem cells differentiated to form new tissue in the liver, skin and gastrointestinal tract. (Martin Körbling, Ruth L. Katz, Abha Khanna *et al*, 'Hepatocytes and Epithelial Cells of Donor Origin in Recipients of Peripheral-Blood Stem Cells', *New England Journal of Medicine*, vol. 346, no. 10, 7 March 2002, pp. 738–46.)

43 Naohiro Terada, Takashi Hamazaki, Masahiro Oka *et al*, 'Bone marrow cells adopt the phenotype of other cells by spontaneous fusion', *Nature*, vol. 416, no. 6880, 4 April 2002, pp. 542–5; Qi-Long Ying, Jennifer Nichols, Edward P. Evans, Austin G. Smith, 'Changing potency by spontaneous fusion', *Nature*, vol. 416, no. 6880, 4 April 2002, pp. 545–8.

44 Anon, 'All too human', *New Scientist*, vol. 165, no. 2228, 4 March 2000, p. 19.

45 Jason A. Barritt, Carol A. Brenner, Henry E. Malter and Jacques Cohen, 'Mitochondria in human offspring derived from ooplasmic transplantation', *Human Reproduction*, vol. 16, no. 3, March 2001, pp. 513–6.

46 Amelia Hill, 'Horror at "three parent foetus" gene disorders', *Observer*, 20 May 2001.

47 A.W.S. Chan, K.Y. Chong, C. Martinovich *et al*, 'Transgenic Monkeys Produced by Retroviral Gene Transfer into Mature Oocytes', *Science*, vol. 291, no. 5502, 12 January 2001, pp. 309–12. In October 2001 Randy Prather at the University of Missouri announced that his research team had cloned five piglets, four of which successfully expressed the jellyfish genes, displaying yellow snouts and hooves which glow under fluorescent light. The purpose of this experiment was apparently to demonstrate that the genetic modification of pigs to yield organs suitable for transplantation into humans is a viable project.

48 Antinori is the publicity-seeking gynaecologist who pioneered the technique that became intracytoplasmic sperm injection, and who enabled a post-menopausal sixty-three-year-old Italian woman to become a mother in 1994. In March 1999 he also claimed at a conference in Venice that he had enabled four sub-fertile men – three Italians and one Japanese – to father children by maturing their sperm for three months in the testes of rats. (Philip Cohen and Michael Day, 'Never say die', *New Scientist*, vol. 161, no. 2179, 27 March 1999.) In April 2002 Antinori is reported to have told a journalist for the *Gulf News*, following a conference in the United Arab Emirates, that a woman he was 'treating' was carrying an embryo which he had produced by nuclear transfer, and he later said on an Italian talk show that he had three such pregnant women in his programme. These claims were not taken seriously by the scientific community, and by this time even fellow members of the original cloning consortium – including Panos Zavos – said they had severed their links with him.

49 Richard Dawkins, 'Thinking Clearly About Clones: How Dogma and Ignorance Get in the Way', *Free Inquiry*, Summer 1997, pp. 13–14. Facetious he may be, but Dawkins's comment evokes an appropriate comment from Leon Kass: 'should we not assert as a principle that any so-called great man who *did* consent to be cloned should on that basis be disqualified, as possessing too high an opinion of himself and of his genes? Can we stand an increase in arrogance?' (Leon R. Kass, 'Making Babies', in *Towards a More Natural Science: Biology and Human Affairs*, New York: Free Press, 1985, p. 68.)

50 Richard Dawkins, 'What's Wrong with Cloning', in Martha C. Nussbaum and Cass R. Sunstein, eds, *Clones and Clones: Facts and Fantasies about Human Cloning*, New York: W.W. Norton, 1998, pp. 54–66.

51 The determination of scientists in the late 1950s and early 1960s to replicate in animals the increasingly obvious teratogenic effects of thalidomide on human foetuses is a classical example of this mentality, the cost of which was a delay in the recall of the drug by five years, allowing over 10,000 babies to be born with serious developmental defects and skeletal deformities. Thanks to numerous epidemiological studies, the connection between smoking and cancer had also been confirmed by the mid-1960s, yet today scientists still receive funding to produce definitive evidence that tobacco smoke causes cancer in animals, while rats, mice, rabbits and dogs are continually experimented on to satisfy somebody's morbid curiosity of the full genetic, biochemical, physiological and cardiovascular effects of a substance which we know is harmful to human health. (See C. Ray Greek and Jean Swingle Greek, *Sacred Cows and Golden Geese: The Human Cost of Experiments on Animals*, New York: Continuum, 2000, pp. 44–7, 143–8.)

52 Cited by John O'Neill, *Ecology, Policy and Politics: Human Well-Being and the Natural World*, London: Routledge, 1993, p. 159.

9 Genetic Discrimination

1 Antony Barnett and Gaby Hinsliff, 'Fury at plan to sell off DNA secrets', *Observer*, 23 September 2001. Provoking fierce resistance from the church and human rights groups, arrangements similar to the deCODE deal are being pursued by the Australian biotech firm Autogen in Tonga (Patrick Barkham, 'Faraway Tonga cashes in on its gene pool secrets', *Guardian*, 23 November 2000), while the Estonian government is preparing to compile a national DNA database with a view to raising money via its sale to the biotech industry

(Lone Frank, 'Estonia Prepares for National DNA Database', *Science*, vol. 290, no. 5489, 6 October 2000, p. 31). In the UK, the Wellcome Trust and the Medical Research Council have funded a £60 million project to establish the world's biggest medical research gene bank, which aims to collect samples from half-a-million volunteers recruited through doctors' surgeries. Though the backers claim the information collected will not be available to insurance companies, employers or the police, there is no plan to prevent commercial exploitation of the material through patenting. This has led to calls from an independent watchdog for the development of the gene bank – called BioBank UK – to be frozen until proper safeguards are in place. (James Meikle, 'Biggest gene bank seeks 500,000 volunteers', *Guardian*, 17 April 2002.)

2 What little research that has been conducted in this area – in this case examining the strategic responses of women who have been identified as carrying the BRCA1 mutation – suggests that such self-seeking behaviour is in fact uncommon. (Cathleen D. Zick, Ken R. Smith, Robert N. Mayer, and Jeffery R. Botkin, 'Genetic Testing, Adverse Selection, and the Demand for Life Insurance', *American Journal of Medical Genetics*, vol. 93, no. 1, 2000, pp. 29–39.)

3 The alternative outcome predicted by more optimistic analysts, is that the difficulties faced by private insurance companies, and the level of discrimination required if they are to retain their monopoly over provision, will strengthen the argument that basic health insurance should be provided for everyone, using funds from community-rated taxes. (See Ronald Dworkin, 'Playing God', *Prospect*, no. 41, May 1999; and Dworkin, *Sovereign Virtue: The Theory and Practice of Equality*, Cambridge, Mass.: Harvard University Press, 2000, pp. 445, 452.)

4 Already over forty-five million people in the US have no public or private health insurance. (Lori Andrews and Dorothy Nelkin, *Body Bazaar: The Market for Human Tissue in the Biotechnology Age*, New York: Crown, 2001, p. 89.)

5 E. Virginia Lapham, Chahira Kozma, Joan O. Weiss, 'Genetic Discrimination: Perspectives of Consumers', *Science*, vol. 274, no. 5287, 25 October 1996, pp. 621–4.

6 Lawrence Low, Suzanne King, Tom Wilkie, 'Genetic discrimination in life insurance: empirical evidence from a cross sectional survey of genetic support groups in the United Kingdom', *British Medical Journal*, vol. 317, 12 December 1998, pp. 1632–5.

7 Joanna Marchant, 'Bush plans to bar firms from using gene tests', *New Scientist*, vol. 170, no. 2297, 30 June 2001, p. 5.

8 Dorothy Nelkin and M. Susan Lindee, *The DNA Mystique: The Gene as a Cultural Icon*, New York: W.H. Freeman and Company, 1995, p. 166.

9 Ruth Hubbard and Elijah Wald, *Exploding the Gene Myth*, Boston: Beacon Press, 1999, p. 142.

10 Paul R. Billings, Mel A. Kohn, Margaret de Cuevas, *et al*, 'Discrimination as a Consequence of Genetic Testing', *American Journal of Human Genetics*, vol. 50, 1992, pp. 476–82.

11 Many commentators also foresee a situation in which genetic tests become so extensive and precise, that people given a clean bill of genetic health will feel confident enough to withdraw their subscriptions to collective insurance programmes, thus undermining the operating principle of mutuality, and forcing up premiums for those at risk. With huge amounts of money to be made through the discovery of new genetic risks, susceptibilities and abnormalities, however, immunity to the predatory warnings of the insurance industry is unlikely to be enjoyed by many.

12 James Meek, 'Insurers insist on wider gene tests', *Guardian*, 13 October 2000.

13 James Meek, 'Insurers "broke code on gene information"', *Guardian*, 2 May 2001.

14 E.J. Calabrese, 'Genetic predisposition to occupationally-related diseases: current status and future directions', in Philippe Grandjean, ed., *Ecogenetics: Genetic Predisposition to the Toxic Effects of Chemicals*, London: Chapman and Hall, 1991, p. 26.

15 Cited in Andrews and Nelkin, *Body Bazaar*, p. 90.

16 Andrews and Nelkin, *Body Bazaar*, pp. 109–14.

17 Hubbard and Wald, *Exploding the Gene Myth*, p. 134.

18 The results of a genetic test sold in UK outlets of the health and beauty chain, the Body

Shop, for example, are analysed by the firm Sciona at the cost of an annual subscription of £120. The cell samples are tested for polymorphisms in nine different genes, most of which are associated with enzymes which help metabolise vitamins and remove toxins and free radicals from the body. Positive results lead to the recommendation that the client should avoid excessive consumption of things like meat and alcohol, and increase their intake of vegetables, vitamins and antioxidants. (James Meek, 'Public "misled by gene test hype"', *Guardian*, 12 March 2002.)

19 This has already been attempted in cases of work-related physical injury. In May 2002, Burlington Northern Santa Fe Railway Company, facing a lawsuit from the US Equal Employment Opportunity Commission, agreed to pay up to $2.2 million in compensation to employees suffering from work-related carpal-tunnel syndrome. Although the company denied it had violated the Americans with Disability Act or engaged in employment discrimination, it admitted that it had secretly tested workers who submitted compensation claims, hoping to find a rare genetic mutation which some scientists believe predisposes individuals to the painful musculo-skeletal condition. (Leigh Strope, 'Railroad to pay workers over secret DNA testing', *Boston Globe*, 9 May 2002.)

20 Chronic Obstructive Pulmonary Disease (COPD) is today estimated to be the fourth leading cause of death in the US. Its increasing prevalence is partly a consequence of an ageing population. (G. Viegi, A. Scognamiglio, S. Baldacci *et al*, 'Epidemiology of Chronic Obstructive Pulmonary Disease (COPD)', *Respiration*, vol. 68, no. 1, January–February 2001, pp. 4–19).

21 Teresa Brady, 'Genetic testing: Medical and Legal Issues, and DuPont's Program', *Employment Relations Today*, vol. 20, no. 3, Autumn 1993, pp. 256–66.

22 Calabrese, 'Genetic predisposition to occupationally-related diseases', in Grandjean, ed., *Ecogenetics*, pp. 22, 27.

23 Calabrese, 'Genetic predisposition to occupationally-related diseases', in Grandjean, ed., *Ecogenetics*, p. 22.

24 Troy Duster, *Backdoor to Eugenics*, London: Routledge, 1990, pp. 47–8.

25 Dorothy Roberts, *Killing the Black Body: Race, Reproduction, and the Meaning of Liberty*, New York: Vintage, 1999, p. 257.

26 Elaine Draper, *Risky Business: Genetic Testing and Exclusionary Practices in the Hazardous Workplace*, Cambridge: Cambridge University Press, 1991, pp. 130–1.

27 Duster, *Backdoor to Eugenics*, p. 26.

28 Andrews and Nelkin, *Body Bazaar*, pp. 82–4.

29 Jeremy Webb, 'A fragile case for screening?', *New Scientist*, vol. 140, no. 1905/6, 25 December/1 January 1994, pp. 10–11.

30 A.J. Zametkin, T.E. Nordahl, M. Gross *et al*, 'Cerebral Glucose Metabolism in Adults with Hyperactivity of Childhood Onset', *New England Journal of Medicine*, vol. 323, no. 20, 15 November 1990, pp. 1361–6. For a summary of criticisms of this article, see Richard DeGrandpre, *Ritalin Nation: Rapid-Fire Culture and the Transformation of Human Consciousness*, New York: W. W. Norton, 1999, pp. 141–3.

31 Brian Vastag, 'Pay Attention: Ritalin Acts Much Like Cocaine', *Journal of the American Medical Association*, vo. 286, no. 8, 22/29 August 2001, pp. 905–6.

32 DeGrandpre, *Ritalin Nation*, p. 181.

33 Peter Breggin, *Talking Back to Ritalin: What Doctors Aren't Telling You About Stimulants and ADHD*, Cambridge, Mass.: Perseus, 2001, p. 13.

34 Cited in Joseph T. Coyle, 'Psychotropic Drug Use in Very Young Children', *Journal of the American Medical Association*, vol. 283, no. 8, 23 February 2000, pp. 1059–60.

35 Breggin, *Talking Back to Ritalin*, p. 3.

36 Julie Magno Zito, Daniel J. Safer, Susan dosReis *et al*, 'Trends in the Prescribing of Psychotropic Medications to Preschoolers', *Journal of the American Medical Association*, vol. 283, no. 8, 23 February 2000, pp. 1025–30.

37 Anthony Browne, 'Ritalin made my son a demon', *Guardian*, 9 April 2000.

38 Barry Meier, 'Suits Charge Conspiracy by Maker and Doctors' Group to Expand Ritalin Use', *New York Times*, 14 September 2000. In 1991 the lobbying efforts of CHADD to have ADHD officially classified as a disability were successful, though its campaign to have Ritalin redefined as a Schedule III drug, which would have relaxed prescription guidelines and removed state control over the supply of the product, failed when Novartis was found to have made undisclosed donations to CHADD of nearly $900,000.

39 See Francis Fukuyama, *Our Posthuman Future: Consequences of the Biotechnology Revolution*, New York: Farra, Straus and Giroux, 2002, p. 46.

40 Paul Virilio, *Polar Inertia*, London: Sage, 2000, p. 70.

41 Stephen V. Faraone and Joseph Biederman, 'Neurobiology of Attention-Deficit Hyperactivity Disorder', *Biological Psychiatry*, vol. 44, no. 10, 1998, pp. 951–8.

42 Gregory Stock, *Redesigning Humans: Our Inevitable Genetic Future*, New York: Houghton Mifflin, 2002, pp. 106–7.

43 Edward M. Hallowell and John R. Ratey, *Driven to Distraction: Recognising and Coping with Attention Deficit Disorder from Childhood Through Adulthood*, New York: Touchstone Books, 1995, pp. 191–2.

44 Achim Kramer, Fu-Chia Yang, Pamela Snodgrass *et al*, 'Regulation of Daily Locomotor Activity and Sleep by Hypothalamic EGF Receptor Signaling', *Science*, vol. 294, no. 5551, 21 December 2001, pp. 2511–5.

45 See, for example: Michael W. Young, 'The tick-tock of the biological clock', *Scientific American*, vol. 282, no. 3, March 2000, pp. 64–71; Karen Wager-Smith and Steve A. Kay, 'Circadian rhythm genetics: from flies to mice to humans', *Nature Genetics*, vol. 26, no. 1, September 2000, pp. 23–7; Takashi Ebisawa, Makoto Uchiyama, Naofumi Kajimura *et al*, 'Association of structural polymorphisms in the human *period3* gene with delayed sleep phase syndrome', *European Molecular Biology Organisation Reports*, vol. 21, no. 4, 2001, pp. 342–6.

46 Andrew Kimbrell, *The Human Body Shop: The Cloning, Engineering, and Marketing of Life*, Washington: Gateway, 1997.

47 Andrews and Nelkin, *Body Bazaar*.

48 Andrews and Nelkin, *Body Bazaar*, p. 61.

49 Seth Shulman, *Owning the Future*, Boston: Houghton Mifflin, 1999, pp. 43–50.

50 Jon F. Merz, Antigone G. Kriss, Debra G.B. Leonard and Mildred K. Cho, 'Diagnostic testing fails the test', *Nature*, vol. 577, no. 6872, 7 February 2002, pp. 577–9.

51 Kimbrell, *The Human Body Shop*, pp. 7–13.

52 Kimbrell, *The Human Body Shop*, pp. 41–7.

10 Making Babies

1 See: <www.who.int/whr/1999/en/annex1.htm>

2 The most famous of these is certainly the Herman J. Muller Repository for Germinal Choice, colloquially known as the 'Nobel Prize Winners' Sperm Bank'. This was founded in California in 1976 by the millionaire inventor and eugenicist, Robert K. Graham, who saw himself as performing a public service in spreading superior genes. He expected donors to share his charitable enthusiasm (they were not in fact paid for their contributions), and did not charge clients directly for the sperm. After disastrous publicity greeted his announcement that he had received the sperm of three Nobel Prize winners in 1980 – including William Shockley, father of the silicon chip industry and a passionate believer in racial explanations for intelligence who advocated voluntary sterilisation of people with modest IQs – Graham was forced to cast his net wider, searching out future Nobel laureates, successful athletes, artists, businessmen and scientists. The bank closed in 1999, two years after Graham's death. A spin-off from the project – in 1984 the director, Paul Smith, left to found his own sperm bank, Heredity Choice, taking all the regular donors with him – is still running in the California desert.

3 For a critical review of UK surrogacy legislation and practices in the decade or so after the Warnock report, see Margaret Brazier, Alastair Campbell, Susan Golombok, *Surrogacy: Review for Health Ministers of Current Arrangements for Payments and Regulation. Report of the Review Team*, London: Stationary Office, 1998.
4 Cited by George J. Annas, 'Using Genes to Define Motherhood – The California Solution', *New England Journal of Medicine*, vol. 326, no. 6, 6 February 1992, pp. 417–20.
5 Scott B. Rae, *The Ethics of Commercial Surrogate Motherhood: Brave New Families?*, Westport: Praeger, 1994, p. 142.
6 Cited by Mark Rose, 'Mothers and Authors: *Johnson v. Calvert* and the New Children of Our Imagination', *Critical Inquiry*, vol. 22, no. 4, Summer 1996, p. 617.
7 Cited by Rose, 'Mothers and Authors', p. 617.
8 Joseph Fletcher, *The Ethics of Genetic Control: Ending Reproductive Roulette*, Buffalo, New York: Prometheus Books, 1988, p. 178.
9 Marjorie Maguire Schultz, 'Reproductive Technology and Intent-Based Parenthood: An Opportunity for Gender Neutrality', *Wisconsin Law Review*, vol. 1990, no. 1, 1990, pp. 378, 380, 384, 386–7.
10 See Stephen A. Marglin, 'What do Bosses do? The Origins and Functions of Hierarchy in Capitalist Production', *Review of Radical Political Economics*, vol. 6, no. 2, Summer 1974, pp. 60–112. This article is reprinted in André Gorz, ed., *The Division of Labour: The Labour Process and Class Struggle in Modern Capitalism*, Hemel Hempstead: Harvester Wheatsheaf, 1976, pp. 13–54.
11 No doubt the Calvert's, like most professionals in the US, also had the resources to intend to hire a nanny to ease the anticipated burden of child care.
12 According to Kimbrell, citing figures from the US Office of Technology Assessment, '64 percent of clients have incomes in excess of $50,000. Most brokers report that from half to 80 percent of their clients have had graduate school education. By contrast, the OTA found that most surrogate mothers earn just above the poverty line, and less than 4 percent of surrogate mothers are reported to have received graduate school education. Over 40 percent of surrogates are unemployed, receiving financial assistance, or both.' (Andrew Kimbrell, *The Human Body Shop: The Cloning, Engineering, and Marketing of Life*, Washington: Gateway, 1997, p. 127.) Most women who agree to be a surrogate of course do so for a variety of reasons, but the evidence suggests that non-economic motives are alone insufficient to persuade women to make this commitment. Even the advocates of commercial surrogacy argue that without financial incentives the supply of willing women would dry up, and a valuable service would therefore be lost. It should also be pointed out that surrogacy is unlikely to be utilised by people on low incomes because of the high costs of assisted conception, as well as the agency, surrogate and medical fees, which together may amount to over $100,000.
13 Gena Corea, *The Mother Machine: Reproductive Technologies from Artificial Insemination to Artificial Wombs*, London: Women's Press, 1988, pp. 214, 245.
14 Rory Carroll, 'Conceived, delivered and sold: a baby for $6,000', *Guardian*, 9 December 2000.
15 Cited by Corea, *The Mother Machine*, p. 229.
16 Kimbrell, *The Human Body Shop*, pp. 130–1; Rae, *The Ethics of Commercial Surrogate Motherhood*, pp. 93–4.
17 Elizabeth S. Anderson, 'Is Women's Labour a Commodity', *Philosophy and Public Affairs*, vol. 19, no. 1, Winter 1990, pp. 85–6. The description of the birth mother as a 'surrogate' is of course a linguistic attempt to sever our association of childbearing with motherhood. Though many feminists disavow the term for this reason, being aware of the ideological function of this conceptual ruse is adequate here.
18 Rae, *The Ethics of Commercial Surrogate Motherhood*, pp. 88–91, 22. Elizabeth Anderson makes a similar point, but stresses that the attachment between mother and baby that is violated by commercial surrogacy is founded not on physiological experiences – which of

course may vary – but on the normative social practices of pregnancy. 'Pregnancy is not simply a biological process but also a social practice', she writes, and 'the social norms surrounding pregnancy are designed to encourage parental love for the child'. The labour of the surrogate mother is alienated 'because she must divert it from the end which the social practices of pregnancy rightly promote – an emotional bond with her child. The surrogate contract thus replaces a norm of parenthood, that during pregnancy one create a loving attachment to one's child, with a norm of commercial production, that the producer shall not form any special emotional ties to her product.' (Anderson, 'Is Women's Labour a Commodity', pp. 81–2.) Given that the social norms of parenthood are historically variable – until fairly recently, for example, unwed mothers in modern industrial societies were expected to surrender their children at birth – Anderson's reliance on the norms of social practice is not entirely convincing.

19 In an advisory booklet, *Information for Surrogates*, produced by the British surrogacy organisation Childlessness Overcome Through Surrogacy (COTS), surrogates are reminded: 'Finally and most importantly it would be the final blow for your [sic] parents-to-be if you kept your child. *You would have robbed them of all hope they have placed in you*. You have your own children to go home to, they are *empty handed*. It is a tremendous trust that your couple have placed on your shoulders . . . *DO NOT BETRAY THAT TRUST*.' Cited in Brazier, Campbell, and Golombok, *Surrogacy*, p. 25 (original emphasis).

20 Rae, *The Ethics of Commercial Surrogate Motherhood*, p. 63.

21 Anita Stuhmcke, 'For Love or Money: The Legal Regulation of Surrogate Motherhood', *Murdoch University Electronic Journal of Law*, vol. 3, no. 1, May 1996, §47–9.

22 This hypothetical case is one of eight offered by John Harris as strong candidates for cloning. John Harris, 'Clones, Genes, and Human Rights', in Justine Burley, ed., *The Genetic Revolution and Human Rights*, Oxford: Oxford University Press, 1999, pp. 86–88.

23 Leon R. Kass, 'The Meaning of Life – in the Laboratory', in *Towards a More Natural Science: Biology and Human Affairs*, New York: Free Press, 1985, p. 113.

24 Sarah Boseley, 'US woman has test tube baby with her brother', *Guardian*, 14 July 2001.

25 Jon Henley, '"It was a favour – he helped me become a mother and I helped him become a father"', *Guardian*, 21 June 2001.

26 Angelique Chrisafis, 'Parenthood postponed', *Guardian*, 20 February 2001.

27 Yvonne Roberts, 'Women at the mercy of the foetal police', *Guardian*, 17 December 1999; Sarah Boseley, 'Ban on births from frozen embryos lifted', *Guardian*, 26 January 2000.

28 Maria Mies, 'From the Individual to the Dividual: In the Supermarket of "Reproductive Alternatives"', *Reproductive and Genetic Engineering*, vol. 1, no. 3, 1988, pp. 225–37.

29 See Barbara Katz Rothman, *The Tentative Pregnancy*, New York: Viking, 1986, p. 114.

30 Maria Mies, 'From the Individual to the Dividual', p. 235.

31 Even Carole Pateman, whose analysis of surrogacy is otherwise faultless, argues that 'references to baby-selling completely fail to meet the defence of surrogacy contracts derived from contract theory . . . In the surrogacy contract there is no question of a baby being sold, merely a service'. (Carole Pateman, *The Sexual Contract*, Cambridge: Polity, 1988, p. 212.)

32 Cited by Janice G. Raymond, *Women as Wombs: Reproductive Technologies and the Battle over Women's Freedom*, North Melbourne, Victoria: Spinifex, 1995, p. 33.

33 Lori B. Andrews, 'Control and Compensation: Laws Governing Extracorporeal Generative Materials', *Journal of Medicine and Philosophy*, vol. 14, no. 5, October 1989, pp. 550–1.

34 Christine T. Sistare, 'Reproductive Freedom and Women's Freedom: Surrogacy and Autonomy', *Philosophical Forum*, vol. 19, no. 4, Summer 1998, p. 237.

35 Carmel Shalev, *Birth Power: The Case for Surrogacy*, New Haven: Yale University Press, 1989, pp. 18–9, 12, 159–60.

36 Pateman, *The Sexual Contract*, p. 216.

37 Margaret Jane Radin, 'Market-Inalienability', *Harvard Law Review*, vol. 100, no. 8, June 1987, p. 1913.

38 Richard Titmuss, *The Gift Relationship: From Human Blood to Social Policy*, Harmondsworth: Penguin, 1977.
39 Barbara Katz Rothman, 'Surrogacy: A Question of Values', in Dianne M. Bartels, Reinhard Priester, Dorothy E. Vawter, and Arthur L. Caplan, eds, *Beyond Baby M: Ethical Issues in New Reproductive Technologies*, Clifton, New Jersey: Humana Press, 1990, pp. 240–1.
40 Sara Ann Ketchum, 'Selling Babies and Selling Bodies', *Hypatia*, vol. 4, no. 3, Fall 1989, p. 123.
41 Rae, *The Ethics of Commercial Surrogate Motherhood*, pp. 30–8.
42 Rae, *The Ethics of Commercial Surrogate Motherhood*, p. 51.
43 Ketchum, 'Selling Babies and Selling Bodies', pp. 116–27.
44 Julien S. Murphy, 'Is Pregnancy Necessary? Feminist Concerns About Ectogenesis', *Hypatia*, vol. 4, no. 3, Fall 1989, p. 69.
45 Jeremy Rifkin, 'The end of pregnancy', *Guardian*, 17 January 2002.
46 Peter Hadfield, 'Japanese pioneers raise kid in rubber womb', *New Scientist*, vol. 134, no. 1818, 25 April 1992; Anon, 'Here's looking at you, kid', *New Scientist*, vol. 155, no. 2092, 26 July 1997; N. Unno, K. Baba, S. Kozuma *et al*, 'An evaluation of the system to control blood flow in maintaining goat foetuses on arteriovenous extracorporeal membrane oxygenation: A novel approach to the development of an artificial placenta', *Artificial Organs*, vol. 21, no. 12, December 1997, pp. 1239–46.
47 Anil Ananthaswamy, 'Brave new babies', *New Scientist*, vol. 170, no. 2292, 26 May 2001, pp. 4–5.
48 Peter Singer and Deane Wells, *The Reproductive Revolution: New Ways of Making Babies*, Oxford: Oxford University Press, 1984, p. 133.
49 Fletcher, *The Ethics of Genetic Control*, p. 103.
50 Cited in Anil Ananthaswamy, 'Brave new babies', *New Scientist*, vol. 170, no. 2292, 26 May 2001, p. 4.
51 Fletcher, *The Ethics of Genetic Control*, pp. 164–5 (his emphasis).
52 Shulamith Firestone, *The Dialectic of Sex: The Case for Feminist Revolution*, London: Women's Press, 1979.
53 André Gorz, *Capitalism, Socialism, Ecology*, London: Verso, 1994, p. 62.
54 'Through the struggle with father and mother as personal targets of love and aggression, the younger generation entered societal life with impulses, ideas, and needs which were largely *their own*. Consequently, the formation of the superego, the repressive modification of their impulses, their renunciation and sublimation were very personal experiences. Precisely because of this, their adjustment left painful scars, and life under the performance principle still retained a sphere of private non-conformity.' Herbert Marcuse, *Eros and Civilisation*, London: Abacus, 1972, p. 78 (his emphasis).
55 Edmund Husserl, *The Crisis of European Sciences and Transcendental Phenomenology: An Introduction to Phenomenological Philosophy*, Evanston, Ill.: Northwestern University Press, 1970, p. 52.
56 André Gorz, *Critique of Economic Reason*, London: Verso, 1989, p. 124.
57 For an extensive analysis of the relationship between the Left and eugenic thinking, see Diane Paul, 'Eugenics and the Left', *Journal of the History of Ideas*, vol. 45, no. 4, October–December 1984, pp. 567–90. See also Allen Buchanan, Dan W. Brock, Norman Daniels, Daniel Wikler, *From Chance to Choice: Genetics and Justice*, Cambridge: Cambridge University Press, 2000, chapter 2.
58 In Sweden, for example, 60,000 young women were compulsorily sterilised between 1935 and 1976. (David Galton, *In Our Own Image: Eugenics and the Genetic Modification of People*, London: Little, Brown, 2001, pp. 100–1.) Over 60,000 Americans also fell victim to forced sterilisation programmes, which in Virginia continued until 1979. (Matthew Engel, 'State says sorry for forced sterilisations', *Guardian*, 4 May 2002.)
59 Charles Murray, 'Genetics of the Right', *Prospect*, no. 51, April 2000.
60 In Israel, where most pregnant women undergo transvaginal sonography (a form of internal ultrasound) screening at fourteen to sixteen weeks after conception, a study found that in

twenty-four cases where cleft lip was detected, all but one of the women chose to terminate their pregnancies as a result. (Zeev Blumenfeld, Israel Blumenfeld, Moshe Bronshtein, 'The Early Prenatal Diagnosis of Cleft Lip and the Decision-Making Process', *Cleft Palate-Craniofacial Journal*, vol. 36, no. 2, 1999, pp. 105–7.)

61 After the Paris high court in December 2001 awarded £200,000 damages to a child with Down's syndrome in compensation for the medical 'error' of doctors who failed to detect the chromosomal defect, French medical staff threatened the opposite course of action – to refuse to carry out prenatal tests in order to eliminate the risk of being sued for faulty diagnoses. The French parliament subsequently approved a bill which prevents children born with disabilities from suing doctors or their own parents on the grounds that they should not have been born. (Paul Webster, 'Doctors in threat to end screening', *Observer*, 2 December 2001; Jon Henley, 'France limits the right of those born disabled to sue doctors', *Guardian*, 11 January 2002.)

62 For a discussion of the way cost-benefit analysis has been used to justify the testing and termination of foetuses with Down's syndrome, see Thomas E. Elkins and Douglas Brown, 'The Cost of Choice: A Price Too High in the Triple Screen for Down Syndrome', *Clinical Obstetrics and Gynecology*, vol. 36, no. 3, 1993, pp. 532–40.

63 The same can of course be said of the selective abortion of female embryos in countries where women are seriously disadvantaged – socially disabled – in terms of life chances, health and opportunity. 'An Indian woman who knows what faces her third daughter is not making a morally different decision, it seems to me, than an American woman who knows what faces her child with Down syndrome.' (Barbara Katz Rothman, *Genetic Maps and Human Imaginations: The Limits of Science in Understanding Who We Are*, New York: W.W. Norton, 1998, p. 201.)

64 Tom Shakespeare, 'Back to the Future? New Genetics and Disabled People', *Critical Social Policy*, vol. 15, no. 1, Summer 1995, pp. 22–35.

65 Hans Jonas, 'Against the Stream: Comments on the Definition and Redefinition of Death', in *Philosophical Essays: From Ancient Creed to Technological Man*, Englewood Cliffs, New Jersey: Prentice-Hall, 1974, pp. 139–40.

66 Hilary Putnam, 'Cloning People', in Justine Burley, ed., *The Genetic Revolution and Human Rights*, Oxford: Oxford University Press, 1999, p. 13.

67 Joel Feinberg, 'The Child's Right to an Open Future', in William Aiken and Hugh LaFollette, eds, *Whose Child? Children's Rights, Parental Authority, and State Power*, Totowa, NJ: Rowman and Littlefield, 1980.

68 Axel Kahn, 'Clone mammals . . .clone man?', *Nature*, vol. 386, no. 6621, 13 March 1997, p. 119.

69 Axel Kahn, 'Cloning, dignity and ethical revisionism', *Nature*, vol. 388, no. 6640, 24 July 1997, p. 320.

70 Leon Eisenberg, 'Would Cloned Humans Really Be Like Sheep', *New England Journal of Medicine*, vol. 340, no. 6, 11 February 1999, p. 474.

71 Richard Lewontin, 'The Confusion over Cloning', *New York Review of Books*, vol. 44, no. 16, October 1997, p. 20.

72 Rothman, *Genetic Maps and Human Imaginations*, pp. 187–8.

73 As many as 100,000 women are thought to be 'missing' from the world's population, many as a result of the abandonment of female babies and abortion of female foetuses in regions of Asia where a girl is considered to be a heavy economic burden for the parents. In one study of a clinic in Bombay in the late 1980s, for example, 7999 out of 8000 aborted foetuses were found to be female. (Cited in R.K. Sachar, J. Verma, V. Prakash *et al*, 'Sex Selective Fertility Control – An Outrage', *Journal of Family Welfare*, vol. 36, no. 2, June 1990, pp. 30–35.) 'Missing women' is a term coined by Nobel Prize laureate Amartya Sen to describe the abnormal sexual demographic of many Third World countries. (See Amartya Sen, 'Missing women', *British Medical Journal*, vol. 304, 7 March 1992, pp. 587–8.)

74 Dorothy C. Wertz and John C. Fletcher, 'Ethical and Social Issues in Prenatal Sex Selection: A Survey of Geneticists in 37 Countries', *Social Science and Medicine*, vol. 46, no. 2, 1998, pp. 255–73.

75 See Ruth Schwatz Cowan, 'Genetic Technology and Reproductive Choice: An Ethics for Autonomy', in Daniel J. Kevles and Leroy Hood, eds, *The Code of Codes: Scientific and Social Issues in the Human Genome Project*, Cambridge, Mass.: Harvard University Press, 1993.

76 A legal loophole currently allows this service to be offered by private unlicensed clinics in the UK. (Kamal Ahmed, 'Baby "gender clinics" to face investigation', *Observer*, 4 November 2001.)

77 David McCarthy, 'Why sex selection should be legal', *Journal of Medical Ethics*, vol. 27, no. 5, October 2001, pp. 302–307.

78 See Leon R. Kass, 'Making Babies', in *Towards a More Natural Science: Biology and Human Affairs*, New York: Free Press, 1985, pp. 68–9. This ethical critique is no less relevant to the case, mentioned in the Introduction, of the deaf woman who gave birth to a deaf child after choosing, with her deaf female lover, a deaf man as a sperm donor. Despite their celebration of alternative deaf culture, one could say that the couple are more victims of biological reductionism – culture is, for them, premised on the possession of specific genes – than crusaders against the prejudices of the able-bodied. Today's obsession with cultural separatism and 'difference' thus finds its perfect ally and apologist in the view that genetic variation invalidates the universalism of humanity.

79 Hans Jonas, 'Biological Engineering – A Preview', in *Philosophical Essays: From Ancient Creed to Technological Man*, Englewood Cliffs, New Jersey: Prentice-Hall, 1974, pp. 161–2 (his emphasis).

80 Jürgen Habermas, *Die Zukunft der menschlichen Natur*, Frankfurt/Main: Suhrkamp Verlag, 2001, p. 111. Translated as *The Future of Human Nature*, Cambridge: Polity Press, forthcoming.

81 Habermas, *Die Zukunft der menschlichen Natur*, pp. 103–4.

82 Lee M. Silver, *Remaking Eden: Cloning, Genetic Engineering and the Future of Humankind?*, London: Phoenix, 1999, pp. 281–93.

83 Francis Fukuyama, *Our Posthuman Future: Consequences of the Biotechnology Revolution*, New York: Farrar, Straus and Giroux, 2002, p. 157.

84 Buchanan *et al*, *From Chance to Choice*, p. 178.

85 Edward O. Wilson, *Consilience: The Unity of Knowledge*, London: Abacus, 1999, pp. 305–10.

11 The Cyborg Solution

1 Hans Jonas, 'Cybernetics and Purpose: A Critique', in *The Phenomenon of Life: Toward a Philosophical Biology*, Evanston, Ill.: Northwestern University Press, 2001, p. 110.

2 Lee M. Silver, *Remaking Eden: Cloning, Genetic Engineering and the Future of Humankind?*, London: Pheonix, 1999, pp. 278–80.

3 Peter Sloterdijk, 'Anthropo-Technology', *New Perspectives Quarterly*, vol. 17, no. 3, Summer 2000, p. 18.

4 Sloterdijk, 'Anthropo-Technology', p. 19.

5 Sloterdijk, 'Anthropo-Technology', pp. 18–9.

6 Gregory Stock, *Redesigning Humans: Our Inevitable Genetic Future*, New York: Houghton Mifflin, 2002, p. 34.

7 Stock, *Redesigning Humans*, pp. 65–72; Michaeline Bunting, Kenneth E. Bernstein, Joy M. Greer *et al*, 'Targeting genes for self-excision in the germ line', *Genes and Development*, vol. 13, no. 12, 15 June 1999, pp. 1524–8.

8 Donna Haraway, 'A Cyborg Manifesto: Science, Technology, and Socialist-Feminism in the Late Twentieth Century', in *Simians, Cyborgs, and Women: The Reinvention of Nature*, London: Free Association Books, 1991, p. 180.

9 Hans Moravec, *Mind Children: The Future of Robot and Human Intelligence*, Cambridge, Mass.: Harvard University Press, 1988, p. 108.

10 Ray Kurzweil, *The Age of Spiritual Machines*, London: Pheonix, 1999, p. 163 (his emphasis).

11 Kurzweil, *The Age of Spiritual Machines*, pp. 129–32.
12 Kurzweil, *The Age of Spiritual Machines*, p. 193.
13 Kevin Warwick, 'Cyborg 1.0', *Wired*, vol. 8, no. 2, February 2000; Polly Curtis, 'Scientist becomes world's first cyborg', *Guardian*, 22 March 2002.
14 Craig Holdrege, *A Question of Genes: Understanding Life in Context*, Hudson, NY: Floris Books, 1996, p. 144–7.
15 Erwin W. Straus, 'The Upright Posture', in *Phenomenological Psychology*, London: Tavistock, 1966, p. 141.
16 Straus, 'The Upright Posture', p. 141.
17 Straus, 'The Upright Posture', p. 143.
18 Straus, 'The Upright Posture', p. 162.
19 Erwin W. Straus, 'Awakeness', in *Phenomenological Psychology*, London: Tavistock, 1966, p. 116.
20 Straus, 'The Upright Posture', p. 165.
21 Leon R. Kass, 'Thinking About the Body', in *Towards a More Natural Science: Biology and Human Affairs*, New York: Free Press, 1985.
22 Straus, 'The Upright Posture', p. 142.
23 Immanuel Kant, 'Conjectural Beginning of Human History', in Lewis White Beck, ed., *Kant On History*, Indianapolis: Bobbs-Merrill, 1963, p. 57.
24 Michel Foucault, *The History of Sexuality: vol. 1*, Harmondsworth: Penguin, 1990.
25 Anthony Browne, 'New surgery may be a sweat but it can save the lady's blushes', *Guardian*, 21 January 2001.
26 Aside from the regenerative potential of stem cells, Geron is also interested in – and indeed has patented – the gene for telomerase, an enzyme which rebuilds the tips of the chromosomes. As the cell divides and ages, the ends of the chromosomes, which are comprised of a single six-nucleotide sequence repeated thousands of times, get shorter and shorter until, after fifty or more divisions, the cell can divide no further and becomes senescent. Since offspring must of course be born with readily dividing cells, the telomerase gene is activated in the egg and sperm cells, serving to restore the tips of the chromosomes – called telomeres – to their original length. Geron's belief that telomerase offers a means of immortalising human cells must be tempered by the recognition that, because most cancerous cells exhibit an *active* telomerase gene, the natural shortening of the telomeres and resulting finite life-span of cells is a built-in defence against uncontrolled cell growth (cancer). (See Nicholas Wade, *Life Script: The Genome and the New Medicine*, London: Simon and Schuster, 2001, pp. 135–8.)
27 Francis Fukuyama, *Our Posthuman Future: Consequences of the Biotechnology Revolution*, New York: Farrar, Straus and Giroux, 2002, pp. 65–7.
28 Max More is the president of the Californian 'Extropy Institute', which holds conferences and disseminates ideas on future technology and its relevance to human beings. Selling himself as a 'corporate philosopher' (which presumably means 'management consultant'), More's has yet to publish his work with a commercial publisher. His lectures and articles are available at <www.maxmore.com> and <www.extropy.org>.
29 Max More, 'Beyond the Machine: Technology and Posthuman Freedom', New York, 1997.
30 As Nietzsche wrote in 1885–6: 'the possibility has been established for the production of international racial unions whose task will be to rear a master race, the future "masters of the earth"; – a new, tremendous aristocracy . . . [who will] work as artists upon "man" himself'. (Friedrich Nietzsche, *The Will to Power*, New York: Vintage, 1968, §960.)
31 Max More, 'Technological Self-Transformation: Expanding Personal Extropy', *Extropy*, no. 10, Winter/Spring 1993.
32 Hans Jonas, *The Imperative of Responsibility: In Search of an Ethics for the Technological Age*, Chicago: University of Chicago Press, 1984, p. 33.
33 John Armitage, 'From Modernism to Hypermodernism and Beyond: An Interview with Paul Virilio', in Armitage, ed., *Paul Virilio: From Modernism to Hypermodernism and Beyond*, London: Sage, 2000, p. 30.

34 Jonas, *The Imperative of Responsibility*, pp. 200–1.
35 Jonas, *The Imperative of Responsibility*, p. 99.
36 Kevin Warwick, *In the Mind of the Machine: The Breakthrough in Artificial Intelligence*, London: Arrow, 1998, p. 261.
37 Cited in Kurzweil, *The Age of Spiritual Machines*, p. 225.
38 Bill Joy, 'Why the Future Doesn't Need Us', *Wired*, vol. 8, no. 4, April 2000. <www.wired.com>
39 Moravec, *Robot*, pp. 3, 7.
40 Alvin Toffler and Heidi Toffler, 'More Technology, Not Less', *New Perspectives Quarterly*, vol. 17, no. 3, Summer 2000, p. 8.
41 George Dyson, *Darwin Among the Machines*, Harmondsworth: Penguin, 1997, p. 209.
42 Paul Virilio, *The Art of the Motor*, Minneapolis: University of Minnesota Press, 1995.
43 Kurzweil, *The Age of Spiritual Machines*, p. 123.
44 Manfred E. Clynes and Nathan S. Kline, 'Cyborgs and Space', in Chris Hables Gray, ed., *The Cyborg Handbook*, New York: Routledge, 1995, pp. 29–33.
45 Edmund Husserl, *Experience and Judgement: Investigations in a Genealogy of Logic*, London: Routledge and Kegan Paul, 1973, p. 43.
46 Edmund Husserl, 'Philosophy and the Crisis of European Man', in *Phenomenology and the Crisis of Philosophy*, New York: Harper and Row, 1965, p. 186.
47 Edmund Husserl, *The Crisis of European Sciences and Transcendental Phenomenology: An Introduction to Phenomenological Philosophy*, Evanston, Ill.: Northwestern University Press, 1970, p. 22.
48 Maurice Merleau-Ponty, 'Eye and Mind', in John O'Neill, ed., *Maurice Merleau-Ponty: Phenomenology, Language and Sociology*, London: Heinemann, 1974, p. 281. 'L'Oeil et l'espirit' was the last work Merleau-Ponty saw published. It appeared in the inaugural issue of *Art de France* in January 1961, and was republished after his death by *Les Temps Modernes* in the same year.

Index